Changing Images in Mathematics

Studies in the History of Science Technology and Medicine
Edited by John Krige, CRHST, Paris, France

Studies in the History of Science, Technology and Medicine aims to stimulate research in the field, concentrating on the twentieth century. It seeks to contribute to our understanding of science, technology and medicine as they are embedded in society, exploring the links between the subjects on the one hand and the cultural, economic political and institutional contexts of their genesis and development on the other. Within this framework, and while not favouring any particular methodological approach, the series welcomes studies which examine relations between science, technology, medicine and society in new ways e.g. the social construction of technologies, large technical systems.

Other Titles in the Series

Changing Images in Mathematics

From the French Revolution to the New Millennium

Edited by

Umberto Bottazzini
University of Palermo, Italy

and

Amy Dahan Dalmedico
National Centre of Scientific Research, Paris, France

Routledge
Taylor & Francis Group

LONDON AND NEW YORK

First published 2001
by Routledge
2 Park Square, Milton Park, Abingdon, Oxon OX14 4RN (UK)

Simultaneously published in the USA and Canada
by Routledge
711 Third Avenue, New York, NY 10017 (US)

First issued in paperback 2013

Routledge is an imprint of the Taylor & Francis Group, an informa business

Typeset by Expo Holdings, Malaysia

British Library Cataloguing in Publication Data
A catalogue record for this book is available from the British Library

ISBN: 0–415–27118–5
ISBN: 978-0-415-86827-3 (Paperback)

CONTENTS

LIST OF CONTRIBUTORS

Tom Archibald, Department of Mathematics and Statistics, Acadia University, Wolfville, Canada

Michel Armatte, Department of Applied Economy, University of Paris IX-Dauphine, Paris, France

David Aubin, Max Planck Institute, Berlin, Germany

Bruno Belhoste, National Institute of Pedagogical Research, Paris, France

Umberto Bottazzini, Department of Mathematics, University of Palermo, Palermo, Italy

Aldo Brigaglia, Department of Mathematics, University of Palermo, Palermo, Italy

Leo Corry, Cohn Institute for the History and Philosophy of Science and Ideas, Tel Aviv University, Tel Aviv, Israel

Amy Dahan Dalmedico, Koyré Centre of the History of Science, National Centre of Scientific Research, Paris, France

Jean-Paul Gaudillère, INSERM Paris, France

Hélène Gispert, IUFM of Versailles, University of Paris XI, Orsay, France

Jeremy Gray, Department of Mathematics, Open University, Milton Keynes, UK

Gabriele Lolli, Department of Mathematics, University of Turin, Turin, Italy

David E. Rowe, Department of Mathematics, Mainz University, Mainz, Germany

INTRODUCTION

Umberto Bottazzini & Amy Dahan Dalmedico

LESSONS FROM RECENT HISTORIOGRAPHY OF SCIENCE

Over the last few decades, the classical approach to the history of science and technology has been challenged profoundly. One might take T.S. Kuhn's celebrated book [1962] as the starting point of a 'new' historiography whose central issues were concepts such as revolutions in science, 'normal' science, and scientific paradigms. In the context of the 1960s, it was the notion of revolution which was the most striking. With hindsight, however, we can see that it was not this which had the greatest long-term impact, but rather the insertion of the concept of *scientific community* into the ancestral confrontation of Subject vs. Nature, characteristic of the philosophy of science since the 17th century. Although anticipated by Alexandre Koyré, Kuhn's notion of 'paradigm' stimulated attempts at describing the inner workings of scientific communities in terms of 'normal science', as well as their mental frameworks and habits of training which led them to tackle problems using the same approaches and methods and even to overlook anomalies. Finally, Kuhn also drew attention to the theme of *consensus* and the processes of persuasion among scientists.

For some twenty years now, the social studies of science have gone further, and have shown that the separation of cognitive aspects from social ones in scientific practice was reductive. This approach has shown how fruitful it is to conceive of science as a complex product of a human and social activity—a product like others bearing the marks of the place and time of its development—raising at the same time new types of questions. The separation between internalist studies of scientific knowledge and externalist analyses of institutional, cultural, or political contexts, which emerged in the 1950s, has therefore been forcefully questioned. On the one hand, this recent historiography of science has underlined the intimate relationship between cultural aspects, forms of scientific practice, and contents of knowledge. On the other hand, it has questioned the traditional conception of scientific activity as totally specific and endowed with its own logic and methodology, and separate from other human activities. For the empirical sciences, the emphasis was first put on the precise study of scientific controversies and on the way they were resolved. It was shown that these processes mobilized intellectual and theoretical aspects as much as the (never self-evident) discussion and interpretation of experiments, the examination of instrumental practices and tacit knowledge, mental and cultural traditions or training.[1] Numerous historical and sociological studies have confirmed that several basic aspects of a prevalent epistemology, such as the notion of experimental replication, or the establishment of a proof, could not be taken for granted. Notions of *proof* or *truth* of theories have been laden with a human collective meaning: *intersubjectivity* (which had already been emphasized by philosophers of science) and *consensus* among scientific communities could no longer be ignored by the historian.

The history of mathematics has remained at the margins of this renewal. Traditionally based on erudition and bibliography, the 'positivistic historiography' of mathematics—as it was sometimes called at the turn of the twentieth century—maintained that direct knowledge of sources and the elimination of interpretations not supported by documented evidence sufficed to enable the historian to arrive at the ultimate factual truth. Because of the apparently cumulative character of mathematics, its history was mainly considered in terms of the collection of scientific achievements through the ages, and often reduced to a museum of documents and results. History being the search for continuity in the progress of mathematical ideas, its leading philosophy was an implicit (or sometimes explicit) form of continuism—in other words, a belief in the steady progress of science from Antiquity to the present day.

Written by and for mathematicians, this historiography accordingly focused on the search for unpublished documents allowing one to 'fill the gaps' thus preserving the ideal continuity with the past. It led to the desperate search for precursors, past theorems and results that went unnoticed in their time. Moreover it focused on the 'longue-durée' development of mathematical concepts and theories. Growth in the corpus of the discipline was perceived as a progressive, continuous accumulation of results. Never questioned, since always true—this being at the very core of the specificity of mathematics—these results seemed to insert themselves spontaneously within global, rich and complex theories.

Because of its dramatic impact on the history of science, Kuhn's approach was likewise considered, in the 1970s, as being applicable to the history of mathematics. In particular, the question which was at stake was whether Kuhnian revolutions could occur in mathematics.

This was debated from opposing points of view.[2] Carrying a strong political meaning, the concept of revolution itself provided different reactions on emotional and political levels. Indeed, the intended reference to revolutions in political history carried over by the term was transferred to the history of science when used as a metaphor. One might argue that revolution is not an appropriate concept for the long-term processes of change taking place in the historical development of mathematics. Be that as it may, this issue has now lost most of its interest. As a result of these discussions, however, the image of the development of mathematics as a continuous and cumulative process no longer enjoys an undisputed position. Unity and continuity in the progress of ideas no longer provide useful guidelines for historical research. The history of mathematics has increasingly turned towards multiple visions which leave room for conjectures, difficulties, dead ends, and which, passing beyond formalisms, look at concrete problems and analogies, attempting more accurately to reconstitute the many ways and mazes of mathematical practice. Today it seems commonplace to claim that the domain of mathematics has not always looked the same. New branches of mathematics were born and developed; others have lost their prominence; still others have disappeared. Crossbreeds and methodological transfer have given birth to diverse subdisciplines. Moreover, whether conceptual or concerning institutional prestige, the implicit *hierarchy* of mathematical branches has also changed. The same is true of the images that mathematicians have cast of their science, its methods and status, with respect to the general organization of knowledge as much as to society, by excluding or putting aside entire branches of activity from the academic corpus.

Over the last two decades, the question of the relation of the history of mathematics to general history, and especially the history of science, has been raised several times and by various authors. On occasions, it has been discussed with the variety of approaches and methods which it deserves. In the preface of the book titled *The Social History of Nineteenth-Century Mathematics*, published in the early 1980s, it was noted that 'some scholars have begun to apply the methods of historical sociology of knowledge to mathematics'.[3] The interest in understanding the development of mathematics in a broad context led to the study of such factors as the formation of mathematical schools, changes in the professional situation of mathematicians, and the role of social and cultural milieus. Upheavals and reorganization in the mathematical corpus may indeed be due to a dynamic evolution in mathematical problems as much as to many other causes: institutional changes, political and social factors, technological demands, interactions with other disciplines, rivalry among mathematical schools, and so on. Concurrently, distinct mathematical schools and different nations may have privileged different hierarchies of disciplines and embraced distinct conceptions of mathematics and of the mathematician's trade. Accordingly, the historical perspective on 'mathematics in society', as well as the questions it raises, has been considerably extended. As H. Mehrtens [1992, 42] aptly noted, however, 'although there is a substantial body of literature on the social history of mathematics, no integrative history of mathematical knowledge and mathematical practices inside or outside academia has been achieved'.

Our aim in *Changing Images in Mathematics* is to develop these innovations with respect to the traditional historiography of mathematics in two distinct directions: on the one hand, to consider mathematics as belonging to the whole of human knowledge in interaction with numerous disciplines (physics and natural science, engineering science, the social sciences, biomedecine, etc.); on the other hand, to inscribe the history of mathematics even more in the contemporary, properly historical approaches to the history of science. The notion of *image* or *representation* (of mathematics) upon which this whole book is built has appeared to us as a privileged size for the articulation of several levels of analysis: (1) a properly disciplinary level; (2) a social and institutional level; (3) finally, a more diffuse cultural level which concerns the place of mathematics in the general configuration of knowledge, the prestige it enjoys, and the values it embodies. For example, how is one to understand the undertaking of Hilbert's 1899 *Grundlagen der Geometrie* and its later impact on twentieth-century mathematics without bringing together these three levels: (1) a study of the precise forms of axiomatization at play in the mathematical text and its repercussions on the development of geometry and, then, of all mathematics; (2) an articulation of this program with the establishment of the Göttingen school; and (3) an evaluation of the impact of the values of abstraction and universality that would be associated with it ?

ON THE NOTIONS OF REPRESENTATION AND IMAGE

Let us deal with the notion of representation. To avoid any misunderstanding, let us say at once that by promoting this notion in the domain of the history of science we certainly do not want to reintroduce the notion of appearance and with that the oldest metaphysical opposition in the occidental tradition, that is, the Greek distinction between appearances and reality. Extended by Descartes and Kant, this dichotomy went hand in hand with the distinction between the role of the subject and the object in the constitution of knowledge. Its long history conditioned philosophical attempts at grasping the intrinsic nature of reality. Clearly, our usage of the notion of representation (or of image) is opposed to this essentialist point of view. Since the beginning of the 1950s, moreover, by emphasizing the relation between truth and justification, the pragmatist and neopragmatist currents in philosophy (Peirce, James, Dewey, then Quine, Goodman, Putnam, Rorty, etc.) and their diverse variants had already cast doubt on the appearance/reality and subject/object dichotomies. From this perspective, language is seen more as a set of tools used to gain access to objects and to manipulate them (including in thought) than as faithful correspondents to these objects. In other words, the relationship between language tools and manipulated objects is dependent on an operative criteria of usefulness rather than a theory of reflection or correspondence.

With this possible misunderstanding laid to rest, let us now come to historical and sociological uses of the notion of representation. Several historians have emphasized that the notion of representation, which was central in Old Regime societies, could serve to study how structures, institutions, and (political or revolutionary) powers were represented in language and images (i.e. literature, art, and rites) and how, conversely, these various structures, institutions, and powers were determined by their symbolic representations [Chartier, 1989]. Concerning sociological approaches, one might refer to Boltanski's book on managerial executives [*cadres*], where the author describes social mechanisms through which a group is unified, acquires an identity, projects it onto others, and achieves its stability through the recognition of its institutional existence [Boltanski, 1982]. In this analysis, the accent is put on the importance of denomination—here, the term 'cadres'—which contributes to collective representations and group identities. Finally, other historians have introduced the idea that a representation may exhibit a dual aspect: a transitive one which is projective—according to the case, exhibiting a presence or making visible an absence—and a reflexive one. Similarly, when various groups construct representations of the same other group, the heterogeneity and nonuniformity of images can reveal differences among groups, and add complexity and nuance to the understanding of a global picture. These misfits and mismatches therefore suggest the importance of struggles over representations and their significance or possible interpretation in a given social, historical context.

For the history of science, it may be fruitful to take up the notion of representation as an analytic concept in order to understand, socially and culturally, the relations among various groups. The analogy with disciplinary, or subdisciplinary, scientific communities is indeed natural. For instance, it can be helpful in establishing a cartography of the mathematical milieu in a given country, or else to grasp the importance of a mathematical school (or place) with respect to others. Thus for a long time the analysis of late nineteenth century French mathematics was exclusively associated with the thesis of decline—a feeling expressed by Darboux and Hermite, who told Mittag-Leffler that by coming to Paris, one would be led astray and would be better visiting Berlin. But recent historical studies, Gispert's work[1991] in particular, show that this was just *one* representation that coexisted with others at the same time. Selected by historiography as the only true image of mathematics, this dominant representation was put forward by a group of elite mathematicians who clearly shared that feeling of decline. The mathematical community should be disaggregated into several groups however. Actors and authors of mathematical articles are distributed over several networks (*Normalien* scientists, engineers from the École polytechnique, high school [*lycée*] teachers, etc.). The whole set of published papers is divided up in two types (research, diffusion and the advancement of science). Institutional locations and teaching establishments must be distinguished according to various categories. Mathematical publishing must be divided between research and intermediary levels. Foreign countries themselves are distributed over distinct groups, with three or four great mathematical powers, and a less developed periphery.

Assuming the possibility of multiple representations, the notion of representation can therefore capture various units of interaction among a social professional subgroup (engineers or the group of great university professors can be considered for instance), such as the way in which it represents itself through the mathematical papers produced by its members (e.g., whether it prefers elementary geometry or higher analysis), the way these are in turn perceived, and the people they interact with in foreign countries. Whether envisioned in its restricted sense as great names who have filtered through history and memory, or in an extended sense which includes all mathematics teachers, members of the Société Mathématique de France, or authors of articles in certain journals, there is not one unified, homogeneous mathematical milieu. Rather, this milieu is stratified, made up of various groups which, in a reasonable period, are engaged in more or less the same type of mathematical activity. The notion of representation can enhance this logic of stratification: the unity of social professional aspects and of types of content produced is therefore constitutive of the various components of the milieu. There are various other locations excluded from official representations of the discipline, and still studied by historians, where other units of interaction little are at play (such as actuaries and artillerymen who often exhibited quite creative mathematical practices and production in the fields of statistics and probability theory). We have mentioned the case of late nineteenth-century France, but similar 'cartographic' studies might be rather useful if undertaken for other European scientific communities.

The heterogeneity of representations is obviously invoked to enrich our knowledge of the sources of these representations. Let us mention two examples among many other possible ones appearing in this book. Firstly the images of the Berlin school headed by Weierstrass, Kronecker and Kummer that can be reconstituted, respectively, from the correspondence of Hermite and other French mathematicians' as well as that of various Italians mathematicians such as Casorati, Dini, Pincherle. These tell us as much about the state of French and Italian mathematics than, reflexively, about Berlin mathematicians themselves and their scientific conceptions. Similarly, in the nineteenth century the reciprocal images constructed by mathematicians and statisticians about each other are highly significant, not only concerning the borders and conception of each domain, but also of the circulation of techniques, ideas and concepts from one to the other. In fact, the notion of representation goes hand in hand with that of reappropriation as for both, active, critical practices such as reading, translating, understanding, and interpreting, intervene, serving to question, adapt, and make operative what is the object of the reappropriation. Representation and reappropriation seem more useful to us than the simple notion of diffusion which assumes the circulation and unchanged preservation of ideas and concepts.

What is striking about mathematics is that a representation or image has the capacity of being projected as the single, true image for the whole community. This distorted projection may be the work of some of the actors themselves and it can also be produced or reproduced by historians. In fact, mathematics seems endowed with a reflexive character, perhaps a very specific one, namely the capacity for a dominant portion of the discipline to absorb local images and to turn them into a predominant, sometimes single, image. This was spectacularly the case for the conception of mathematics as the *science of structures*. As has been shown by Corry,[4] this image was produced by van der Waerden for algebra. In the 1950s, the Bourbaki group was responsible for the forceful attempt at turning this image and representation into the single, true conception of mathematics as a whole. A few dozen years earlier, this was not the case and the secular competition between algebra and geometry seemed to have been definitely settled by the development of algebraic function theory.

Today it can be said that the image of mathematics ruled by structures and the axiomatic method is past history. In a new technological environment characterized by the omnipresence of computers, a different set of domains—precisely those excluded by Dieudonné [1970], [1971] from the Bourbakist program ['*le choix bourbachique*']—have acquired an increasing importance: discrete mathematics, finite-difference methods, algorithms, recursive logic and functions, coding theory, probability theory and statistics, applied mathematics and so on.[5] When the disciplinary hierarchy is upset in such a way, the role of past mathematicians is reevaluated and mathematics' own methodology is reexamined. With the recent advent of the science of chaos, Poincaré, for example, has regained an important position which was hidden by the Hilbertian heritage. Similarly, the experimental method no longer seems an antinomy to mathematics. While in the 1950s and 1960s, structure was the emblematic notion of science, the notion of model seems to be typical of today's knowledge, even if it might be premature to say which of the contemporary images of mathematics will endure. We merely wish to suggest that, at given times, images and representations decisively contribute to the determination of a whole hierarchy of thriving questions, open problems deserving attention, and privileged methods of solving them. Inside the mathematical community, they are strongly linked to values. In particular, images and representations are responsible for 'constructing exclusion' from the mathematical world.[6]

Once more, let us emphasize the changing character of images and representations. Not only, do several contemporary representations coexist and compete with one another but historians also produce, over time, distinct representations of the same historical moment which are superimposed and blurred. We cannot ourselves escape fashions and dominant ideologies, and we easily grant that our interpretative frameworks are imprinted by the present we live in and the questions it privileges. For this reason our choice of the theme of representation, as far as the validity and stability of the analyses suggested here are concerned, necessarily leads us to favour comparative approaches and a reflexive attitude.

THE CONTENT OF THE BOOK

This book contains thirteen studies dealing with the nineteenth and twentieth centuries. Once again, let us emphasize that our aim has not been to present a linear, univocal account of a development spanning two centuries, but rather, multiple contrasted images of mathematics depending on countries, schools, and historical moments, and therefore to show how the conception and practice of mathematics depend on the cultural and national contexts in which they take place.

The first group of seven papers deals with nineteenth-century mathematics. By and large, in this field the dominant countries were France and Germany, and by the end of the century, Italy. With a balance oscillating between pure and applied mathematics, these countries provided contrasting images of the discipline. The role played by several centers and the different emphasis put on applications, in contrast with pure mathematics, provide guidelines through the relevant articles. In the mathematics of the first half of the nineteenth century, the dominant image was provided by French mathematicians trained at the École polytechnique. As Belhoste argues in his paper, the interest for application dominated academic life in France even prior to the Revolution. The polytechnician's training contributed to the construction of a disciplinary matrix understood as a set of values shared by the French mathematical community. A lasting interest in application was combined with research activity in pure mathematics. Not specific to French mathematicians, this dual orientation was also common among leading German mathematicians, like Gauss, Dirichlet, and Jacobi.

In the 1860s this image started to change profoundly. Bottazzini's paper shows how the establishment of German leadership in mathematics with the emergence of the Berlin school headed by Kronecker and Weierstrass placed a new image of pure mathematics at the fore. A deliberate effort was made to place pure mathematics at the top of a hierarchy of values. In this regard, the hackneyed story according to which Crelle's journal was jokingly re-named 'Journal für reine unangewandte Mathematik' (Journal for Pure Unapplied Mathematics) is quite enlightening. Arithmetical rigor became analysts' major concern, and analytic function theory acquired a leading role among mathematical theories. A contrasting image is offered by Archibald's essay on the community of German mathematicians who practiced 'rational physics'. By analogy with rational mechanics, he suggests using this term as a way of describing the set of values and practices specific to mathematicians occupied with modeling aspects of physical reality. Generally, their problems were formulated in terms of differential equations, and their solutions enriched the domain with a wealth of new methods and results. A closer look at the relationship of mathematics with the natural sciences is provided by Gray's paper. The contributions of Poincaré, Volterra, and Hadamard to the description of the physical world mathematically, notably in terms of partial differential equations, are seen from an original perspective that embraces the ups and downs of mathematics and mathematical physics at a time when physics was entering a period of dramatic change. From these articles, the emerging images of nineteenth-century mathematics are clearly partial, shifting, even contradictory.

At a time when the balance in Germany shifted increasingly toward theory, Felix Klein endeavored to associate the University of Göttingen mathematics department with technology and engineering. Some aspects of the unique career of the leader of the Göttingen school, and in particular his role as *Wissenschaftspolitiker*, are highlighted by Rowe's paper. Considering the various mathematical schools in German universities, he details the strategic goals which motivated Klein's efforts at Göttingen. Among Klein's long-term projects was the promotion of the *Encyklopädie der mathematischen Wissenschaften*. The contrasted images of mathematics emerging from this wide-ranging enterprise, particularly with respect to applied mathematics and the question of the existence of national 'schools', are discussed by Gispert in his comparison of the German and French editions of the Encyclopedia.

The last five articles are concerned with the twentieth century. In a sense, they all position themselves with respect to the structural axiomatic image of mathematical modernity which played a central role in the history of the century. Having questioned the very notion of the image of mathematics and its articulation with a body of knowledge, Corry carefully studies the emergence of the structural image first in van der Waerden's book *Moderne Algebra*, then for the field of algebra as a whole. He then considers the role of Hilbert and Bourbaki in the rise of this image, which was associated progressively with increasingly diverse branches of mathematics and which soon was to be identified with the discipline as a whole. Concerning the Italian school of algebraic geometry, Brigaglia's article shows that, whatever its hegemonic pretensions, the structural image and the tendency toward algebraization always coexisted with very different representations and practices. Considering the work of Segre, Peano, Enriques, Castelnuovo, and others, the author questions the nature of a 'school' and whether its unity rests on shared methods, problems, or other extra-mathematical elements. Whatever its persistence over several decades could be explained, often treated as marginal, this tradition has nevertheless been extremely fruitful, so much so that modern algebraic geometry sought to reclaim this entire corpus because of its capacity of allowing research on many topics. In terms of image, Gödel's incompleteness theorems had a paradoxical consequence. Arguing that his results changed nothing in practice, mathematicians (and even logicians) did indeed try to downplay their importance. Hilbert's foundation program and his metamathematics thus ruined, Gödel's theorems gave birth to a very different representation among an audience more or less removed from technical issues. They came to believe that the ambition of entirely formalizing (and indeed of mechanizing) the whole of mathematics had suddenly come to a halt.[7] Linking definition theory with the technique of paradoxes and the coded representation of syntactical operations, Lolli's article analyzes the mutation—which flowed from Gödel's results—of a precise mathematical concept, that is definability.

The way in which the structural image of pure mathematics, already in the process of being crystallized in the 1930s, was shaken up as a result of World War II is studied in Dahan's paper. In the United States, the war indeed stimulated the development of applied mathematics and the redefinition of the boundaries of the discipline. Moreover it favored the insertion of mathematicians in networks of increasingly diverse actors. It also prompted the appearance of a new cultural figure for the mathematician symbolized by John von Neumann. Emphasizing the resistance to this different representation of mathematics by pure mathematicians (whose ideology in its extreme form was embodied by Bourbakism), the author describes the unparalleled competition that, up to the 1970s, divided the two images of mathematics. As argued in Aubin's article, the same tension also underlies the evolution of a specific mathematical branch—namely topology. Evoking the role and work of 'applied topologists', such as Thom, Zeeman, Smale, or the physicist Ruelle, Aubin shows that all bore the mark of the abstract mathematics of the Bourbaki era. Despite differing widely in their viewpoints, approaches, and philosophies, all of them were inspired by differential topology to develop models aimed at providing forms of explanation of natural phenomena, from embryological development to linguistics, via economics. Suggesting a classification for these hetero-geneous models, Aubin analyzes a few mediations in the physicists' adoption and adaptation of topological modeling practices.

Two articles look at the development and the images of statistics by emphasizing its relation, firstly, to mathematics in the course of the entire nineteenth century, and, secondly, to medicine in the second half of the twentieth century. As Armatte writes, statistics is not merely a branch of mathematics linked to notions of probability and chance, and to write its history from this standpoint alone would be reductive. Intimately linked with the State—as its etymological root bears witness—statistics provides the tools for its description. It constructs social links and constitutes a public space. Called 'statistical thinking' by the author, who in this follows Alain Desrosières's [1993] concept ('la raison statistique'), the global rationality of this discipline located at the crossroads is therefore both social and cognitive. Armatte distinguishes three phases whose different images he analyzes over the course of the nineteenth century: (1) a quantification of serial facts; (2) a construction and interpretation of the significant representations of their organization; (3) a phase relative to action and the measurement of efficiency with respect to its social goals and uses. Finally through an analysis of various treatises of statistics, he proposes a synthesis of self-images. Along similar lines, Gaudillière's article explores the relation-ships between mathematics, statistics, and medicine in the second half of the twentieth century. He likewise considers medical statistics as an interactive process between actors belonging to different worlds: one concerned with health and sickness management, another constructing computational tools for decision-making. In particular, he analyzes the organization of randomized clinical trials and the establishment of inference rules in the case of tobacco and cancer.

Despite the long time period covered, the variety of topics and countries, this book strives to make an underlying unity in the history of mathematics. To achieve this, the contributors have paid attention throughout to the interaction of mathematics with themes (practice, institutional involvement, relations with other scientific disciplines, cultural and political images and so on) whose proper treatment has been neglected by the traditional historiography of the discipline.

ACKNOWLEDGMENTS

This book is the result of an international conference held at the Centre de Rencontres Mathématiques (Marseille-Luminy) in September, 1997, and seminars held in Paris in 1998 and 1999. We thank our colleagues who participated, the institutions which provided support (Centre Alexandre-Koyré and Association Henri-Poincaré), particularly Gyl Meillon for her help in preparing the manuscript, David Aubin for translations and comments, and John Krige for his support in completing the project.

NOTES

1 The first of these controversy studies was Collins [1985] on gravitational waves and the most noteworthy was probably Shapin and Schaffer [1985]. For a general introduction to this current and its major research topics, see Pestre [1995], [1996].
2 See for instance Crowe [1975], Mehrtens [1976], and later, Dauben [1992].
3 Mehrtens and al. (eds) [1981].
4 See Corry ([1992], [1996] and below).
5 See Dahan Dalmedico [forthcoming].
6 Due to the mathematician Jean-Pierre Bourgignon, this expression—'fabriquer de l'exclusion'—is especially striking for its political undertones in the French context. We interpret it as meaning that if the hegemonic representation of mathematics is structural, this image excludes applied mathematics (which do not fit this framework) for the realm of mathematics.
7 On this topic, see Nagel and al. [1989], and in particular J.-Y. Girard's postface.

REFERENCES

Birkhoff, G., (ed.) [1975] 'Proceedings of the American Workshop on the Evolution of Mathematics', *Historia Mathematica* 2.
Boltanski, L., [1982] *Les Cadres. La formation d'un groupe social*. Paris, Éditions de Minuit.
Chartier, R., [1989] 'Le monde comme représentation', *Annales ESC* 6 (November–December), 1505–1520.
Collins, H.M., [1985] *Changing Order: Replication and Induction in Scientific Practice*, London, Sage.
Corry, L., [1992] 'Nicolas Bourbaki and the Concept of Mathematical Structure', *Synthèse*, 92, 315–348.
    [1996] *Modern Algebra and the Rise of Mathematical Structures*, Basel, Birkhäuser.
Crowe, M., [1975] 'Ten "Laws" Concerning Patterns of Change in the History of Mathematics', *Historia Mathematica* 2, 161–166; repr. in Gillies [1992], 15–20.
Dahan Dalmedico, A., [forthcoming] 'Chaos, Disorder, and Mixing: A New "Fin-de-Siècle" Image of Science ?', *Growing Explanations: Historical Perspective on the Sciences of Complexity*, ed. M.N. Wise, Princeton, N.J., Princeton University Press.
Dauben, J. [1992] 'Are there Revolutions in Mathematics?' in Echevarria J. and al (eds.), *The Space of Mathemics*, Berlin, Walter de Gruyter
Desrosières, A., [1993] *La Politique des grands nombres. Histoire de la raison statistique*, Paris, La Découverte.
    [1998] *The Politics of Large Numbers: A History of Statistical Reasoning*, transl. By C. Naish, Cambridge (Mass), Harvard University Press.

Dieudonné, J., [1970] 'The Work of Nicolas Bourbaki', *American Mathematical Monthly* 77, 134–145.

[1977] *Panorama des mathématiques pures. Le choix bourbachique*, Paris, Gauthier-Villars.

Gillies, D., (ed.) [1992] *Revolutions in Mathematics*, Oxford, Oxford University Press.

Ginzburg, C., [1980] *Les batailles nocturnes. Sorcellerie et rituels agraires en Frioul, XVIᵉ-XVIIIᵉ siècles*, Paris, Lagrasse-Verdier (Origin. Italian ed. 1966).

Gispert, H., [1991] *La France mathématique. La Société Mathématique de France (1872–1914)*, Paris, Société Française d'Histoire des Sciences et des Techniques and Société Mathématique de France.

Gooding, D., Pinch, T., and Schaffer, S., (eds.) [1989] *The Uses of Experiments: Studies in the Natural Sciences*, Cambridge, Cambridge University Press.

Heering, P., and Sibum, O., [1994] 'The Replication of the Torsion Balance Experiment: The Inverse Square Law and its Refutation by Early 19th-Century German Physicists', *Restaging Coulomb*, ed. C. Blodel and M. Dörries, 47–66, Florence, Leo S. Olschki.

Kuhn, T., [1962] *The Structure of Scientific Revolutions*, Chicago, The University of Chicago Press.

Mehrtens, H., [1976] 'T.S. Kuhn's Theories and Mathematics: A Discussion Paper on the "New Historiography" of Mathematics', *Historia Mathematica* 3, 297–320; repr. in Gillies [1992], 22–31.

[1992] 'Appendix: Revolutions Reconsidered', in Gillies [1992], 42–48.

Mehrtens, H., Bos, H., Schneider, I. (eds) [1981] *Social History of Nineteenth Century Mathematics*, Boston, Birkhäuser.

Nagel, E., Newman, J., Gödel, K. and J.Y. Girard, [1989] *Le théorème de Gödel*, Paris, Le Seuil.

Pestre, D., [1995] 'Pour une histoire sociale et culturelle des sciences. Nouvelles définitions, nouveaux objets, nouvelles pratiques', *Annales, Histoire, Sciences sociales* 3, 487–522. [1996] 'Les 'Social Studies of Science' et leurs effets sur le travail historique', *Raison présente* 119, 35–46.

Shapin, S., and Schaffer, S., [1985] *Leviathan and The Air-Pump. Hobbes, Boyle and the Experimental Life*. Princeton, Princeton University Press.

Chapter 1

THE ÉCOLE POLYTECHNIQUE AND MATHEMATICS IN NINETEENTH-CENTURY FRANCE*

Bruno Belhoste

AUGUSTE COMTE'S MATHEMATICS AND THE 'POLYTECHNICIENS'

In Auguste Comte's classification of the sciences, mathematics took up a special place at the opening of the encyclopedic series, both as the first term and as the basis of his *Cours de philosophie positive* [Comte 1830–1842, vol. 1, 2nd lesson]. Indeed it constituted 'the most powerful instrument that the human mind can use in its study of natural phenomena.' Through its 'rigorous universality,' mathematics provided the best model for what a science should be. The other fundamental sciences in the series—astronomy, physics, chemistry, physiology, and sociology—were themselves classified according to the degree of abstraction of the phenomena they sought to describe and the degree of mathematization of their resulting laws. According to Comte, therefore, mathematical science ought to form 'the true starting point of any scientific training, either general or specialized.' Moreover, Comte divided mathematics internally between calculus (*calcul*) (its abstract part) and geometry and mechanics, which represented its concrete part. Under its various (arithmetic, algebraic, or analytic) guises, calculus was the science of relations *par excellence,* while geometry and mechanics, to which calculus could be applied, were natural sciences based on observations, just as astronomy and physics.

For Comte, fundamental sciences in general, and mathematics in particular, had a speculative nature even if they found practical application. So, he wrote, 'the human mind must proceed to theoretical research, while disregarding any practical consideration.' But, he added, 'these days, an intermediary class, that of engineers [*ingénieurs,* Comte's emphasis], is being formed, whose special purpose is to organize the relationships of theory with practice.' Comte cited Monge's descriptive geometry as an example of the 'intermediary doctrines between pure theory and direct practice' concerning the engineers. Whatever the importance of these applied sciences, they remained, in Comte's view, subordinate to the fundamental sciences, i.e. to purely theoretical sciences.

In mathematics, Auguste Comte was a conservative. He referred to authors from the beginning of the century (Laplace, Monge, Fourier, and Lagrange above all), but generally ignored his contemporaries' work. Moreover, his relationships with the mathematicians of his day soured for reasons having as much to do with principles as personal matters. He blamed the greatest (Poisson, Cauchy, Liouville) for their excessive specialization. However, far from being a fantasy, the image of mathematics emerging from the *Cours de philosophie positive* reflected, admittedly with a few distortions, the spirit of the discipline in nineteenth-century France. With his very own originality and intensity, Auguste Comte expressed the dominant conceptions of the mathematical milieu he associated with.

Comte himself was a *polytechnicien* who became a mathematics professor by necessity [Pickering, 1993]. He taught mathematics in a preparatory institution for the École polytechnique, and then at the École itself as an analysis assistant [*répétiteur*]. In addition he served for seven years as an entrance examiner covering for a long time one of the two professorships in analysis and mechanics. This experience nourished his vision of mathematics. Like Auguste Comte, most French mathematicians in the nineteenth century were trained, and many taught, at the École polytechnique. Consider for example the mathematicians elected to the Academy of Sciences [Crosland, 1992, pp. 228–229]: out of 25 academicians from the geometry section between 1816 and 1900, fourteen were *polytechniciens,* and, among the eleven that were not, four were elected before 1816 (Laplace, Legendre, Lacroix, Ampère) and three after 1880 (Darboux, Appell, Picard). In the mechanics section, ten academicians out of the twelve whose activity was partly mathematical were also *polytechniciens,* among whom were Cauchy and Poncelet. In the astronomy section, one also finds seven *polytechniciens* out of the 14 academicians working in mathematical astronomy, including Liouville and Leverrier, and out of the seven non-*polytechniciens,* five were elected before 1816 (Messier, Cassini, Lalande, Bouvard, Burckhardt). Finally, in the physics section, the *polytechniciens* Poisson and Duhamel were known mainly for their mathematical work. Thus, out of the fifty-odd academicians who worked on mathematics between 1816 and 1900, two thirds were *polytechniciens.* At the middle of the century this domination was overwhelming, and only after 1880 did it perceptibly decline.

Until the end of the 1860s, it seemed therefore that the École polytechnique was by far the main training place for mathematicians in France, and that, at least until the 1880s, it played a decisive structuring role for the mathematical milieu. For a long time, universities [*facultés des sciences*] had no true students, their essential role consisting in granting degrees, principally the *baccalauréat* (a high school degree). Only the Sorbonne in Paris provided high-level teaching in mathematics, given by renowned professors, but attracting only a small audience. This situation was just starting to change after 1870. As for the École normale supérieure, initially it exclusively trained high school [*lycée*] teachers. One had to wait for the appointment of Cauchy's disciples—Briot, Bouquet, Puiseux, and most notably Hermite—as assistant professors [*maîtres de conférence*] in 1850–1860, to see the level of mathematics teaching rise in universities [Tannery, 1895].

After 1880 the expansion of university education led to a profound reorganization of the French mathematical milieu. At the end of the century, the main French mathematicians went through the École normale supérieure and the influence of the École polytechnique quickly dwindled—Henri Poincaré's case being an exception—until it became nearly negligible after 1914. Hélène Gispert's [1991] study of the Société Mathématique de France, founded in 1872, confirms this decline: at its foundation, *polytechniciens* accounted for nearly 60% of its members; in 1914, for less than 30%.

What was the effect of this polytechnician hegemony for nearly a century on the representations and practices of French mathematicians? Beyond a diversity of men and works, was there a common thread linking such diverse mathematicians as Poncelet, Cauchy, and Poinsot? When drawing attention to the eminent position of mathematics in Auguste Comte's system, I claimed above that this position could well reflect the conceptions of the French mathematical milieu trained at the École polytechnique. More precisely, I would like to show here that the polytechnician training contributed to the construction of a 'disciplinary matrix,' understood as a set of values and interests dominating the French mathematical community in the nineteenth century.

I shall first sketch the institutional framework—strongly affected by the intervention of the State—in which mathematical activities took place since the end of the eighteenth century. I shall then try to bring out characteristics of the mathematical teaching at the École polytechnique. This study will lead me to consider two complementary aspects of the polytechnician matrix concerning mathematics, namely, the interest for useful applications and the predilection for general theories.

ENGINEERS AND MATHEMATICS: THE HERITAGE OF THE ENLIGHTENMENT

In 1794, the École polytechnique was founded mainly as a means to insure the initial training of the technical fraction of the civilian, and above all military, State bureaucracy [Belhoste, Dahan Dalmedico, Pestre, Picon, 1995]. At this time, that is, during the Revolution, this technocracy was already benefiting from a long historical experience. Having started to be put in place under the reign of Louis XIV a century earlier, it subsequently acquired a specific organization, culture, and identity. Organized in hierarchical, self-administered professions [*corps*] (*corps du génie* [military engineering], *corps de l'artillerie* [artillery], *corps des ponts et chaussées* [civil engineering], and *corps des mines* [mining engineering], to cite the main ones), this technocracy invented or perfected methods of modern administration—surveys and evaluations, decision and financing procedures, control and management of building sites and factories, etc. These methods mobilized many intellectual and material resources with a constant concern for precision and quantification. The very organization of the *corps* was inspired by these principles: recruitment and promotion of men according to objective criteria, standardization of their salaries, creation of schools for the training of young recruits and constant follow-up of their careers with the constitution of personal files and inspections. Thus was this technocracy established as a model of enlightened administration.

After having played a major role in the reorganization of the State apparatus (a role symbolized by the names of Lazare Carnot from the *corps du génie,* and Napoléon Bonaparte from the *corps de l'artillerie*), the technical elite survived intact, but transformed, from the revolutionary crisis. On the one hand, the creation of the École polytechnique reinforced its homogeneity and its intellectual prestige, while recruitment based solely on entrance examinations [*concours*] with no condition of birth gave it a democratic legitimacy unique among the senior civil service. On the other hand, the fragile compromise struck among different fractions of the landed class, on which the succeeding regimes rested after 1815, put technocrats aside in a France ruled by *notables* and oscillating between conservatism and political liberalism, where the model of the enlightened State was no longer in fashion. Under the restored monarchy, the frustration created by such a situation among many State engineers favored the success of technocratic utopias, such as Saint-Simonism, celebrating the alliance between *savants* and industrialists in the conduct of society. But while *Saint-Simoniens* became the new apostles of industry, the technical *corps* mistrusted private initiatives that eluded them, like the building of the first railway lines: much more than that of State engineers, the development of new industries was the work of entrepreneurs and civilian engineers, i.e. civil engineers employed by the private sector.

The technocracy's authority rested mainly on its mastery of the exact sciences, for which the École polytechnique was the warrant. This cultural trait itself stemmed from a complex process going back to the seventeenth century. The relationship between the engineer's theoretical knowledge and his technical art had for long remained blurred. While it was commonly claimed that theory could shed light on practice, the engineer's art remained firmly rooted in field experience. Only in the last third of the eighteenth century, with the development of technical schools (whose best model was Mézières), did the idea take hold that this art consisted chiefly in applying theoretical knowledge to the practical domain. The State engineer then forfeited his status as an 'artist' to become some kind of 'practical *savant*' whose science was all about application.

Without the existence of an academic milieu always concerned with its public usefulness, this change could never have taken place. As we know, the problems posed by mechanics and astronomy were a major source of inspiration for seventeenth-century mathematicians who cared to deal with technical questions. In France, the Academy of Sciences received a mission of official expertise in arts and trades, which prompted *savants* to be interested in machines, war and navigation techniques, and so forth [Hahn, 1971]. However, academic sciences for a long time remained foreign to the culture of the mechanical arts they pretended to guide. At the close of a slow evolution marked by the growth of mathematical physics, social mathematics, and engineering science, the rapprochement was only accomplished at the end of the eighteenth century [Gillispie, 1980]. Only then did academic knowledge recompose itself around the notion of application.

In the new conception defended by Condorcet [Belhoste, 1997] and—with different modalities—illustrated on the eve of the Revolution by the work of the three main mathematicians of the Academy (Lagrange, Laplace, and Monge), the ultimate justification of theoretical endeavor rested on the importance of its applications, rather than, as before, on the speculative search for truth. A science could thus be applied to another, for example, analysis to geometry, mechanics, and physics; probability theory to celestial mechanics, and to the moral and political sciences; geometry and mechanics to engineering science.

Moreover, a science could be applied to an art, like geometry to drawing, as tronomy to navigation, mechanics to machine building, and the calculus of probabilities to the social art. The notion of application therefore erased the hitherto insurmountable barrier between the theoretical and the practical spheres.

In this context, mathematics became, in France, a determining element of engineering culture. To measure, draw, and build, the field worker used to have at his disposal instruments and recipes, a whole body of practical mathematics. In place of this art, school-trained engineers of the Enlightenment substituted a mathematical science to be applied to problems encountered in practice. To illustrate this transformation, the best example already cited by Comte was provided by the 'method of projections,' or descriptive geometry, taught by Monge at the Mézières school instead of the engineers' traditional techniques of drawing [Sakarovitch, 1998]: inspired by stone-cutting processes, this method consisted in representing surfaces and curves in ordinary space by their cylindrical projections on two reference planes pictured on a same sheet of paper, and in the construction of their intersections and differential elements (tangent lines at one point, curvature lines, etc.).

This role ascribed to mathematics and its applications contributed to the fashioning of the technical *corps'* new identity. Henceforth, they would not be corporations of tradesmen anymore, but scientific *corps* represented at the Academy of Sciences. In 1786, out of the 24 ordinary academicians in the mathematical classes, six were directly linked to the State technocracy: the three examiners Laplace, Bossut, and Monge, the captain [*capitaine de vaisseau*] Borda, and the *ingénieurs du génie* Coulomb and Meusnier. And each *corps* hosted young *ingénieurs-savants*, who sometimes were mathematicians like Lazare Carnot and Prony. In the recruiting and training of engineers, the increasing importance granted to mathematics simultaneously denoted a change in social status. Considered a military matter, mathematics indeed belonged to a noble's education by tradition, just like fencing and dance. By requiring mathematical knowledge for admittance into the technical *corps,* the royal power showed its willingness to promote them in the social hierarchy. In truth, in the course of the eighteenth-century, the State technocracy had been recruiting less and less from the reservoir of the trades and more and more among the educated elite. Progressively, access to artillery and military engineering (the *armes savantes*) had been shut to commoners, to the point that at the dawn of the Revolution these branches recruited exclusively among noblemen [Blanchard, 1979].

To this social mutation corresponded the adoption of a new value system. Henceforth an engineer's legitimacy rested less on the recognition of his practical competence than that of his personal merit. A value of aristocratic origins, merit associated quality of birth with personal excellence. In the *armes savantes* the mathematical exam became the main tool for measuring this excellence. By eliminating birth conditions for granting access to civil services, the Revolution democratized this meritocratic ideal without questioning it, since the *concours* was what granted access to the École polytechnique and, beyond that, to the various technical *corps*. Therefore, the role it played in meritocratic selection gave mathematics a special status in France: it dominated the arts and sciences not only because of the effects of its application, but also because of its purported capacity of measuring intelligence.

MATHEMATICS AT THE ÉCOLE POLYTECHNIQUE

The neologism 'polytechnique' used to name the École underscored its vocation to train general, versatile engineers. Indeed, its founders had envisioned the suppression of specialized technical *corps* and their replacement by commissions given to independent, polyvalent techni-cians to whom the public power would appeal in case of need. At this time, this project of dismantling the State technocracy to the benefit of self-employed engineers was part of the Jacobine struggle against corporations and their defense of small, independent producers. This seemed even more justified by the fact that a vast pyramidal system was envisioned for the scientific and technical training of tradesmen ['*hommes de l'art*'], whose summit would have been the École polytechnique [Belhoste, Dahan Dalmedico, and Picon, 1994]. All of these projects having been abandoned almost at once, the *corps* demanded that their schools be maintained and the École polytechnique reformed, if not suppressed. In the end, the École subsisted as a preparatory school for future State engineers, while its curriculum was reduced from three to two years. The *corps* schools were transformed into application schools ['*écoles d'application*'] where graduates of the École polytechnique acquired specialized training. By ensuring a uniform recruitment and scientific training for all the technical *corps*, the École polytechnique therefore became the matrix of the State technocracy.

This strategic role accounts for the fact that, far from being a simple pedagogical and scientific affair, the definition of its curriculum, which was dominated by mathematics, was also an important political issue. The École's founder, Gaspard Monge, organized the students' curriculum with a dual aim: to bind as tightly as possible the study of theories with their applications so as never to lose sight of the practical purposes of teaching; and to adopt as general as possible a viewpoint so as to embrace all engineering specialties. Thus in the initial program of the École, mathematics was always thought of in relation to applications. On the one hand, engineering drawing practice called for the creation of a course of descriptive geometry associating the study of figures in space with applications to the representation of shadows, the art of sketching, and the drawing of machines. On the other hand, the practice of the engineer-builder demanded a course of mechanics associating the study of analysis with its applications to the mechanics of solids, fluids, and machines. At the same time, in mathematics as in other disciplines of teaching, Monge's program was characterized by the determination to provide students with general viewpoints able to be very widely applied. The initial choice of professors was wholly shaped by this consideration, since two famous mathematicians of the Academy were selected: Monge in descriptive geometry, and Lagrange in analysis and mechanics.

Monge and Lagrange delivered very ambitious courses, but going in opposite directions: the former favored the role of geometrical intuition, the latter that of algebraic calculus. A shared concern for generality, however, led both to disentangle the exposition of methods from their applications. From the very outset therefore, the École witnessed a tendency towards greater abstraction in mathematical teaching. This is evident from a quick comparison between Monge's course of descriptive geometry at the École polytechnique[1] and the study of stereotomy as practiced at the Mézières school before the Revolution [Belhoste, Picon, and Sakarovitch, 1990].

Nearly all problems studied at the Mézières *École du génie* emerged directly from practice. Mandatory sketches were thus almost exclusively sketches of stone and wood cuts. At the end of the course, each student had to consider a particular problem. Rare, however, were those given geometrical problems: in 1780, out of 13 students, ten were asked to solve stone-cut problems. At the École polytechnique, on the other hand, the theoretical part of descriptive geometry dealt exclusively with geometry in space to which 42 sketches were devoted. Moreover, students considered in parallel the same problem (or at least some of them) by means of analysis. Synthetic and analytic, this dual approach to geometry which characterized Monge's teaching at the École polytechnique was rich with potential. It incited students to reflect on their respective advantages in terms of simplicity, generality, and applicability.

In analysis and mechanics, whose study expanded with the arrival of Joseph Fourier in charge of a first-year course [propédeutique], Lagrange's influence was decisive, even if his course was optional (the mandatory course being taught by Prony). In his lessons, Lagrange introduced his calculus of functions in which he adopted a uniform, simple, and general approach allowing him, in his own terms, to 'link all parts of analysis in a simple body of sciences,' from arithmetic to mechanics which, according to him, merely was 'a four-dimensional geometry' [Pepe, 1986].

Thus, from the very first years of the École, the tone was set in mathematics: even if the broader aim remained that of application, the teaching was resolutely theoretical. The goal was to provide students with abstract methods susceptible by their very nature of the most general applications. When the écoles d'application were put in place, this tendency was but reinforced. While some application courses had at first been retained by the École polytechnique, they were progressively reduced and finally disappeared in 1816, at the end of a very fierce struggle between the écoles d'application and the direction of the École polytechnique [Fourcy, 1828]. This amounted to more than simply a general evolution in the curriculum: in all courses, some applications were similarly excluded. In mathematics, as elsewhere, the program was focused on the study of general methods, with the understanding that students would later have the occasion to apply them to particular problems.

Institutional dynamics therefore followed the direction laid down by Monge and Lagrange at the creation of the École. 'More abstraction for better application' was the motto. In an 1800 introductory report about the programs in analysis presented to the Conseil de perfectionnement by the Conseil de l'École, Lacroix, himself a professor of analysis, claimed that 'as soon as one has passed the first elements, one must elevate oneself very high in order to be able to find truly useful subjects for applications in analysis.' The increase in the level of entrance examinations, the augmentation in the number of lessons, and the creation of new professorships and teaching assistantships [répétiteurs], provided resources for the adoption of ambitious programs in analysis and mechanics. The course in analysis was centered, on the one hand, on differential and integral calculus and the study of differential equations and, on the other, on analytic and infinitesimal geometry. Tightly linked to it, the course in mechanics was purely theoretical, the study of applied mechanics being wholly entrusted to the *écoles d'application*. As for the course in descriptive geometry whose importance had begun to decline after 1815, it was devoted almost exclusively to the graphical study of ruled surfaces, surfaces of revolution, and their intersections. Stone cuts and other applications occupied only a small part of the program.

Thus was put in place a mathematical cannon which would hardly change in the course of the nineteenth century. Analysis occupied an hegemonic position, while arithmetic and algebra, whose elements had been learnt by students in secondary and preparatory schools, were almost entirely neglected, and geometry only appeared under the forms of analytic geometry and descriptive geometry. The domain of applications itself was essentially reduced to rational mechanics, and marginally to engineering drawing. Even a domain to which nineteenth century French mathematicians devoted numerous studies, mathematical physics, was more or less ignored by the École. True, some attempts were made to enlarge this cannon. For example, under the Restoration, Cauchy introduced some elements of his theory of functions of one complex variable with applications to mathematical physics in mind [Belhoste, 1991]; fifty years later, Hermite had the study of elliptic functions (to which he gave some applications in number theory) adopted in the program [Gispert, 1995] and [Belhoste, 1996]. But no matter how interesting, these relatively marginal innovations did not upset the cannon of mathematical teaching at the École.

It was in the framework sketched above that Cauchy developed his teaching of analysis. Without reviewing this course here—it has been the topic of numerous historical studies (see for example [Gilain, 1989], [Grattan-Guinness, 1990], [Belhoste, 1991], [Bottazzini, 1992])— I would like to hint briefly at what Cauchy's enterprise owed to his institution's concerns. As he himself attested, it was through a concern for rigor, and in direct reaction to a certain mathematical laxity, that in his course he cared to define as often as possible the domain of validity for a definition or theorem. This led him to introduce new notions in analysis, such as converging series and continuous functions, and to give existence proofs, for example for summations of series, definite integrals, and solutions to differential equations. On ideological as well as mathematical levels, this exigency incontestably marked a rupture with respect to Lagrange's formalism and Lacroix's empiricism. Still, rigor was construed neither as a limitation—even if Cauchy now gave up the purported 'generality of analysis'—nor as a goal in itself: it appeared rather as the condition for the development of general methods with a view to applications.

THE INFLUENCE OF APPLICATIONS AND THE PRIMACY OF THEORY

Many a mathematician graduating from the École polytechnique in the nineteenth century was a State engineer: Cauchy and Saint-Venant were *ingénieurs des ponts et chaussées,* Lamé and Jordan *ingénieurs des mines,* Poncelet a *génie* officer, and Laguerre an artillery officer. Others avoided integration in a *corps* or quickly quit, like Poinsot, Poisson, Chasles, Liouville, Hermite, and Poincaré, often in order to embrace a teaching career. All were to various degrees influenced by their passage through the École. However, a rapid comparison with the mathematical communities across the Rhine or the Channel reveals a typically polytechnician trait of mathematical activity in nineteenth-century France, namely, the concern for applications.

Two preliminary remarks need to be made. First, a persisting interest in applications can hardly be explained by the mere existence of the École polytechnique. As it developed at the start of the nineteenth century, mathematical activity was indeed inscribed in an applied science tradition which, as we said, dominated academic life even prior to the Revolution. The École polytechnique, however, played a decisive role in consolidating this tradition. Second, one can hardly deny that during this period a research activity existed in France that was centred on pure mathematics. A few mathematicians, such as Legendre and Hermite, showed a predilection for questions of this kind, and rare were those who entirely ignored them. For that matter, the development of teaching of elementary mathematics provided an institutional and intellectual basis for this activity. Under Lagrange's influence, the theory of algebraic equations, for instance, became an important part of high school algebra between 1800 and 1830, and this was probably how the young Galois was attracted by the problem of their solvability by radicals. But the practice of mathematics for its own sake—as one would speak of art for its own sake—was not much valued in France, especially in comparison with Germany. Known for their interest in pure mathematics, Liouville and Hermite thus directed their efforts at Dirichlet and Jacobi's Germany after 1840.

'The time for the great applications of the sciences has come,' proclaimed the perpetual secretary of the Academy of Sciences Joseph Fourier while introducing the mathematical work of 1824 in a public séance [Fourier, 1825, p. xxxvi]. Whatever their approach, the historians of mathematics who study this period cannot but agree (see for example [Dhombres J. and N., 1989] and [Grattan-Guinness, 1990]). But, as it is obviously impossible and perhaps vain to review each and every example confirming Fourier's claim, I shall limit myself to a few derived from the first half of the century.

Let me begin with mechanics, which at the École polytechnique and at the *écoles d'application* constituted the main domain of application of the analysis course. Here the impulse mainly came from the study of mechanical artifacts, machines, and structures by *ingénieurs-savants*, such as Navier, Coriolis, Poncelet, and Saint-Venant, who taught what was called 'applied' or 'industrial' mechanics. The strong development of machine theory therefore led to the introduction of the notion of mechanical work, paving the way for the physics of energy and a clearer distinction between kinetic and dynamic properties in mechanics. The study of structures forced them to analyze more realistically the behavior of solid and fluid bodies, taking their properties of elasticity, viscosity, and even plasticity, into account. In order to deal with these questions, *ingénieurs-savants* often mobilized sophisticated mathematical means, though it must be to admitted that the development of applied mechanics did not produce major advances later on.

The situation was very different in mathematical physics, which with celestial mechanics constituted French mathematicians' main field of research in the nineteenth century [Dahan Dalmedico 1992]. Essentially they strove to establish and study the equations governing physical phenomena in fluid mechanics, electrostatics and electrodynamics, heat theory, light theory, etc. Thus did Fourier's analysis emerge from his research on heat diffusion; potential theory from Lagrange's, Laplace's, and Poisson's research on the effect of mass and electric charge distributions; Sturm-Liouville theory from the study of second-order linear partial differential equations; Cauchy's complex analysis (at least in part) from an even more general study of linear partial differential equations. Obviously, the domain was immense, and French mathematicians were not the only ones interested in it.

In his study of French mathematics in the first third of the nineteenth century [1990], the historian I. Grattan-Guinness distinguishes two principal groups: theoreticians concerned with analysis and mathematical physics, including Cauchy and Liouville, and applied mathematicians concerned with mechanics and engineering science, including Navier and Poncelet. Indeed, there was at this time a rather clear disciplinary separation between, on the one hand, mathematical physics and celestial mechanics, which were natural sciences and, on the other, mechanics, which dealt with artefacts. Belonging to academic traditions, and being taught neither at the École polytechnique nor in application schools, the former endeavored to deduce phenomena from the mathematical expression of physical laws. On the contrary, developed by working engineers or professors at the engineering schools, applied mechanics aimed at the invention of methods directly useful in practice. Nevertheless, to systematically oppose applied mechanics on one side, and physics and astronomical mathematics on the other would be misguided. Actually, problems and methods circulated between these different domains. In his work an *ingénieur-savant* such as Navier used the whole analytical arsenal developed by Lagrange, Laplace, and Fourier. Cauchy developed elasticity theory in between the studies of solids and light. Lamé passed without difficulty from problems of applied mechanics to questions of mathematical physics. Rather than a specialization in a domain of application, each mathematician had a dominant, but nonexclusive, orientation, favoring the physical sciences in Poisson's case, or engineering science in Navier's.

In fact, the classification suggested by Grattan-Guinness suffers from its neglect of geometry whose development still constituted one of the major characteristics of early nineteenth-century French mathematics. The renewal of geometry itself found its sources directly in applications. The initial interest came from Monge's teaching of descriptive geometry at the École polytechnique [Taton, 1950]. Thus did geometrical optics, stone cutting, mechanisms, construction, and surveying inspire many geometers, such as Malus, Hachette, Dupin, Olivier, Servois and Brianchon. True, the growth of projective geometry was independent of such concerns, but the subtitle of Poncelet's *Treatise on projective properties of figures: A work useful to those who deal with descriptive geometry and geometric operations in the field*[2] showed clearly that these geometers were interested in applications. Some of them, like Poncelet himself, also dealt with applied mechanics.

While mathematical research was mainly inspired by applications, its orientation remained theoretical. It is commonly acknowledged that the domain of application was as general as the viewpoint was elevated. We have seen that this principle, borrowed from eighteenth-century academic science, underlay the mathematical teaching at the École polytechnique. On this point, Lagrange's influence was decisive. His theory of analytic functions and his analytical mechanics provided universally admired models for a science wholly proceeding from one general idea (the principle of derivative functions in analysis, the principle of virtual velocities in mechanics).

The same spirit led Fourier, against Lagrange's advice, to base his analysis on the simple idea that a function can always be represented by a trigonometric series. So, opposed to Lagrange on other matters, Cauchy himself did not hesitate to link his residue calculus to the differential calculus, because he too always looked for the most general viewpoint; with his complex function theory, he built a framework for the study not only of analytic functions (locally expandable in Taylor series), but also meromorphic series (locally expandable in Laurent series). While the unity of inspiration characterizing the work of Lagrange, Fourier, and Cauchy was lacking from that of Liouville, one recognizes in his work, as shown by J. Lützen [1990], the same concern for the adoption of the highest viewpoint, be it in elliptic function theory, potential theory, or the stability theory of a rotating fluid mass.

The search for the greatest generality was also found in the works of the French geometers belonging to Monge's school [Chemla, 1998]. These mathematicians were looking for transformation principles which allowed them to reduce the study of all cases to a single simple case or a general configuration [Chasles, 1837]. Thus were stated and used the 'principle of figure correlation' by Carnot, the 'principle of projection' by Brianchon and Poncelet, the 'principle of continuity' by Poncelet, the 'principle of duality' by Poncelet and Gergonne and the 'principle of homography' by Chasles. Poncelet's extension of real projective geometry to complex projective geometry provided the framework for dealing with projective properties of conics and quadrics with the greatest generality.

Far from being opposed to applications, this theorizing movement found its source and inspiration in them. A characteristic example of this was provided by mechanics, where Cauchy and Saint-Venant elaborated a theoretical framework for analyzing kinetic and dynamic behaviors in non-rigid bodies which underlay continuum mechanics. To solve this problem, Cauchy generalized notions and methods invented by Euler for the study of perfect fluids. While more abstract and more general than that of flow and pressure, the notions of infinitesimal deformations and stresses provided a more 'realistic' model for the study of (elastic) structures in construction and mechanical engineering, as well as material fluids (viscous fluids) and immaterial ones (the ether) in physics.

The interest for applications, on the one hand, and the search for ever more abstract, ever more general theories, on the other, constituted a dual orientation that, obviously, was not specific to the French mathematicians of the nineteenth century, since it can be found in the work of mathematicians such as Gauss, Riemann, or Cayley. But what undoubtedly characterized mathematical activity in France at this time was an original way of linking the theoretical with the applied, by attributing to applications themselves the virtue of engendering theoretical generalizations. This 'disciplinary matrix' found an institutional realization in the École polytechnique, the breeding ground of French mathematicians for more than sixty years. In this respect, however, the remarkable point is less that one had to be an engineer to become a mathematician than that one needed to be a mathematician (at least to some degree) to become an engineer. From this configuration, emerged this strange conviction, so widespread among *polytechniciens* and echoed in Auguste Comte's philosophy, that mathematics should be the basis for all knowledge and action.

NOTES

1. Monge did not publish his course at the École polytechnique, but it is known that he more or less followed the exposition published in his 1799 *Géométrie descriptive*, which is a reproduction of the lessons he delivered in 1795 at the École Normale of Year III [Belhoste and Taton 1992].
2. *Traité des propriétés projectives des figures: ouvrage utile à ceux qui s'occupent de la géométrie descriptive et d'opérations géométriques sur le terrain.*

REFERENCES

Belhoste, B., [1991] *Augustin-Louis Cauchy : a Biography*, New York, Springer.
[1996] 'Autour d'un mémoire inédit: la contribution d'Hermite au développement de la théorie des fonctions elliptique', *Revue d'histoire des mathématiques*, 2, 1–66
[1997] 'Condorcet, les arts utiles et leur enseignement' in *Condorcet homme des Lumières et de la Révolution*, Fontenay-aux-Roses, ENS Éditions.
Belhoste, B., Dahan Dalmedico, A., Picon, A., (eds.), [1994] *La Formation polytechnicienne, 1794–1994*, Paris, Dunod.
Belhoste, B., Dahan Dalmedico, A., Pestre, D., Picon A., (eds.), [1995] *La France des X. Deux siècles d'histoire*, Paris, Economica.
Belhoste B., Picon, A., Sakarovitch, J., [1990] 'Les Exercices dans les écoles d'ingénieurs sous l'Ancien Régime et la Révolution', *Histoire de l'éducation*, n 46, 53–109.
Belhoste, B., Taton, R., [1992] 'L'invention d'une langue des figures' in Dhombres, J. (ed.), *L'École Normale de l'an III. Leçons de mathématiques*, Paris, Dunod, 269–317.
Blanchard, A., [1979] *Les 'Ingénieurs du Roy' de Louis XIV à Louis XVI. Étude du corps des fortifications*, Montpellier, Université Paul-Valéry.

Bottazzini, U., [1992] 'Geometrical Rigour and "Modern Analysis", an Introduction to Cauchy's *Cours d'analyse*' in A.L. Cauchy, *Cours d'analyse de l'École royale polytechnique, 1ère partie, Analyse algébrique,* (reprint, Bologna, Editrice CLUEB, Instrumenta Rationis, vol. 7, 1990).

Brezinski, C., [1990] 'Charles Hermite, père de l'analyse mathématique moderne', *Cahiers d'histoire et de philosophie des sciences,* 32.

Chasles, M., [1837] *Aperçu historique sur l'origine et le développement des méthodes en géométrie, particulièrement de celles qui se rapportent à la géométrie moderne, suivi d'un mémoire de géométrie sur deux principes généraux de la science: la dualité et l'homographie,* Bruxelles

Chemla, K., [1998] 'Lazare Carnot et la généralité en géométrie. Variations sur le théorème dit de Ménélaüs', *Revue d'histoire des mathématiques,* 4, 163–190

Comte, A., [1830–1842] *Cours de philosophie positive,* 6 tomes, Paris, (reprint Paris, Hermann, 1975).

Crosland, M., [1992] *Science under Control: the French Academy of Sciences, 1795–1914,* Cambridge, Cambridge University Press.

Dahan Dalmedico, A., [1992] *Mathématisations. Augustin-Louis Cauchy et l'École Française,* Paris, A. Blanchard.

Dhombres, J., Dhombres, N., [1989] *Naissance d'un pouvoir. Science et savants en France: 1793– 1824,* Paris, Payot.

Fourcy, A., [1828] *Histoire de l'École polytechnique,* Paris (reprint Paris, Belin, 1987).

Fourier, J., [1827] Rapport lu dans la séance publique de l'Institut du 24 avril 1825, *Mémoires de l'Académie royale des sciences,* t. 7.

Gilain, C., [1989] 'Cauchy et le cours d'analyse à l'École polytechnique', Bulletin de la Société des Amis de la Bibliothèque de l'École polytechnique, t. 5.

Gillispie, C. C., [1980] *Science and Polity at the End of the Old Regime,* Princeton, Princeton University Press.

Gispert, H., [1991] 'La France mathématique. La Société Mathématique de France (1870–1914)', *Cahiers d'histoire et de philosophie des sciences,* 34.

[1995] 'De Bertrand à Hadamard, quel enseignement d'analyse pour les polytechniciens' in Belhoste, B., Dahan Dalmedico, A., Pestre, D., Picon A., (eds.), [1995], 181–196.

Grattan-Guinness, I., [1990] *Convolutions in French Mathematics, 1800–1840,* 3 vols, Basel-Boston-Berlin, Birkhäuser.

Hahn, R., [1971] *The Anatomy of a Scientific Institution: the Paris Academy,* Berkeley, London, University of California Press.

Lützen, J., [1990] *Joseph Liouville 1809–1882. Master of Pure and Applied Mathematics,* New York, Springer.

Pepe, L., [1986] 'Tre "prime edizioni" ed un'introduzione inedita della *Théorie des fonctions analytiques* di Lagrange', *Bollettino di Storia delle Scienze Matematiche,* 6, 17–44.

Pickering M., [1993] *Auguste Comte: an Intellectual Biography,* vol. 1, Cambridge, Cambridge University Press.

Sakarovitch, J., [1998] *Épures d'architecture. De la coupe des pierres à la géométrie descriptive, XVIe-XIXe siècles,* Basel-Boston-Berlin, Birkhäuser.

Tannery, J., [1895] 'L'enseignement des mathématiques à l'École' in *Le Centenaire de l'École Normale,* Paris, 387– 394 (reprint Paris, Presse de l'École Normale Supérieure, 1994)

Taton, R., [1950] *L'Œuvre scientifique de Monge,* Paris, Presses Universitaires de France.

Chapter 2

FROM PARIS TO BERLIN: CONTRASTED IMAGES OF NINETEENTH-CENTURY MATHEMATICS

Umberto Bottazzini

PRELUDE

Towards the end of 1864 the young Italian mathematician Ulisse Dini, a graduate from the Scuola Normale Superiore in Pisa, went to Paris in order to complete his studies, following the advice of his supervisor Eugenio Beltrami. Introduced to outstanding Parisian mathematicians through Scuola Director Enrico Betti, Dini began to attend their lecture courses. Betti had been known to his French colleagues since the fall of 1858 when, together with Francesco Brioschi and Felice Casorati, he undertook a journey through Germany and France in order to get in touch with leading European mathematicians. In Göttingen, they had met with Dirichlet, Dedekind, and Riemann, in Berlin with Kummer, Kronecker, and Weierstrass, and in Paris with Hermite and Bertrand.

A few months after his arrival in Paris, Dini informed Betti that the only courses new to him were those of Chasles and Bertrand. He had much appreciated the former, the latter being a course on elliptic functions—in Dini's words 'a material calculation which is rather tiring' [Bottazzini, 1994, p. 140]. To his disappointment, Dini realized that Italian mathematical journals were completely lacking from French libraries. Even series of French journals were 'incomplete,' and the library of the Scuola in Pisa apparently held more of them. Nonetheless, during his stay in Paris he was able to produce some ten papers on differential geometry inspired by the works of Bonnet, Serret, and Beltrami. But, as he did not hesitate to write to Betti, he gave up following lectures at the university because working at home was much more productive. After a short visit to London and Berlin, Dini went back to Pisa at the end of the summer. He would be the last Italian graduate student to spend a period of research in Paris for some considerable time.

Since the early 19th century, Paris—and the École polytechnique in particular—had attracted Italian students. However, Dini's letters provide factual evidence that in the 1860s the French capital was not the leading center of mathematical research any longer. German universities were acquiring a dominant position on the international scene. Needless to say, at this time Germany was not a scientific and political monolith centered around its capital as France was. And Germany was still waiting for its political unification. The 'Deutsche Bund' included some 39 states, and Berlin was merely the main town of Prussia, other centers being Königsberg, Göttingen, Leipzig, and Munich, to mention only a few.

From the 1870s onward, German universities—rather than the French Grandes Écoles—became the privileged destination for gifted Italian students who wanted to get acquainted with the more recent advances in mathematics. For instance, Luigi Bianchi and Gregorio Ricci-Curbastro spent some time attending Klein's lectures in Munich, while Giuseppe Veronese went to Leipzig, and Salvatore Pincherle attended Weierstrass's and Kronecker's lectures in Berlin.

The emergence of Berlin as the center of the mathematical world in the second half of nineteenth century was the result of a double, almost contemporary move: first, the transfer of the leadership in mathematics from France to Germany, and secondly the dominant position achieved by Berlin with respect to other German universities, in particular Königsberg and Göttingen. Consequently, a new image of mathematics prevailed for some thirty years dominated in the main by the theory of functions established according to Weierstrass's principles. Indeed, after Steiner's death in 1863, geometry virtually disappeared from the agenda of Berlin mathematicians, and the search for arithmetic rigor in analysis became their major concern. Combined with their lack of interest in applied mathematics, the Berlin mathematicians heralded an image of 'pure unapplied mathematics', borrowing this expression from a joke popular at that time.[1] Although, thanks to both Felix Klein's political and academic abilities and Weierstrass's retirement, the 1890s saw the balance swing back in favor of Göttingen, the Weierstrassian conception of mathematics—the 'arithmetization' of mathematics, as Klein labeled it—exerted a lasting influence well into the 20th century.

IMAGES OF FRENCH MATHEMATICS: CAUCHY AND HIS HERITAGE

In the first half of the nineteenth century, French mathematics was dominated largely by the figure of Augustin-Louis Cauchy. Exceptional as he was as a mathematician, Cauchy was a typical *ingénieur-savant* (see [Belhoste 1984] and his article in this volume). Trained at the École polytechnique, he served as an ingénieur for some years before devoting himself completely to an academic career. After the fall of Napoleon and the subsequent Restoration of the Bourbons, he was named a member of the Paris Academy of Sciences in 1816, and in the following year began teaching analysis as a professor at the École polytechnique.

From the very beginning of Cauchy's scientific activity, complex analysis—in his words the 'passage from the real to the imaginary'—was one of his favorite research topics, and the need for 'geometric rigor' in dealing with imaginary quantities one of his major concerns. In a number of papers and books, in particular his *Cours d'analyse* (1821), Cauchy presented a new image of analysis—'modern analysis' as he himself did not hesitate to call it. His *Cours* was devoted to algebraic analysis that was designed as an introduction to infinitesimal calculus proper. Although calculus constituted the core of the course Cauchy taught first-year students at the École polytechnique, the concepts introduced in the *Cours* supplied, in his opinion, the whole of analysis built on rigorous foundations. There, he defined such basic notions as infinitesimals, limits, continuity of functions, convergent series, and related criteria for both real and 'imaginary' quantities and expressions. The real and 'imaginary' parts of the *Cours* culminated with, respectively, the binomial theorem and the fundamental theorem of algebra.

In contrast to an alledgedly loose treatment of foundations by eighteenth-century mathematicians, Cauchy's *Cours d'analyse* is commonly held as the source of a modern concept of rigor in mathematical analysis. This image was first nourished by Cauchy himself, who claimed that, instead of resorting to the 'generality of algebra', 'geometric rigor' was needed in analysis. This criticism was especially directed against the work of Joseph-Louis Lagrange. According to Lagrange (1799, p. 328) 'à proprement parler, l'algèbre n'est en général que la théorie des fonctions. [...] Dans l'algèbre proprement dite, on ne considère que les fonctions primitives qui résultent des opérations algébriques ordinaires; c'est la première branche de la théorie des fonctions. Dans la seconde branche on considère les fonctions derivées, et c'est cette seconde branche que nous désignons simplement par le nom de théorie des fonctions analytiques, et qui comprend tout ce qui a rapport aux nouveaux calculs'.[2] Indeed, assuming that any function can be expanded in a power series, Lagrange was able to define the derivative of a function by means of 'purely algebraic operations'.

In 1821 Cauchy was pleased to find, in non-analytic functions such as $\exp(-1/x^2)$, counterexamples to Lagrange's assumption. Rejecting

Lagrange's approach, he stated in a celebrated passage from the introduction to his *Cours*: 'Quant aux méthodes, j'ai cherché à leur donner toute la rigueur qu'on exige en géométrie, de manière à ne jamais recourir aux raisons tirées de la généralité de l'algèbre.'[3] This approach was applied in particular to 'the passage from convergent to divergent series and from real quantities to imaginary expressions' [Cauchy, 1821, p. ii].

By emphasizing the need for rigor in analysis, Cauchy met opposition not only from students at the École polytechnique, but also the Council of Instruction, its institutional authority. One might even wonder whether Cauchy himself ever used the *Cours* at all as a textbook for his students. Indeed, as Gilain [1989, p. 12] convincingly argued, it 'already seemed obsolete as a textbook for the students of the École polytechnique before it even appeared', due to changes in official programs. During the 1820s Cauchy himself increasingly reduced the weight of algebraic analysis in his course. He was officially invited by the Council of Instruction of the École to conform to the program, and not waste the time scheduled for teaching the application of calculus (of the highest importance for future engineers) by dealing with abstract and abstruse questions of rigor, 'a luxury of analysis which was undoubtedly suitable for papers to be read at the Institute, but superfluous for the instruction of the students of the École' [Belhoste, 1984, p. 36]. As a result, he announced that he was giving up 'completely rigorous proofs' in his courses.

Thus, more than a textbook, Cauchy's *Cours d'analyse* apparently served as the *manifesto* for a new attitude in research. For some thirty years Cauchy himself hardly missed an opportunity to refer to it in his papers, thus turning his *Cours* into the *manifesto* for a new image of rigor in mathematics. This attitude was first shared by Niels Abel, who in his 1826 paper on the binomial theorem recommended that Cauchy's *Cours* should be studied by 'everyone who loves rigor in mathematical analysis' and who in private letters lamented 'the prodigious obscurity that one incontestably finds today in analysis' [Bottazzini, 1986, p. 86].

Eventually, the image of the alleged looseness of eighteenth-century mathematics in comparison with Cauchy's 'modern' analysis has generally been accepted by mathematicians and historians of mathematics. One century after Abel's statement, Klein wrote in his *Lectures on mathematics in the 19th Century* [1926, I, p. 84] : 'in all critical points [Cauchy's *Cours*] provides an arithmetical foundation, free of objections; starting from this fundamental work the so-called 'arithmetization' of the whole of mathematics begins'.

Although questioned by authoritative mathematicians like Caratheodory and Hardy (see Bottazzini 1992, pp. xiv–xv), this deeply rooted image remained almost unquestioned until recent times. In a very influential book, the *Éléments d'histoire des mathématiques*, Bourbaki [1960, p. 224] for instance claimed that 'weary of this [earlier] unbridled formalism devoid of foundations, mathematicians at the beginning of the nineteenth century put analysis back on the tracks of rigor'.

In the same vein Morris Kline [1972, p. 947] wrote : 'By about 1800 the mathematicians began to be concerned about the looseness in the concepts and proofs of the vast branches of analysis'. Accordingly, 'several mathematicians resolved to bring order out of chaos'.

A more accurate historical analysis however shows that the plot is far more intricate than this view seems to suggest [Bottazzini, 1992]. Mathematical rigor is itself an historical concept and therefore it changes. Consequently, from the eighteenth century to Cauchy's work and, beyond, up to Weierstrass's 'arithmetical' rigor, standards of rigor have evolved. In this connection, it is interesting to note that Bourbaki's rather crude statement was later revised by Jean Dieudonné [1986, p. 21], one of the leading figures of the Bourbaki group: 'In the nineteenth century, [eighteenth-century analysts] were condemned too hurriedly, their faulty language being solely taken into account while the context was not closely examined'.

Cauchy taught at the École polytechnique until 1830, when, because of the July Revolution which overthrew his beloved Bourbons, he left Paris for Italy in self-exile. He returned to Paris in 1838, but was unable to get his position back at the École. He kept his seat at the Academy and filled up the *Comptes Rendus* with a stream of notes on various subjects, and in particular complex analysis. He clarified, refined, and extended theorems and results he had obtained earlier, such as the integral theorem and the integral formula both bearing his name, the theory of residues, and the 'calculus of limits' (or, in modern terms, the method of majorants for expanding a complex function in power series). Combined with contemporary studies by Liouville, Laurent, Puiseux, and Hermite, Cauchy's work contributed to the establishment of a 'French school' of complex analysis.

However, Cauchy did not publish a systematic treatise on the theory of complex functions. The first exposition of complex analysis according to his methods—which was to become the standard reference in the French school of complex analysis—was provided by the first part of a volume published in 1859 by Briot and Bouquet. Translated into German in 1862, their treatise also included Liouville's 'beautiful theory' of doubly periodic functions and was enriched by 'Mr. Hermite's wonderful works on the same topic'.

Despite being tremendously influential for French mathematics, Cauchy's scientific heritage prevented the French mathematicians from appreciating new developments taking place abroad. In addition, for most of them, including Briot and Bouquet, Liouville, Bertrand, and Chasles, their more productive and original period was over, the only remarkable exception being Hermite, teacher at both the Sorbonne and the École Normale. After Cauchy's death in 1857, 'the French milieu worked in autarchy', as Hélène Gispert [1996, p. 401] has aptly remarked.

In the 1860s, the temporary decline of French mathematics was reflected in that of the École polytechnique. There is no doubt that in the first half of the nineteenth century this École played a decisive role in training French mathematicians. But according to Hélène Gispert, the glorious heritage of the École polytechnique turned out, in the 1860s, to be a handicap for the development of new mathematics. As she writes [1996, p. 400], 'excluded for some thirty years from research in geometry as well as in analysis, [the École polytechnique] perpetuated an ossified heritage, centered around a course on descriptive geometry and stereotomy, as well as a course of analysis notable for a plethora of developments in the infinitesimal elementary geometry of curves and surfaces'.

In 1870 Chasles openly recognized that new branches of mathematics like invariant theory and the theory of transcendent functions (notably elliptic functions and Abelian integrals) were flourishing outside France, in Germany and England and also, for a few years already, in Italy. This very same year a highly interesting report on the status of French mathematics was jointly written by two young German mathematicians, Felix Klein and Sophus Lie, during their stay in Paris. They addressed it to the academic mathematical union (*Verein*) of Berlin University. 'It is not only our personal view', they wrote, 'but also a widespread belief in all mathematical circles here that mathematical studies in France nowadays are by no means at the same level as they were about fifty years ago'.

In Klein and Lie's opinion the essential reason for the decline of French mathematics lay in the awareness of its own remarkable success, which in the course of time led to self-sufficiency combined with isolation and closure with respect to the new developments of mathematics occurring in other countries. Klein and Lie were pleased to add that in recent times this view was more and more widely shared even in France, to the point that the French minister of education 'had been sent to Germany' in order to get acquainted with the system of education in German universities.

In Klein and Lie's report there is no further hint to the political climate of the time. However, one has to keep in mind that they were visiting Paris in the spring of 1870, just a few weeks before the Franco-Prussian war was declared. According to Hélène Gispert [1996, p. 403] the 1860s 'were the gloomy years of the end of the Second Empire, synonymous with political and ideological authoritarianism, with censorship in political life and the university. The place of mathematics was then ambiguous. The leading science at the École polytechnique, mathematics fell victim to its enviable position in the French education system, which was divorced from research, and which promoted above all the "useful", 'practical', and 'concrete' side in the training of future engineers, and so distrusted theoretical and abstract developments'.

Just the opposite was going on across the Rhine.

INTERLUDE

By taking a contemporary Italian mathematicians' perspective, one can perhaps better appreciate the changes in the picture of mathematics which took place around the middle of the nineteenth century. By the end of the 1850s Italy was experiencing a process of independence and political unification, the Risorgimento, culminating in 1861 with the establishment of the Italian Kingdom. Most young Italian mathematicians belonging to the Risorgimento generation, like Beltrami, Betti, Brioschi, Casorati, Cremona, and Genocchi, conceived of mathematical research as intertwined with political activity. They took part in the wars of independence and, subsequently, in the political life of the new State. At the same time, they were deeply concerned with the simultaneous process of the rebirth of Italian mathematics.

For this purpose, they looked at the models provided by France and Germany, by far the most mathematically developed countries of the time. In 1858, at the initiative of Brioschi, Betti, and Genocchi, a new journal of mathematics was founded, the *Annali di matematica pura e applicata*, whose name was even modeled after Crelle's *Journal* in Germany and Liouville's in France. In the fall of that year, Betti,

Brioschi, and Casorati undertook a journey through Europe which marked a turning point in the history of Italian mathematics, and which marked its entry on the modern international stage.

Upon going back to Italy, Betti drew his Italian colleagues' attention to Riemann's work by translating and publishing in the *Annali* one of his most seminal papers, his *Inauguraldissertation* on the foundations of a general theory of complex functions, with which Riemann graduated from the University of Göttingen in 1851. As Ahlfors [1953, p. 3] has stated, 'very few mathematical papers have exercised an influence on the later development of mathematics which is comparable to the stimulus provided by Riemann's dissertation. It contains the germ of a major part of the modern theory of analytic functions, it initiated the systematic study of topology, it revolutionized algebraic geometry, and it paved the way for Riemann's own approach to differential geometry.'

Following a geometric approach, the German mathematician introduced the concept of Riemann surfaces, and by means of the Dirichlet principle proved both the existence of a complex function with prescribed singularities and the mapping theorem named after him.[4] However impressive his results were, Riemann's ideas and methods in complex function theory were not widely accepted until much later. As Betti said in his inaugural lecture for the academic year 1860–61, they were 'almost entirely a magnificent work of pure thought. But so great is the force of the mind, so great is the concision and obscurity of this eminent geometer's style, that at the moment it is as if his work did not exist in the scientific world' [Bottazzini 1986, p. 280].

As for Brioschi, by this time he was mainly interested in the solution of the quintic equation by means of elliptic functions, and he corresponded with Kronecker and Hermite on this matter. Together with invariant and covariant theory, elliptic functions and their applications were among the favorite topics covered by the papers published in the *Annali*, which in a very short time became a recognized internationally journal.

In order to improve the level of mathematical teaching, new chairs were created, for instance, Cremona's and Battaglini's chairs of higher geometry at Bologna and Naples. They taught modern projective geometry as developed by Chasles and Steiner, a topic which was a 'completely new guest' in Italian universities, as Cremona recognized in the opening lecture of his course. In the 1860s the Italian mathematicians were also involved in the reform of national education, including high schools and universities. In November 1862, while Brioschi was serving as the secretary of the Minister of Public Education, the Pisan scientist Carlo Matteucci, a decree established a new polytechnic school—the Higher Technical Institute (*Istituto Tecnico Superiore*)—in Milan. Its intended aim was the instruction of both civil and military engineers as well as secondary technical school teachers. Brioschi was appointed as its Director. His speech at the opening ceremony in November 1863 [Brioschi, 1863] is interesting in many respects. Published in the official journal of the Ministry of Public Education, it significantly received a German translation that appeared the following year in Grunert's *Archiv*.

According to Brioschi, 'the civil history of nations' showed that 'educational institutions have no chance of fulfilling their high mission if their creation and organization do not correspond to both the new needs of science and new social conditions'. Therefore, the newly founded Technical Institute was intended to fulfill the 'intellectual and material need of our country', and its organization was the result of a precise comparison between analogous institutions that shared the same goals 'in the most civil nations'. The comparison with foreign nations and institutions was the guideline of Brioschi's talk. As for the training of secondary school teachers, he referred to the 'enlightening examples' provided by the École Normale in Paris and, above all, by the German seminars with respect to classical and historical studies.

'The greatest political revolutions', Brioschi went on, had always been accompanied by 'either the creation of new institutions or deep modifications in the organization of existing ones'. By comparing the current conditions in Italian politics, economy, and administration with the ones existing in 1859 at the beginning of the war of independence, everybody had to admit that in Italy 'a great political, economical, and administrational revolution has taken place'. That granted, Brioschi asked, could we say that a revolution had also taken place in Italian public education? 'Have we actualized any of those great concepts that accompanied great revolutions abroad, like the École polytechnique, the École Normale, and the National Institute in France, and which spurred the scientific movement of the German universities?'

In Brioschi's opinion, political unification and the creation of a new state represented a 'revolution' for Italy, which should find its expression even in the educational system 'which reflects the culture of a nation'. This was one of Italy's most 'urgent needs' and motivated the creation of the Technical Institute. However, it was modeled after the German *Technische Hochschulen* more than after the French École polytechnique.

Italian mathematicians looked to Germany rather than France not only because of German mathematicians' leading position in Europe, but also because of the similar experience of political unification that Germany was at that time undergoing under Prussian leadership. It is worth recalling that a political and military alliance linked Italy to Prussia by the mid-1860s, which was decisive for the victory in the final Italian independence wars in 1866.

Some of the leading German mathematicians had the same feeling toward Italy as Italians had toward Germany. Thus for instance Weierstrass wrote to Casorati in 1867: 'The happy development of science in your country can nowhere be followed with more interest than in Northern Germany. You also should be convinced that nowhere has Italy more sincere and disinterested friends. Happily, we are ready to extend to science the alliance between you and us, which has been so successful in politics, in order that even in this field more and more barriers, which an unfortunate politics has erected for such a long time between two peoples in many ways spiritually related, can fall down' [Neuenschwander, 1978, p. 72].

Referring to a paper on Riemannian complex function theory that Casorati had sent him, Weierstrass took the occasion of comparing the development of mathematics in different European countries. 'Our scientific efforts', he wrote, 'are better understood and appreciated in Italy than in France and England, especially in the latter country where a stifling formalism threatens completely to choke off the feeling for deeper researches. How significant indeed it is, that our Riemann, whose loss we cannot mourn enough, is studied and appreciated, apart from Germany, only in Italy. In France he seemingly is well recognized but poorly understood, and in England he remains almost unknown'.

In 1863 Riemann settled in Pisa for a couple of years due to his illness. Betti became his 'dearest friend' and indeed, after Riemann's untimely death in 1866, the main reference for all those interested in his last, unpublished work. Riemann's stay in Pisa had a tremendous influence not only on Betti's own work but also on the whole of Italian mathematics. Casorati himself was one of the most convinced followers of Riemann's methods in complex analysis. In 1867, at the Technical Institute in Milan, he presented them in a celebrated course taught jointly with Brioschi and Cremona, the former lecturing on Jacobi's theory of elliptic functions and the latter on the geometric work of Clebsch and Gordan which had been inspired by Riemann's theory of Abelian integrals. By then Casorati was working on a complex analysis textbook [Casorati, 1868] which Klein [1926, I, p. 274] later ranked first among those inspired by Riemann's ideas. In preparing his book in the fall of 1864, Casorati traveled to Berlin and discussed the most recent advances in analysis with Kronecker, Weierstrass, and their pupils.

THE EMERGENCE OF THE BERLIN SCHOOL

In 1857, when Cauchy, the leading figure of French mathematics died in Paris, the forty-year old Karl Weierstrass had just begun his teaching in Berlin. One might assume that this date marked a turning point in the history of nineteenth-century mathematical analysis, and in fact the symbolic transfer of leadership in mathematics from the French to the Prussian capital. Weierstrass in particular resumed and refined the rigorization program in mathematics that had been launched by Cauchy in the 1820s, and he made a point of establishing the whole of analysis on rigorous, arithmetical foundations.

Perhaps because of his very different training, Weierstrass did not share with Cauchy his broad interests in applied mathematics. Contrary to the French ingénieur, Weierstrass was virtually a self-taught mathematician: his mathematical education was limited to lectures on elliptic function theory he attended for only one semester in 1839 at the Münster Academy where he enrolled in order to get a teacher's diploma. He then taught in secondary schools for some fifteen years. Working in almost complete isolation, Weierstrass reached a number of results which for the most remained unpublished until 1895 when he himself edited the first volume of his collected works.

Weierstrass's mathematical interest was completely dominated by the theory of elliptic and Abelian integrals (and in particular by Jacobi's inversion problem), the topic of a great paper published in Crelle's *Journal* in 1854 which called the mathematicians' attention to this obscure teacher. In 1856, he was hired to teach in Berlin where he joined Kronecker and Kummer in building the celebrated 'trio' which for some twenty years dominated Berlin mathematics. Kronecker had no professorship until Kummer's retirement in the early 1880s, but as a member of the local Academy, he had the right to teach free courses at the University. In 1860, Kummer and Weierstrass founded the Mathematical Seminar, which played a decisive role in creating the

Berlin 'school'. At these meetings, professors lectured on various topics and graduate students presented their own research.

It is worth noticing that at the very beginning of his teaching career at Berlin University Weierstrass had no interest in rigor and foundations. Selected chapters of mathematical physics constituted the topic of his very first course in the winter semester of 1856/57. This was followed by a course on the representation of analytic functions by series, where he expounded his own results, and by a course on elliptic functions. The applications of elliptic functions to geometry and mechanics was the subject of his lectures in the winter semester of 1857/58.

As he stated in his inaugural address in front of the Berlin Academy in 1857 [Weierstrass, *Werke* I, pp. 223–224], elliptic function theory exerted a 'powerful attraction' on him, which had a 'decisive influence' on his mathematical research. In this speech he forcefully emphasized the need for a deep understanding of the link between mathematics and the natural sciences—a matter which 'lay very close to my heart' (*mir allerdings sehr am Herzen liegt*), he said—but he completely refrained to utter a single word about rigor.

Apparently, Weierstrass began to feel the need for greater rigor in analysis by lecturing on analytic function theory. In his view, this provided a foundation for the whole theory of elliptic and Abelian functions. In the winter semester of 1863/64, he gave his first course on this topic. He worked on this theory for some twenty years through continuous refinements and improvements. However, instead of publishing his course, he preferred either to present his discoveries in his lectures, or occasionally to communicate some of his results to the Berlin Academy.

Rumors about new discoveries made by Weierstrass, combined with lack of publication, motivated Casorati's journey to Berlin in 1864. The notes he made after his talks with Berlin mathematicians provide a first-hand account of problems at the forefront of mathematical research there [Bottazzini, 1986, pp. 257-264]. In addition to Kronecker and Weierstrass, some of Riemann's students also attended the meetings.

Riemann's work was continually referred to in Casorati's discussions with his German colleagues. 'Riemann's things are creating difficulties in Berlin', Casorati reported, adding that Weierstrass claimed that 'he understood Riemann because he already possessed the results of his research'. A major difference between Riemann's and Weierstrass's respective points of view emerged in the approach to complex function theory. Weierstrass and Kronecker criticized Riemann's use of the Dirichlet principle and instead emphasized the basic role of the method of power series and analytic continuation, which, according to them, Riemann treated nowhere with the necessary rigor. In addition, Riemann seemed to share the idea that a function could be continued in any part of the plane along a path that avoided critical points (branch-points and singularities). 'But this is not possible', Weierstrass stated, 'and it was precisely while searching for the demonstration of the general possibility that he realized that it was in general impossible'.

In this connection Kronecker suggested the example of a function defined by the series $\theta(q) = 1 + 2 \sum_{n \geq 1} q^{n^2}$. This turned out to be the first example of a lacunary series. It has the unitary circumference as its natural boundary and cannot be analytically continued outside the convergence disk. This series plays a role in the example of a continuous, nowhere differentiable, function which Weierstrass later communicated to the Berlin Academy. This might explain Kronecker's claim, reported by Casorati's, to the effect that he 'knows functions that do not admit differential coefficients, that cannot represent lines etc.'

On the same occasion, Kronecker added that he had learned 'to be more exact by cultivating algebra and the theory of numbers' while 'in the use of infinitesimal analysis he had not found the occasion of acquiring this exactitude' [Bottazzini, 1986, p. 259]. Kronecker's contribution to the research on the foundations of mathematics has generally been underestimated by historians. In the author's opinion, however, his algebraic way of thinking exerted an influence on Weierstrass's work that is likely to have been greater than is commonly acknowledged. Kronecker moreover played a role in interesting Weierstrass to search for an algebraic foundation of analysis. In this regard, the 'confession of faith' Weierstrass made in 1875 in a letter to his former student Hermann Amandus Schwarz is particularly interesting. 'The more I ponder the principles of function theory—and I do so unceasingly', Weierstrass wrote, 'the firmer becomes my conviction that they have to be built upon a foundation of algebraic truths. It is therefore not correct to turn around and, expressing myself briefly, use 'transcendental' notions as the basis of simple and fundamental algebraic propositions—however brilliant may appear for example the consideration by which Riemann discovered so many of the most important properties of algebraic functions (That every path should be allowed to the researcher in the course of his investigations goes without saying; what is at issue here is merely the question of a systematic theoretical foundation)' [Weierstrass, Werke II, p. 235].

THE BERLIN SCHOOL: IMAGE AND NORM IN MATHEMATICS

Among Weierstrass's students, Schwarz was perhaps the closest to his master's ideas [Biermann, 1972]. At his suggestion Schwarz made a point of establishing some of Riemann's results in a rigorous way according to Weierstrass's standards. In particular, he devoted a number of papers—including a short Italian note in the *Annali* (Schwarz 1869)—to Riemann's mapping theorem, which he proved for particular domains without resorting to the questionable use of the Dirichlet principle. Instead of this, he proposed his own method of *'alternierendes Verfahren'* [Schwarz, 1870]. This paper motivated Weierstrass to communicate his celebrated counterexample to the Dirichlet principle to the Berlin Academy [Weierstrass, 1870].

The (still unpublished) correspondence between Schwarz and Weierstrass represents a unique source for the history of the Berlin school of mathematics. In their letters, which span over thirty years, they discussed not only mathematical, but also institutional, matters. They exchanged critical remarks about current mathematical publications as they appeared. For instance, in a letter dated July 1872, Schwarz [Bottazzini, 1992a, p. 78] advised Weierstrass that he should display his counterexample of a continuous, nowhere differentiable, function in front of the Berlin Academy [Weierstrass, 1872]. They also discussed academic policies in German universities, and Weierstrass did not hesitate to criticize Klein's replacement by Lie in Leipzig in 1883 when the latter left for Göttingen. Klein himself met Weierstrass's sharp criticisms for his style in mathematics, which appeared very imprecise with respect to Berlin standards of rigor.

In addition, Schwarz played a decisive role in spreading the image of Berlin mathematics abroad. In this regard, his correspondence with Italian mathematicians—Casorati and Dini in particular—is enlightening. Schwarz's letters to Casorati mainly concerned the search for a rigorous proof for the Dirichlet principle, which Casorati was looking for in vain. In his letters to Dini, Schwarz detailed the main points of Weiestrass's methods, including his rigorous 'method of proof' in analysis which, according to Schwarz, was based on the (Bolzano)-Weierstrass theorem and its extension to several variables [Bottazzini, 1992a, p. 79].

Dini was by then in charge of teaching analysis at the university of Pisa. His 1871/72 lectures on the foundations of analysis according to Weierstrass's principles were at the basis of his treatise titled *Fondamenti per la teorica delle funzioni di variabili reali* [Dini, 1878], one of the most rigorous and influential books in its day, and one which greatly contributed to the diffusion of the new image of rigorous mathematical analysis heralded by the Berlin school and its leader.

In the 1870s the fame of Weiertrass's lectures spread all over the mathematical world, and Berlin became the destination of an increasing number of gifted young students willing to attend the lectures of the 'great legislator of analysis', as Hermite once called him. Regularly, that is every four consecutive semesters, Weiertrass lectured on the introduction of analytic function theory. This was followed by lectures on elliptic functions, Abelian integrals, and the calculus of variations. Because of his notorious refusal to publish his courses, the content of his lectures was only known through the notes taken by his students.

Among these lecture notes, there was the *Saggio* published in 1880 by Salvatore Pincherle, a student of Dini's and Casorati's, who attended Weierstrass's and Kronecker's lectures in the academic year 1877–78. In introducing analytic function theory, Weierstrass rejected both Cauchy's and Riemann's approaches based on 'transcendental' methods. In his lectures there was no place for the Cauchy-Riemann equations or the Cauchy integral theorem and formula. Instead, he stated and proved theorems on power series, including uniform convergence and term-by-term differentiation, and eventually introduced the method of analytic continuation by means of which, from an element of an analytic function (i.e. a power series), he obtained the analytic function in its totality. In short, as has been rightly observed, 'for the Weierstrass school, to define a function is essentially to give a Taylor series, since from this series one can, by the method of analytic continuation, theoretically deduce the value of the function at every point where it is defined' [Boutroux, 1908, p. 2].

Towards the end of the century Weierstrass's arithmetical approach became dominant, and the German expression *Funktionenlehre* was almost synonymous with analytic function theory *according to Weierstrass's principles*. This 'arithmetization of mathematics', as Klein called it in a talk delivered in 1895 on the occasion of Weierstrass's eightieth birthday, together with Kronecker's extreme attitude 'to ban irrational numbers and reduce mathematical knowledge to relations between *whole* numbers alone', best summarized the image of mathematics dominating in Berlin. In contrast to this view, however, Klein went on, 'I do not see the arithmetical form of the evolution of thought as the essence of the discipline.' Consequently, 'I have to maintain and firmly stress that mathematics will never be completed by logical deduction, and that, in relation to this, *intuition* also retains its full specific importance' [Bottazzini, 1986, p. 290].

NOTES

1. In those days, indeed, Crelle's celebrated *Journal für die reine und angewandte Mathematik* (Journal for pure and applied mathematics) was often ironically called 'Journal for pure unapplied (*unangewandte*) mathematics'.
2. 'Properly speaking, algebra is in general nothing but the theory of functions ... In algebra per se, only primitive functions obtained from ordinary algebraic operations are considered; this is the first branch of function theory. In the second branch, derived functions are considered, and this second branch is what we simply call the theory of analytic functions, which includes all that regards the new calculus'.
3. 'As for the methods, I strove to endow them with all the rigor required in geometry, in such a way as never to rely on reasons derived from the generality of algebra'.
4. The Dirichlet Principle states the existence of a minimum for a certain, always positive integral. According to Dirichlet, it was 'evident' that the integral had a minimum 'since it cannot become negative'. Dirichlet's argument was the object of several criticisms until Hilbert 'resurrected it' in 1900 [see Bottazzini 1986, p. 295–303]

REFERENCES

Ahlfors, L.V., [1953] 'Development of the theory of conformal mapping and Riemann surfaces through a century', *Annals of Mathematical Studies*, 30, 3–13.

Belhoste, B., [1984] *Cauchy 1789–1857. Un mathématicien legitimiste au XIXeme siecle*, Paris (Engl. transl., New York, Springer, 1991).

[1994] 'Un modèle à l'épreuve. L'École Polytechnique de 1794 au Second Empire', in Belhoste, B., Dahan Dalmedico, A., Picon, A., (eds.) *La formation polytechnicienne, 1794–1994*, Paris, Dunod, 9–30.

Biermann, K.R., [1973] *Die Mathematik und ihre Dozenten an der Berliner Universität, 1810–1920*, Berlin, Akademie Verlag.

Bottazzini, U., [1986] *The Higher Calculus. A History of Mathematical Analysis from Euler to Weierstrass*, New York, Springer.

[1992] 'Geometrical Rigour and 'Modern Analysis'. An Introduction to Cauchy's Cours d'Analyse', in A.L. Cauchy, *Cours d'Analyse algebrique*, Paris, 1921, reprint Bologna, CLUEB, xi–clxvii.

[1992a] 'The influence of Weierstrass's analytical methods' in Italy, in: S. Demidov, M. Folkerts, D. Rowe, C. Scriba (eds.), *Amphora. Festschrift für H.Wussing* , Basel, Birkhäuser, 67–90.

[1994] *Va' pensiero. Immagini della matematica nell'Italia dell'Ottocento*, Bologna, Il Mulino.

Bourbaki, N., [1960] *Eléments d'histoire des mathématiques*, Paris, Hermann.

Boutroux, P., [1908] *Leçons sur les fonctions définies par les équations différentielles du premier ordre*, Paris, Gauthier-Villars.

Brioschi, F., [1863] 'Accademia scientifico-letteraria e l'Istituto Tecnico Superiore', *Rivista Italiana di scienze lettere ed arti colle Effemeridi della Pubblica Istruzione*, anno IV, n° 169, 13 Dicembre 1863 (Germ. Transl. : Rede gehalten bei der feierlichen Eroeffnung der Accademia scientifico-letteraria und des Istituto tecnico superiore zu Mailand, *Archiv der Math. u. Physik*, 47 (1864), 42–54).

Casorati, F., [1868] *Teorica delle funzioni di variabili complesse*, Pavia, Fusi editore.

Cauchy, A.L., [1821] *Cours d'Analyse algebrique*, Paris, De Bure (= *Œuvres* (2), 3) (reprint Bologna, CLUEB, 1992).

Dieudonné, J., (ed.) [1986] *Abregé d'histoire des mathématiques*, Paris, Hermann.

Dini, U., [1878] *Fondameuti per la teorica delle funzioni di variabili reali*, Pisa, Nistri & C. (reprint U.H.I. 1990).

Gilain, C., [1989] 'Cauchy et le cours d'analyse à l'École Polytechnique', *Bulletin de la societé des amis de l'Ecole Polytechnique*, n° 5.

Gispert, H., [1996] 'Une comparaison des journaux français et italiens dans les années 1860–1975', in Goldstein, C., Gray, J., Ritter, J., (eds.), *L'Europe mathématique/Mathematical Europe*, Paris, Éditions de la Maison des Sciences de l'Homme, 391–408.

Klein, F., [1926] *Vorlesungen ueber die Entwicklung der Mathematik im 19. Jahrhundert*, Courant, R., and Neugebauer, O., (eds.), 2 vols., Berlin, Springer.

Kline, M., [1972] *Mathematical Thought from Ancient to Modern Times*, New York, Oxford University Press.

Lagrange, J.L., [1799] 'Discours sur l'objet de la Théorie des Fonctions analytiques', *Œuvres 7*, 323–328.

Laugwitz, D., [1996] *Bernhard Riemann, 1826–1866*, Birkhauser, Basel.

Neuenschwander, E., [1978] 'Der Nachlass von Casorati in Pavia', *Archive Hist. Exact Sciences*, 19, 1–89.

Riemann, B., [1990] *Gesammelte mathematische Werke und wissenschaftlicher Nachlass*, Narasihman, R. (ed.), New York, Springer.

Schubring, G., [1996] 'Changing cultural and epistemological views on mathematics and different institutional contexts in nineteenth century Europe', in Goldstein, C., Gray, J., Ritter, J., (eds.), *L'Europe mathématique/Mathematical Europe*, Paris, Editions de la Maison des Sciences de l'Homme, 363–390.

Pincherle, S., [1880] 'Saggio di una introduzione alla teoria delle funzioni analitiche secondo i principi del prof. C.Weierstrass', *Giornale di matematiche*, 18, 178–254; 317–357.

Schwarz, H.A., [1869] 'Notizia sulla rappresentazione conforme di un'area ellittica sopra un'area circolare', *Annali di Matematica pura e applicata*, II serie, 3, 166–170 (= *Gesammelte Mathematische Abhandlungen* II, 102–107)
    [1870] 'Über einen Grenzübergang durch alternierendes Verfahren, *Vierteljahrschrift der Naturforschenden Gesellschaft in Zürich*' 15, 272–286 (= *Gesammelte Mathematische Abhandlungen* II, 133–143)

Weierstrass, K., [1870] 'Über das sogenannte Dirichlet'sche Prinzip', *Mathematische Werke* II, 49–54.
    [1872] 'Über continuirliche Functionen eines reellen Arguments, die für keinen Werth des letzteren einen bestimmten Differentialquotienten besitzen', *Mathematische Werke*, II, 71–76
    [1894–1927] *Mathematische Werke*, 7 vols., Berlin, Mayer & Mueller.

Chapter 3

IMAGES OF APPLIED MATHEMATICS IN THE GERMAN MATHEMATICAL COMMUNITY

Tom Archibald

INTRODUCTION

In the course of the nineteenth century, the image of mathematics shifted profoundly. Around 1800, theoretical and practical mathematics shared the project of representing nature mathematically, and most mathematicians working in the 'high culture' tradition associated with theoretical mathematics saw themselves as part of the same heritage as Newton, Euler, and others. By the end of the century, such activities were no longer central to the newly-formed professional community of mathematicians, and internationally a pure mathematical discourse stood at the top of the hierarchy of values that was expressed in a variety of ways by mathematicians. Discussions of nature were more particularly the property of theoretical physicists, following the dramatic rise of this professional and disciplinary group in the period from 1850 to 1890, most notably in Germany.

This new image of mathematics as being primarily pure, and only secondarily concerned with 'application', proved remarkably durable. It is likewise the image which has dominated much historical writing of the first three quarters of the twentieth century, while efforts to incorporate mathematics into the study of the natural world have been taken in hand mostly by historians of physics. However, most historians of this science have retained a view of the discipline of physics which projects the disciplinary structure and values of the mid-twentieth century onto the nineteenth, and in part for this reason the mathematical affiliations and activities of many researchers have received little attention.

This paper will attempt to approach some aspects of the disciplinary split between mathematics and physics in nineteenth-century Germany by examining one population of authors of mathematico-physical work, namely the authors of papers on mechanical and physical subjects in the *Journal für die reine und angewandte Mathematik* (Crelle-Borchardt) in the period between 1855 and 1875. This work proceeds almost without exception along the lines of eighteenth century rational mechanics.

Research which incorporates the new values of precision measurement and experimental verifiability, certainly on the rise during this period, appears in other journals and (mostly) issues from the hands of other writers. It is a mark of the coming to dominance of this new experimentally based theoretical physics that treatises on physical subjects largely vanish from the *Journal für die reine und angewandte Mathematik* following this period. It is in this later time that the journal was often jokingly described as the *Journal für die reine, unangewandte Mathematik*, corresponding to the shift in priorities.

We shall concentrate on forming an image of this kind of activity, its intellectual debts and research priorites, based on a few examples which seem to me typical. The image we shall obtain, an image which I think is close to the representation that the writers themselves might have expressed, is obtained from the viewpoint of the Berlin mathematical community. The editor of the journal, C.W. Borchardt, had taken over from Crelle in 1856, and the editorial team remained largely stable: Kummer, Kronecker, Weierstrass and Steiner were the core of the editorial board, with the participation of the pedagogue Schellbach. Steiner died in 1863 and was not replaced until Helmholtz joined the team in the 1880s. Despite the overwhelmingly pure-mathematical affiliations of all these men, their continued publication of 'applied' work reflects the activity of the community as a whole.

The term applied mathematics is a problematic one, used as it is today to describe things which may or may not be literally applied or even applicable questions. Likewise, while it would be possible to use the term mathematical physics (as distinct from theoretical physics) to label a certain category of work, I see no prospect for convincing the rest of the world to adopt any unequivocal definition I might propose. In discussing the German context, I will use the term theoretical physics as it has been used by Jungnickel and McCormmach [1986], to denote the mathematically *and* experimentally based work done primarily by professors of physics at universities or *Technische Hochschule* as it was practised circa 1880. These authors emphasize the German development of theoretical physics as a discipline and profession closely associated with a body of practice which includes experiment as a fundamental part of its activity, both as a subject of instruction and as a body of research, while nevertheless making extensive use of mathematical analysis to comprehend the natural world. This newly-formed discipline thus distinguished itself from earlier (say, late eighteenth century) university physics, rooted in natural philosophy, which had as its agenda the production, cataloguing and mastery of diverse natural effects either in the world at large or in the physical 'cabinet', the precursor of the laboratory. By the end of the nineteenth century, despite considerable local variation, a mathematically-based theoretical physics based on (and frequently guiding) experimental work was the international norm in scientifically developed countries. The formation of this discipline, as Jungnickel and McCormmach clearly express, is intimately linked with the creation of professorships in theoretical physics at universities and technical colleges, which is linked in turn to the expansion of physics education in the secondary schools as well as for future engineers.

The main available model in the early nineteenth century for the combination of mathematical analysis and mechanics with detailed observation was theoretical astronomy, or celestial mechanics in Laplace's terminology. It is thus no surprise to find astronomers— particularly Gauss and Bessel—at the heart of the enterprise to extend the realm of mathematical and mechanical analysis to other areas of physics, as both Jungnickel and McCormmach [1986], and Olesko [1991], have described. Olesko also points out the importance of mathematical tools in dealing with precision measurement and error analysis, though these tools are in the main distinct from those used in modelling the phenomena and obtaining predictions. Indeed it is precisely at this point that we may draw a line between the work of mathematical authors such as (most of) those in Crelle-Borchardt and work in the nascent field of theoretical physics: the mathematicians develop the models and the tools for their analysis, especially partial differential equations and potential theory, while stopping short of work with data. Furthermore they did so without sharing Gauss's fears of the 'howls of Boeotians', well-known concerning non-Euclidean geometry, but also expressed to his astronomical colleague Encke in connection with the idea that he might develop a physical theory without data to back it up. (Gauss to Encke April 2, 1839, cited by [Schaeffer 1924/29, p. 93].

I propose the term 'rational physics' to describe much of this work by mathematicians undertaken in the 1850s and 1860s, waning in the 1870s. I suggest this term by analogy with the term 'rational mechanics', which I feel calls to mind a set of values and practices which conveys the flavour of much of this work. This is a label which may be applied to a paper, or to a portion of an author's work, rather than a disciplinary description. In particular there are certain features shared by these works (with one or two important exceptions) which allow us to see more clearly the nature of the transition accompanying the disciplinary split:

1. They provide a mathematical description of some portion of physical reality which is consciously simplified or idealized. The objects of study are idealized in several senses, the most notably being that only one physical domain is usually involved (so that for example thermodynamic considerations are not mixed in with electrical ones). Further, results may be obtained concerning objects which are completely fictitious (e.g. infinitely thin plates, ellipsoids with vanishingly small mass).

2. The description is based on principles (starting points) which are considered by the writer to be *simple* (according to some criterion which is usually not articulated) and *evident* (in the sense that they are thought of as unarguably true characteristics of the natural world or else as clear hypotheses adopted for the purposes of the exercise). These principles have a mathematical expression of some kind, typically as equations the terms of which are associated with the physical entities being described.

3. The description employs a specified model of some aspect of physical reality (e.g. electric fluids made up of particles treated as geometrical points bearing charge) or else is silent on the physical nature of the entity described but provides a function or functions which characterize the important features of its condition or state at a given place and time (e.g. Fourier's conception of heat transfer).

4. There is little or no interest in producing experimentally testable results, or else such issues are taken up in a separate paper (in a different journal). Likewise, measurement issues are rarely discussed.

5. Overwhelmingly the problems in question are formulated as differential equations, and their solution provides the solution of the problem. Mathematical aspects of the solution method are definitely of interest: new special functions are introduced; series convergence is discussed; uniqueness (though not existence, usually) of solutions is important; whether a solution exists in closed form, whether expressed by quadratures, whether in terms of elliptic functions, etc. are all of interest at least sometimes.

Together, these features constitute a sort of sub-genre of mathematical writing, one which is not restricted to Germany during this period but which is international and which is closely and explicitly linked to work of Euler, Lagrange, and Laplace. One specificity of the German situation is the rise of the discipline of theoretical physics, and the corresponding influence on the practitioners of this genre of 'rational physics'. In particular, later in the century, mathematics was still being related to the solution of physical problems occupying the attention of mathematicians (by profession and disciplinary association). However the nature of the relation to physical work had altered, and the mathematical representation of physical problems came to be undertaken principally by people who were professionally physicists—as did the solution of such problems. Mathematicians in the latter period dealt primarily with mathematical issues related to such problems, and are, however, fully inscribed in the pure-mathematical practices of the time.

A further specifically German aspect is the fact that these authors refer frequently to each others' work, and to a rather restricted list of individuals and works by both German and non-German writers. This is discussed in more detail below.

Note that we are not dealing here with the formation or deformation of a discipline or profession. Many of the writers concerned worked as well in other fields of mathematical endeavour, and there were no chairs, institutes or journals devoted to such work. Any question of 'disciplinary space' for applied mathematics arises only much later, in the twentieth century. However, we shall be occupied with some aspects of the disciplinary history of both mathematics and physics during this period, calling notably on the work of Gerd Schubring [Schubring, 1990] and Kathryn Olesko.

THE CRELLE-BORCHARDT AUTHORS

My choice of the group was initially based on a naïve distinction, that between mathematics which has something to do with the study of the natural world and that which does not. Since this distinction is an evolving one in the time period under study, and since it is precisely this evolution which I wish to examine, some care must be exercised in defining the object of study, the physico-mathematical paper. The papers included consist specifically of work which explicitly claims some sort of direct contact with a physical or mechanical problem. Excluded are works which treat mathematical problems which arose in physical contexts, but where the physical context is completely absent from the paper. Thus a treatment of the Laplace equation is not included unless the physical context of a problem is at least mentioned by the author. In this way I have attempted to take the authors at their word in choosing which papers to examine. Admittedly, at times this distinction seemed strained: some papers which are concerned with the attraction of ellipsoids, for example, concern mathematical manipulations only. Yet the very fact that the author describes them using the physical terminology seems to me to be indicative of their embedding in a certain context. In most cases, the distinction is clear enough and poses no real problem.

This definition therefore includes many papers about mechanics. It should be noted that Crelle himself regarded most mechanical papers as part of pure mathematics, as we may see by looking at the analytical tables of contents of the journal up to his death. This seems to me in no way idiosyncratic, but rather conforms to the prevailing notion of mechanics proceeding from clearly stated assumptions and definitions in a purely mathematical way, and dealing with consciously idealized objects (e.g. point masses, infinitely thin bars, etc.) which have a function and ontological status analogous to the objects of Euclidean geometry. This was not generally seen as physical at the time, though the position of mechanics and geometry with respect to the physical world was to be reexamined later in the nineteenth century by Klein and especially Hilbert, as David Rowe and Leo Corry have recently investigated. [Rowe, 1992]; [Corry, 1997]. Mechanics has a certain claim to professional autonomy in Germany in this period since it is associated with the teaching of basic mechanics in the *Technische Hochschule* (a position held for example by Clebsch for a time), though it has no real claim to autonomous disciplinary status. It occupies a rather uneasy space in this regard, since it was employed as fundamental by theoretical physicists, yet it continued to be explored by pure mathematicians, sometimes in rather abstract directions.

The *Journal für die reine und angewandte Mathematik* was easily the most prestigious of the German journals publishing original work in mathematics, and indeed had no serious competition in this regard until the founding of the *Mathematische Annalen* by Clebsch and Neumann in 1868. Apart from Academy journals (effectively including only those of Berlin, Göttingen and Leipzig) it was also the only venue in which extended mathematical treatments of physical issues could be published, unless as monographs. (The short-lived and subject-specific *Resultate aus den Beobachtungen des magnetischen Vereins* of Göttingen is exceptional.) In contrast, the *Annalen der Physik und Chimie*, Poggendorff's journal, tends to contain only short articles with mathematical content. Furthermore, as such articles became more sophisticated mathematically, Poggendorff became less and less inclined to publish them. This had posed problems already for Riemann in the mid-1850s, Poggendorff finding his paper on Nobili's rings only marginally acceptable [Archibald 1992]. While this decision appears to have been Poggendorff's alone, it is clear that he knew his audience (at least in the 1850s and early 1860s), consisting of physicists and chemists whose mathematical training was often perfunctory and whose interest was in phenomena rather than analysis. One measure of the clarity of his vision is that his journal did not encounter the financial difficulties that plagued Crelle and Borchardt. The situation changed abruptly when Poggendorff was succeeded by Wiedemann in 1876. Though an experimentalist, Wiedemann was keenly aware of and interested in theoretical developments, and understood the transformed nature of the physics community.

In the period between 1860 and 1875, the Borchardt journal published 72 articles on physico-mathematical subjects, by 33 different authors. Half of these appeared between 1860 and 1864, and this is a continuation of an earlier trend of the late 1850s. Between 1865 and 1869 there are only 14 papers of this type. An increase may be noted between 1870 and 1875 with the arrival of Helmholtz in Berlin, and his publication of a number of papers in Crelle. Some of the major figures include Helmholtz himself, Clebsch, Carl Neumann, R. Lipschitz, G.R. Kirchhoff, E. Kummer, E.B. Christoffel, H. Weber, L. Boltzmann, E. Heine, and H.A. Schwarz. The subjects of greatest interest are potential theory (whether applied in electricity or in gravitation); fluid mechanics; and elasticity. There are a number of papers having to do with theoretical issues in analytical mechanics, which I have likewise included in this count provided the papers include specific reference to physical 'applications'.

Only two of the papers have direct reference to experiment: Helmholtz's 1860 organ pipe paper is intimately involved with contemporary efforts at precision measurement in acoustics [Helmholtz 1860]; and Oskar Meyer's thesis on friction in fluids, done at Königsberg under Franz Neumann's direction, refers to an experimental portion published in Poggendorff's *Annalen* [Meyer 1861]. A handful of the writers had at some point positions with physics labels: Helmholtz, Kirchhoff, Boltzmann, Lorberg (at Strassburg), Jochmann, and the Danish physicist Ludvig Lorenz. The rest were mathematics professors at universities, *technische Hochschüle*, or secondary schools.

Disciplinary Evolution in Mathematics and Physics, 1860–1880

Gerd Schubring has called attention to many important features of the disciplinary situation of mathematics in Germany, particularly to many internal variations that must be overlooked in our effort to provide a general picture [Schubring 1990]. At the core of the transformations that occur during the nineteenth century is the basic political transformation from 36 states in 1815 (from up to 10 times that many in the late eighteenth century) to a unified state under Prussian rule in 1871. The German *Sprachgebiet* also includes Austria-Hungary and part of Switzerland, and a few of our authors (Boltzmann, Mertens) are from these areas, but overwhelmingly they are from Germany. Within Germany, there are marked differences between the catholic and the protestant states with regard to the position of the philosophical faculty, where mathematics and the natural sciences were located. In the catholic states, such as Bavaria, this faculty had become marginalized in the course of the seventeenth and eighteenth centuries, a situation similar to that in France. In the protestant states, on the contrary, the philosophical faculty flourished, in particular in Prussia and Hannover. Thus a relatively large number of universities existed where mathematics flourished, not least because of its important role in secondary education, where it was one of the three main subjects. This created a demand for mathematics graduates, normally qualified by a doctorate before acquiring certification as teachers. Several of the most important universities mathematically were in Prussia: Berlin, Bonn, Königsberg, Halle—but mobility between the different states meant that an integrated 'system' was already effectively in place by the time of unification. This system was one in which research was prized and rewarded.

Another important feature of the German situation was the rise of the technical colleges, post-secondary institutions for the training of engineers, architects, etc. These institutions, for the most part founded in the first half of the nineteenth century, became stronger at mid-century and expanded mightily in the 1870s and 1880s, providing positions for mathematicians as well as physicists. While mathematical activity here lacked the autonomy of the universities, there was still room for research. For the physicist, it was often possible to have laboratory facilities as well, in some cases better than those of the universities (this was notably the case for Heinrich Hertz, who moved to Karlsruhe from Kiel in large part because of the availability of laboratory facilities at the former institution). Thus in the period under examination we find a very healthy professional base for mathematics, and mathematical research in particular.

During the same period, physics and mathematics grew apart, so we must look at the shifting situation of the physics community. Recent studies have pointed to the new importance of experiment for this community, as well as of precision measurement. Olesko [1991] has likewise emphasized the role of error analysis. The adoption of these trends was gradual and varied according to the location of practitioners. Certainly the arrival of Helmholtz in Berlin placed him in a key position to direct students in the conduct of this new theoretical physics, and his close involvement with students during this period is attested in many accounts. Gauss' associate Wilhelm Weber, working on the precision measurement of theoretical and material constants, likewise produced students skilled in these tasks. Olesko [1991] has pointed to the importance of an apprenticeship in theoretical physics under Franz Neumann at Königsberg, a training which included an experimental component.

Historical writers have said little about the acquisition of the mathematical tools necessary for the tasks either of old-fashioned 'rational physics' or for the new theoretical physics. Here I believe that claiming the centrality of Neumann ignores certain fundamental features of higher mathematical education in the period, though Neumann's lectures were doubtless important in providing his listeners with a basic understanding of differential equations and of the special functions of mathematical physics (most notably the spherical harmonics essential in attraction theory). But most of our writers show a clear debt to French textbooks of calculus and mechanics, a debt which was enforced by the structure of lecture courses where the professor largely lectured on advanced material close to their current research. Some of these influential books include Lacroix's treatise on the differential and integral calculus, which was translated into German. We find frequent references to Lagrange, Euler, and Laplace, and it is certain that the appropriate works—espcially Euler's *Institutiones*, the *Mécanique analytique* of Lagrange, and Laplace's *Mécanique céleste*— were an important part of the basic education of the generation that began to publish in the 1850s. A special role for our authors was played by French mechanics textbooks, above all that of Poisson's *Traité de mécanique* [Poisson, 1833], the correction and modification of whose formulations was a cottage industry among them. Especially important also were recent works and lectures by German authors: Dirichlet's account of Fourier series in Dove's *Repertorium der Physik* [Dirichlet, 1837], Jacobi's lectures on analytical mechanics [Jacobi, 1842/1865], Dirichlet's emphasis on special definite integrals (some published, others in lectures) and his account of potential theory (in lectures, eventually published) had varying degrees of influence. Likewise important was Green's *Essay*, published (in English) in Crelle in the early 1850s, for providing his "Green's function" method. The work of Gauss was likewise influential, especially the theory of capillarity and potential theory [Gauss 1829 and 1839], though many of Gauss's papers provide fundamental tools. While the basic repertoire varied with the individual, this shared background was acquired through individual study, at the hands of *Privatdozenten*, and in lectures, and informs the work produced. Also, we notice that contemporary English work is occasionally referred to, especially that of Thomson (e.g. the method of electric images, discussed by Lipschitz [1863]) and Stokes (Stokes' theorem, used especially in fluid mechanics).

These techniques—what might be described now as the contents of a course in the methods of mathematical physics—did indeed gel into a lecture course in the hands of Riemann in the early 1860s. His two-part lecture course was published posthumously: *Schwere, Elektricität und Magnetismus* deals principally with potential-theoretic methods, while *Partielle Differentialgleichungen* gathers together results on definite integrals, series, the heat equation, the wave equation, and a number of particular physical problems which Riemann had treated [Riemann, 1856/1876]. The course is clearly based on the lectures of Dirichlet, which Riemann had taken in his 1848 Berlin year. Subsequent revisions of the book by Heinrich Weber augmented the contents as the repertoire expanded, not least by methods devised by Weber himself (Riemann-Weber 1900 is the fourth edition). Some of these new techniques appear among the Borchardt papers, notably a method for writing the equations of hydrodynamics as first-order equations which allows for the easy introduction of the hypothesis that a velocity potential exists for a given flow. Even a casual examination of Weber's paper reveals that a high level of specialized mathematical knowledge was necessary to produce this result, specifically mentioned by Weber as 'not merely an analytical simplification but also a physical one.' [Weber, 1868 p. 286].

This draws our attention to a rather obvious reason for the division of labour between the mathematician and the physicist. Considerable mathematical sophistication was required to create the necessary tools or generalizations, fine-tune them, and understand clearly when they were valid. Such efforts were increasingly a full-time job. Likewise, an increasing level of expertise was needed for experimental work. Initially this expertise did not transmit particularly well by literary methods, consisting as it does not only in carrying out experiments, but in formulating them, designing equipment, and grasping the priorities of the community. Such activities did indeed require craftiness and practical experience, since methods for eliminating extraneous effects often require a detailed knowledge of devices and an ability to manipulate equipment. The passage of time did however reduce the component of 'tacit knowledge' necessary to carry out certain kinds of experimental work, not least because of the availability of professionally produced instrumentation.

With few exceptions, professional mathematicians did not later switch to activity in physics. Aside from cognitive obstacles and barriers of taste to doing so, there were compelling institutional reasons for this: mathematicians are supposed to produce papers in mathematics, not fool about with devices. We have one example of an individual who took up an experimental career after a mathematical one: Julius Plücker, Lipschitz's colleague at Bonn, responding to an institutional dictum that he must do physical research in order to continue to be paid as a physics professor, set aside his mathematical career for twenty years. Interestingly this was almost entirely experimental work, and was in no way associated with the analytical modelling of our physico-mathematical writers.

THE PRACTICE OF RATIONAL PHYSICS

I have set out above several characteristics claimed of the bulk of the work I have examined, and tried to position it with respect to contemporary practices in physics. I now propose to treat in greater detail a few examples to document these proposed shared characteristics more clearly.

Hydrodynamics, Elasticity, and Electromagnetism: some typical instances

The first group of examples, all stemming from the early 1860s, presents us with three papers which in many respects typify approaches to physical problems by this group of authors: a paper by Clebsch on floating bodies, one by Kirchhoff on the distribution of charge on two spherical conductors, and one by C. Neumann on elasticity. In each case, earlier treatments are summarized and criticized, notably those of

Poisson, on either physical or mathematical grounds. A new, more refined model is introduced, a system of differential equations formulated, and a solution under certain circumstances is exhibited. Experiment is not mentioned.

Clebsch was doubtless moved to his mechanical studies by his position teaching mechanics at Karlsruhe, a position he was to leave in 1863, at which point his mechanical research also ceases. In [Clebsch, 1860] he criticizes earlier treatments of the problem of equilibrium of floating bodies due to Poisson and Duhamel on the grounds that their simplifying assumptions are never satisfied in actual fact, thus lacking the necessary evidence which should be a characteristic of such a treatment. (Both Poisson and Duhamel's treatments are elaborated in their textbooks). The physico-mechanical problem is to characterize stable equilibria by a sort of principle of virtual displacements. The Poisson hypothesis involves replacement of hydrodynamic by hydrostatic pressure, which, Clebsch argues on intuitive grounds, is inadmissible because near equilibrium hydrostatic pressures will cancel one another while hydrodynamic pressures, those due to the motion of the fluid induced by the wobbling of the body, won't. In fact the latter are opposed to the direction of motion of the body, yet are of the same order of magnitude as the total hydrostatic pressure. Thus, Clebsch maintains, the usual equations are not only inaccurate, they are completely false in that they neglect terms which will in fact determine the whole motion.

Clebsch's analysis appears not to be particularly innovative as regards mathematical technique. His technique is based on potential theory, the difference between hydrostatic and hydrodynamic pressure satisfying Laplace's equation, so that results from steady-state heat conduction could be adduced to solve the problem. His result after integrating the system of differential equations is a general equation of motion, transcendental in form, and a resulting equilibrium condition for which he claims stability when the equation has only negative roots. While the general solution was difficult to handle, Clebsch achieved some results in a simpler case by restricting the investigation to a floating body which is symmetric about two vertical mutually perpendicular planes.

Kirchhoff's treatment of a standard electrostatic problem, likewise one which Poisson had handled (in his well-known 1811 memoir, [Poisson, 1811]), aims much more at an extension of mathematical methods previously employed. Kirchhoff examines the problem of the distribution of static electricity on two charged conducting spheres [Kirchhoff, 1861].

Kirchhoff identifies two parts to the method for solving the problem. First, one must specify the potential $f(x)$ created by the electricity on one sphere for points along the line joining the centres of the two spheres. Then, by obtaining from f the potential for points outside this line, the density may be deduced. Thus this function f, already known to Poisson, is the critical quantity to which Kirchhoff directs his attention. Kirchhoff notes that several physically interesting functions which depend on f may be expressed in closed form using elliptic functions, which were unavailable to Poisson, for example the density for points on the centre line if the spheres have the same potential. Using this observation Kirchhoff revises the estimate for this value, which had different theoretical values in the work of Poisson and in the subsequent work of Plana. By introducing an asymptotic expansion for the potential, Kirchhoff is able to produce values which would in principle be much more reliable than those of either of his predecessors. Significantly in my view, these values are not mentioned. The piece is a technical tour-de-force drawing on mathematical methods due to Gauss (hypergeometric series) and Lipschitz (series expansions for $\log(\Gamma)$), and is part of a number of works from the period attempting to integrate the elliptic functions of Abel and Jacobi fully into analysis by using them to solve differential equations. Its relation to practical physical concerns is doubtful, since it would seem that the improvements lie beyond the level of accuracy of experiment.

The third paper in this group, by Carl Neumann [1860] is likewise primarily concerned with efficient mathematical methods for attacking a physical problem. Arising perhaps from Neumann's 1858 dissertation on the Faraday effect, this paper examines the method for setting up the differential equations expressing conditions for equilibrium and the motion of an elastic body. Two earlier methods are mentioned, one due to Navier in which one writes differential equations satisfied by the forces on a given molecule of the body exerted by all others; and a second, due to Poisson, expressing the pressure experienced by a surface-element inside the body. Neumann's paper proposes a third way, namely to derive the potential for one molecule exerted by all others, then integrating over the body, taking the variation of this integral and setting it to zero to obtain a condition which is necessary for equilibrium. The method is an adaptation of Gauss's method in his paper on the theory of capillarity [Gauss, 1829], and Neumann's treatment of it simplifies an earlier method of Lamé.

These papers are closely grouped in time, and emphasize a revision or extension of existing methods in order to improve the level of accuracy of the existing method or to introduce more effective mathematical methods. We find similar work throughout the period, but two examples are perhaps worth mentioning. The first, already mentioned is the method of Heinrich Weber for rewriting the equations of hydrodynamics [Weber, 1868], reminiscent of Neumann's efforts. The second is an 1871 paper by O.E. Meyer, on the motion of a spherical pendulum taking the influence of friction of the surrounding medium into account [Meyer, 1871].

Meyer, like Neumann, is concerned in this paper with a problem rather similar to that faced by Clebsch in 1860, but in the context of a compressible rather than an incompressible fluid. Here, experimental issues do come into play, though peripherally, providing a point of contrast with the other papers mentioned. Analogously to the situation with Clebsch's treatment of floating bodies, Bessel had already noted that if we observe a pendulum in air, to know its motion in a vacuum two corrections are needed: an aerostatic factor, (noted already in 1749 by Bouguer), to deal with apparent loss of weight in air; and an aerodynamic factor due to an apparent gain in inertial moment since the pendulum moves air along with it. Bessel had subjected this to experiment. Once again, a previous treatment by Poisson is implicated, since Poisson had addressed this issue theoretically in an 1832 paper and in his 1833 text by integration of the system of differential equations on which the motion of fluids depends. Poisson's theory produced a numerical value too small by half ($k = 1/2$ as opposed to Bessel's $k = 0.95$), a fact which Stokes later explained as due to neglect of internal friction of air, modifying the result to bring it into line with Bessel's experimental results.

Meyer raises instead a more specifically mathematical issue: are the solutions that Stokes obtained unique? Uniqueness questions are important physically and mathematically, since the existence of other solutions might point out unforeseen physical situations, or merely indicate weaknesses of the model. Writing shortly before the vogue for existence proofs, Meyer is likely concerned with such issues rather than with more restricted mathematical questions such as how to generally produce such uniqueness proofs on the basis of algebraic and geometrical understanding of partial differential equations. Meyer's uniqueness argument is nonetheless inscribed in the disciplinary environment of the mathematics professor, ignoring physical issues beyond the brief reference to the history of the question. In this regard it is rather different from an earlier paper of his in Borchardt, one in which experimental issues are more explicitly present, to which we now turn.

Two singularities: Helmholtz and Meyer

I mentioned above that only two papers of the seventy-odd under discussion actually broach experimental issues in an essential way. These are Meyer's dissertation, done under the guidance of Franz Neumann at Königsberg, and a paper of Helmholtz on organ pipes.

Meyer [1861] contains the theoretical portion of his dissertation on friction in fluids, the experimental part appearing in Poggendorff's *Annalen*. Meyer claims a history for his approach going back to Newton (*Principia* Book 2 section 9), distinguishing it from the efforts of Euler, Navier, Poisson and Stokes, and attributing it to an independent rediscovery by his professor F. Neumann. The basic hypothesis is that the friction between adjacent lamina of a fluid is proportional to the difference in their velocities and to the surface area of contact, and independent of pressure. On this basis Meyer gives a theory of some experiments due to Coulomb, calculating friction from the loss in velocity which a plate experiences which is oscillating in the liquid about its midpoint (in its own plane). This is thus a typical rational-mechanical exercise in some respects, yet the explicit mention of experiment and an interest in both precision and error make it atypical for the Borchardt authors.

Helmholtz [1860] is even more thoroughly physical, and we now know it to be one of series of papers which he intended to be models for a new practice of theoretical physics in Germany (Darrigol, personal communication). It is a study of the vibration of columns of air in open-ended pipes. Theoretically rooted in work of D. Bernoulli, Euler, Lagrange, and Poisson, Helmholtz finds their idealizations problematic, since they lead to various contradictions of experience. These are on the one hand qualitatively evident: in the Euler method, for example, infinite vibrations may be produced by finite driving force, the vibrations continuing undamped outside the tube. Poisson's treatment fixes this, but his model produces nodal surfaces which precision measurement (by Hopkins at Cambridge and Wertheim in Berlin) clearly showed to be the wrong shape. Helmholtz reformulates the model, employing Green's theorems to establish general results which he had obtained in 1858 by exploiting the analogy between fluid flow and electricity. He then generalizes earlier models to get a potential function and deduce results about the possible form of the nodal surface. For some cases reasonable agreement with experiment is obtained, though the model also produced problematic results for some cases. Here, once again exceptionally, the question is one of fully integrating experimental work into the picture, modifying the physical model as necessary to avoid over-simplification. A further attribute of the Helmholtz treatment is a deft unification of earlier theoretical treatments which in principle would permit an experimental (or thought-experimental) distinction between different models by the determination of parameters in the model. The approach foreshadows his electrical papers (likewise in Borchardt) of the 1870s, and reminds us of later practice by experimentally-oriented theoretical physicists whose use of mathematical tools was much less linked to the elaboration of research programs in mathematics than with immediate concerns of the practising physicist. In this regard the contrast with Neumann, Kirchhoff, and H. Weber is particularly notable.

These examples could be multiplied. One class of works I have not described has to do with determining potentials of various planar or solid bodies of different shapes. Such problems were for a time popular dissertation topics in Berlin, Göttingen and Königsberg, embodying both important physical questions (attractions) and mathematical ones (solving the Laplace equation in different geometrical configurations, dealing with infinities and non-differentiable points produced for example at corners of polyhedra or the vertices of squares). A fundamental early work of this type is E. Heine's 1854 study of the potential of a homogeneous circle. This was used as a limiting case for comparison purposes by several writers on attraction, notably Lipschitz [1861]. Another, such work is due to Mehler [1862], who provides a method for determining the attraction of two shells bounded by similar quadric surfaces. Mehler's paper extends Dirichlet's 1839 method of 'discontinuous factors' to the cases other than ellipsoids. Dirichlet's method itself generalized Poisson's methods (once again!) for attraction of ellipsoids by allowing a unified treatment regardless of whether the attracted point is inside or outside mass.

CONCLUDING REMARKS

These few examples give an idea of the image held by the Borchardt authors of how mathematical writers might appropriately proceed in treating physical subjects. Multiplying such examples, while clarifying the picture, would not however alter it much. Fully attached to the discipline of mathematics by various institutional and educational bonds, these writers clearly shared approaches and values. Mathematical innovation is prized in these papers, while experimental considerations are all but absent. Yet the papers have in general some relevance to experimental questions, indicating solution methods and extending the range and accuracy of existing models (though no deeply innovative models are found in the time period selected). These applications might be taught to physicists and engineers, their use in those contexts clearly lying outside the realm of interest of the mathematician.

As I have stated but not documented in detail, the papers share a background in the classical mechanics of the eighteenth century. Sharing both literary background and methods, they also refer to one another frequently, both when mathematical methods introduced or elaborated by one writer can be adapted to another situation (as in the case of the use of Green's methods) or when a worked-out model can be analogically transfered to another physical situation. Thus the texts in question can be viewed as a representative of a sub-genre in mathematical writing in the period, sharing as they do not only the stylistic factors already described, but also both subject and audience.

As a genre, it provides us with a fine example of one aspect of the image mathematicians held of their subject during the period in question. Representing mathematics as a tool which had power in the study of nature, this genre was increasingly occupied with issues that were of less and less of interest to those outside mathematics—such as differentiability and convergence conditions (Kirchhoff), or different representations of a given system of partial differential equations and of its solutions (H. Weber). Yet the claim of mathematics to rights over this subject area became more and more tenuous with the increasing demands of experiment. Some of our authors, for example Boltzmann, soon became fully integrated into the newer approach. Others, like Kirchhoff, claimed a kind of autonomy for their field. Writing in 1876 in the preface to his lectures on mechanics (which included accounts of much of the work on elasticity and hydrodynamics that the Crelle authors treated), Kirchhoff expressly disavowed the connection between the basic objects of mechanical theory and the physical world, positing for experimental concordances the status almost of happy coincidence [Kirchhoff, 1876]. Yet such an autonomy was never to materialize, and the image of the essence of mathematical activity as being linked with its ability to describe the natural world without appeals to other natural sciences faded into near-oblivion with the advent of the twentieth century.

ACKNOWLEDGEMENTS

I wish to thank the editors for their thoughtful suggestions concerning earlier versions of this paper, which also benefited from remarks by David Rowe and Jesper Lützen. Thanks are also due to the *Centre de recherche en histoire des sciences et techniques* at the Cité des Sciences et de l'Industrie, La Villette, Paris, which has hosted me as an associate researcher (CNRS) for the academic year 1997–1998; and to Acadia University for according me a sabbatical. Research for this paper has also been supported in part by SSHRC, which assistance is gratefuly acknowledged.

REFERENCES

Archibald, T., [1992] 'Riemann and the Theory of Electrical Phenomena: Nobili's Rings', *Centaurus*, 34, 247–271.

Boltzmann, L., [1871] 'Ueber die Druckkräfte, welche auf Ringe wirksam sind, die in bewegte Flüssigkeit tauchen', *Journal reine angew. Math.*, 73, 111–134.

Clebsch, A., [1860] 'Ueber das Gleichgewicht schwimmender Körper', *Journal reine angew. Math.*, 57, 149–169.

Corry, L., [1997] 'David Hilbert and the Axiomatization of Physics (1894–1905)', *Arch. Hist. Exact Sci.*, 51, 83–198.

Dirichlet, P.G., Lejeune, [1837] 'Ueber die Darstellung ganz willkürlichen Functionen durch Sinus- und Cosinusreihen', *Repertorium der Physik*, 1, 152–174.
[1839] 'Sur une nouvelle méthode pour la détermination des intégrales définies', *Comptes Rendus Acad. Sci.*, 7, 156–180.

Gauss, C.F., [1829] 'Principia generalia theoriae figurae fluidorum in statu aequlilibrii'. *K. Soc. Gött. Comm.*, 7 (in *Werke*, vol. 5, 29–77).
[1839] 'Allgemeine Lehrsätze in Beziehung auf die in verkehrten Verhältnisse des Quadrats der Entfernung wirkenden Anziehungs- und Abstossungskräfte', *Res. Beob. Mag. Verein*, (in *Werke*, vol. 5, 195–242).

Helmholtz, H., [1860] 'Theorie der Luftschwingungen in Röhren mir offenen Enden', *Journal reine angew. Math.*, 57, 1–72.

Jacobi, C.G.J., [1842/1865] *Vorlesungen über Dynamik* (ed. by A. Clebsch), Supplementband to Jacobi's *Gesammelte Werke*, Berlin.

Jungnickel, C., MacCormmach, R., [1986] *The Intellectual Mastery of Nature*, 2 vols., Chicago, University of Chicago Press.

Kirchhoff, G.R., [1861] 'Ueber die Vertheilung der Elektricität auf zwei leitenden Kügeln ', *Journal reine angew. Math.*, 59, 89–110.
[1869] 'Ueber die Kräfte, welche zwei unendlich dünne, starre Ringe in einer Flüssigkeit scheinbar auf einander ausüben können', *Journal reine angew. Math.*, 71, 263–273.
[1876] *Vorlesungen ueber mathematische Physik*, vol. 1, Mechanik, Leipzig, Teubner.

Lamé, G., [1840] 'Mémoire sur les coordonnées curvilignes', *Journal math. pures appl.*, 5, 313–347.
[1841] 'Mémoire sur les surfaces isostatiques dans les corps solides homogènes en équilibre d'élasticité', *Journal math. pures appl.*, 6, 37–60.

Lipschitz, R., [1861] 'Beiträge zur Theorie der Vertheilung der statischen und der dynamischen Electricität in leitenden Körpern', *Journal reine angew. Math.*, 58, 1–53.

[1863] 'Untersuchungen über die Anwendung eines Abbildungsprincips auf die Theorie der Vertheilung der Elektricität', *Journal reine angew. Math.*, 61, 1–21.

[1872] 'Untersuchung eines Problems der Variationsrechnung, in welchem das Problem der Mechanik enthalten ist', *Journal reine angew. Math.*, 74, 116–149.

[1874a] 'Reduction der Bewegung eines flüssigen homogenen Ellipsoiids auf das Variationsproblem eines einfachen Integrals, und Bestimmung der Bewegung für den Grenzfall eines unendlichen elliptischen Cylinders', *Journal reine angew. Math.*, 78, 245–273.

[1874] 'Beweis eines Satzes der Elasticitätslehre', *Journal reine angew. Math.*, 78, 329–339.

Mehler, F., [1862] 'Ueber die Anziehung einer von zwei ähnlichen Flächen zweiten Grades begrenzten Schale', *Journal reine angew. Math.*, 60, 321–342.

Meyer, O.E., [1861] 'Ueber die Reibung der Flüssigkeiten', *Journal reine angew. Math.*, 59, 229–303.

[1871] 'Ueber die pendelnde Bewegung einer Kugel unter dem Einflusse der inneren Reibung des umgebenden Medium', *Journal reine angew. Math.*, 73, 31–68.

Neumann, C., [1860] 'Zur Theorie der Elasticität', *Journal reine angew. Math.*, 57, 281–318.

Olesko, K., [1991] *Physics as a Calling: Discipline and Practice in the Königsberg Seminar for Physics*, Ithaca, Cornell University Press.

[1995] 'The Meaning of Precision: the Exact Sensibility in Early Nineteenth-Century Germany', in Wise N., (ed.), [1995], 103–134.

Poisson, S.D., [1811] 'Mémoire sur la distribution de l'électricité à la surface des corps conducteurs', *Mém. Institut de France*, Paris, 1812.

[1833] *Traité de mécanique*. 2e éd. Paris, Bachelier.

Rowe, D., [1992] 'Introduction', in D. Hilbert, *Natur und mathematisches Erkennen: Vorlesungen, gehalten 1919–1920 in Göttingen*, Boston and Basel, Birkhäuser.

Riemann, G.F.B., [1882] (1876) *Partielle Differentialgleichungen und ihre Anwendungen auf physicalische Fragen*, Braunschweig, Vieweg.

Riemann, G.F.B., Weber, H., [1900] *Die partielle Differentialgleichungen der mathematischen Physik*, 4e ed. Braunschweig, Vieweg.

Schaeffer, C., [1924/29] 'Ueber Gauss' physikalischen Arbeiten' in Gauss, *Werke*, vol. 11, Abt. 2, Berlin.

Schubring, G., [1990] 'Zur strukturellen Entwicklung der Mathematik an den deutschen Hochschule, 1800–1845' in Scharlau, W., (ed.), *Mathematische Institute in Deutschland, 1800–1950*, Braunschweig, Vieweg.

Weber, H., [1868] 'Ueber eine Transformation der hydrodynamischen Gleichungen', *Journal reine angew. Math.*, 68, 286–292.

Wise, N., (ed.), [1995] *The Values of Precision*, Princeton, Princeton University Press.

Chapter 4

FELIX KLEIN AS *WISSENSCHAFTSPOLITIKER*

David E. Rowe

Felix Klein (1849–1925) has been the focus of several recent studies on the history of modern mathematics that have highlighted the importance of turn-of-the-century developments in Göttingen for grasping the complex transition in the structure and conduct of mathematical research. Herbert Mehrtens, for example, has drawn a compelling thumb-nail sketch of the Göttingen *Betrieb*, pictured as a community on the cutting edge of the large-scale modernization process that transformed mathematical and scientific research at the German universities [Mehrtens, 1990]. Various other writers have also dealt with Felix Klein's activities in the interpenetrating spheres of educational politics, big science, and engineering reforms, all of which played a part in Klein's efforts to promote applications of mathematical knowledge (see, for example, [Tobies, 1991], [Manegold, 1970], [Tobies, 1981], [Pyenson, 1983], [Hensel, 1989], [Rowe, 1985]). In this chapter, I would like to consider various factors connected with Prussian educational politics that help to account for some of the particular circumstances that shaped Klein's singular career. In places, I will also appeal to anecdotal information that helps us to better appreciate Klein's personality and the manner in which he adapted to the problems faced by mathematics during the Wilhelmian epoch. In broad terms, I will argue that his career can be analyzed in terms of four principal phases: 1) creative period (1865–1882); 2) leader of school (1883–1892); 3) *Wissenschaftspolitiker* (1893–1914); 4) sage (1915–1925). While all four phases deserve consideration, for the questions raised above the third period as well as the transition from the second to the third were of decisive importance, and hence these will receive special attention here.

EARLY QUESTS AND CONQUESTS

Klein was born and raised in Düsseldorf, a bustling city in the heart of the Rhineland. His father, a stern, dutiful Prussian governmental official, set the tone of discipline that pervaded the household. Klein later thought of him as a typically Prussian personality who had nothing of the *fröhlicher Rheinländer* about him: 'tough determination, never slackening diligence, clear realism, unconditional loyalty, and carefully considered thriftiness: those were the traditional characteristics of this hardened German tribe that my father so authentically embodied' ([Klein, 1923, p. 12], my translation). His mother, who came from Aachen, was a far warmer spirit, with a quick mind and many-sided interests. During their boyhood years she taught her sons how to read and cared for their early intellectual development. Klein's first formal education took place at age six when he entered a private school; two years later he began attending the classical Gymnasium in Düsseldorf, which taught him nothing particularly useful for his later career, but did instill in him the value of hard work and discipline. As the sage-like mathematician later described this Catholic educational institution in the twilight of his life:

> We learned precise grammatical thinking and acquired an all-round knowledge of the dialectics and other peculiarities of ancient languages. We also gained a certain proficiency in the independent use of these foreign languages, but we were never exposed to the living substance of the literary works we studied, their poetry, cultural history, folklore, etc. History was taught in the same manner; an enormous amount of material was presented, but without any vivid clarity or larger viewpoints. The accepted opinion was simply that the pupil would be best educated if forced to work through difficult, opaque material that required great effort to master. Yet, even if we went away empty-handed when it came to creativity and artistic sensibility, and despite the fact that these methods failed to expose us to many things of real educational worth, they did teach us one thing of value: we learned how to work and then work some more. Certainly I could not say that I felt unhappy about all the rote learning associated with this type of formal schooling. Indeed, the great demands it made on diligence and energy were in keeping with my eagerness to learn, just as the emphasis on the logical and grammatical elements suited my intellectual tendencies. Even today, I still remember the proud satisfaction I felt after translating a number of verses from Schiller's *Cranes of Ibykus* into flawless Greek, though I have my doubts whether I really understood the content and literary value of the poem completely [Klein, 1923, p. 13].

In 1865, at age sixteen, Klein began his university studies in Bonn, where he quickly came to the attention of Julius Plücker (1801–1868). Initially intent on pursuing a career as a physicist, Klein was offered the splendid opportunity of serving as Plücker's assistant in his lectures on experimental physics. At the same time, however, Plücker engaged his help in the mathematical research he was then doing in the new field of line geometry. After his mentor's death, Klein completed his dissertation on second-degree line complexes in Bonn under the supervision of Rudolf Lipschitz. Yet despite fairly extensive as well as intensive contacts with Lipschitz, it was not he but the Göttingen mathematician, Alfred Clebsch(1839–1872), who emerged as the next major figure to enter his life.

Clebsch was an avid admirer of Plücker's pioneering work in algebraic geometry. Like Plücker, he practiced geometry with a strong analytic flavor, a style severely criticized by Berlin's Jakob Steiner, the leading synthetic geometer of the period. This circumstance contributed to the estrangement Klein's teachers both felt toward the Berlin establishment. Plücker virtually abandoned algebraic geometry for physics in the 1840s, winning considerable acclaim from Faraday and others for his research on cathode ray emissions. Coincidentally or not, he only returned to geometry after 1863, the year of Steiner's death. Plücker's relations with his colleagues in Bonn were also tense, however, and he received only grudging recognition for his achievements in geometry and physics within Germany. Clebsch, on the other hand, had strong ties with the Königsberg mathematics-physics tradition, and during the 1860s he emerged as the most dynamic figure operating outside the Berlin network. In Giessen he established a new school specializing in algebraic geometry and invariant theory, and in 1868 he was offered the chair in Göttingen that Riemann had occupied until his death in 1866. Clebsch and Carl Neumann also launched the new journal *Mathematische Annalen* in 1868, thereby providing a platform from which their allies could challenge Berlin's *Journal für die reine und angewandte Mathematik*. At the same time, Clebsch was already making plans to mobilize support along the periphery of the Berlin-based network, which dominated the chairs in mathematics at the Prussian universities, in an effort to form a national mathematics organization, an effort that would engage Klein's attention during the early 1870s.

Klein entered this atmosphere, so full of promise, early in 1869 after he took his doctorate at Bonn. Still hoping to take up a career in physics, he quickly found himself drawn by the magic of Clebsch's new approach to algebraic geometry based on Riemann's bold ideas. As Klein later described the scene :

> In Göttingen I was surrounded by a completely different academic life. Certainly Clebsch attracted only a small audience for his special lectures, but the stimulation I gained from these lectures and from my personal contact with Clebsch had all the more effect on me. Soon I found myself

surrounded by an intimate circle of like-minded people ([Max] Noether, [Eduard] Riecke, and others), with whom I enjoyed a period of exhilarating scientific enthusiasm. For this reason physics again took a back seat, even though I was soon introduced to the home of Wilhelm Weber. For I was in no way attracted to the finicky work which was then being pursued by Weber's followers, whose one and only goal was to elaborate and support an established physical worldview in detail without introducing any new points of view [Klein, 1923, p. 15].

Klein's friend and later colleague, Eduard Riecke, remained in Göttingen where he became Weber's leading disciple and, in 1881, heir to his chair in physics. Riecke was later instrumental in bringing Klein back to Göttingen and he proved a bulwark of support when Klein needed it most in carrying out his ambitious plans for science and engineering during the 1890s.

Certainly Klein was a mathematician with unusual gifts and even greater ambitions. His career took many surprising turns as well, indicative of his ability to adapt his strategies to circumstances and to seize fresh opportunities. Throughout all of the tumultuous changes he experienced—including the rise and fall of the Second Reich—he was a man who loved to be at the heart of the action. For an aspiring mathematician and natural philosopher in 1869, the chance to study in Berlin and Paris, the two leading centers of the day, made even the stimulating experience of working with Clebsch look drab by comparison. Klein thus set his eyes on an international study tour that would take him first to the Prussian capital, then to Paris, and finally to England. As it turned out, even more dramatic political events interceded before he could cross the English Channel.

Klein's thirst for power and recognition became especially evident during the crucial third phase of his career, yet even as a young man he wanted to do things that would make prominent people stand up and take notice. A good case in point can be seen in his famous journey to Paris just before the outbreak of the Franco-Prussian war when he and his friend, the Norwegian Sophus Lie, met many of the city's leading mathematicians. Much less well known is the fact that this trip was coordinated in advance with the Prussian Ministry of Culture in what might be called a mission of scientific espionage. Klein's self-appointed effort (the first of many trips abroad) was to obtain information about the state of mathematical affairs in France and England (the trip to England had to be postponed until 1873 owing to the war). Shortly before leaving Paris, he and Lie wrote a report on the mathematical scene there, sending copies to both the Berlin Mathematics Club as well as to the Ministry. Nothing seems to have come of this, and the thread connecting Klein with Prussian officialdom remained quite thin throughout the first two phases of his career when he was preoccupied with building a power base within the discipline.

During these years Klein began to develop an impressive network of scientific contacts by building on the school contacts he inherited from Clebsch after the latter suddenly succumbed to diptheria in November 1872. Klein, who had strengthened his ties to the Clebsch school by returning to Göttingen as a *Privatdozent* in January 1871, already reaped the rewards of this affiliation in the form of an appointment as full professor in Erlangen, an extraordinary coup for a young man of 23 years. He quickly rose to the occasion, however, by writing the famous inaugural *Programmschrift* that has since come to be known as *the* 'Erlangen Program'. In it, Klein proclaimed that geometrical research was essentially the study of the invariants of transformations groups that act on manifolds. Clebsch, whose work had clearly inspired Klein to think of geometric properties in terms of invariants and covariants, died just one month before the 'Erlangen Program' came off the presses.

Following Clebsch's death, several of his advanced students left Göttingen to study under Klein in Erlangen, a university which prior to this had been mathematically moribund. Thus, almost overnight, Klein went from being a fresh-baked Ph.D. to assuming the leadership of one of Germany's most important mathematical schools. Within two years, Klein took over the reins of *Mathematische Annalen,* which under his leadership became the preeminent mathematical journal of the day. In the meantime he was offered a new position in Bavaria in 1875 as successor to Otto Hesse at the *Technische Hochschule* in Munich. This marks the beginning of Klein's most productive years when he could churn out a new paper nearly every month. No one else in Germany was even nearly this prolific during the late 1870s and early 1880s. While hoping to find his way back into the Prussian system, Klein accepted a new chair in geometry created in 1880 by the Saxon Ministry at Leipzig.

The next six years saw the high water mark of his creative endeavors in the well- known 'competition' with Henri Poincaré, soon followed by a dramatic collapse in Klein's productivity that brought about the second phase of his career. Recognizing that he could no longer attain the same high level of output as a writer, he consciously chose to curtail his work on research articles while devoting most of his energy to the preparation of lectures and monographs. He did not need to 'refashion' himself radically in order to make this change since it mainly required shifting the focus of his efforts from research to teaching and mentoring, activities he had been cultivating ever since his days in Erlangen. Moreover, there was an excellent precedent for such a shift: the most influential mathematician of the 1870s, Karl Weierstrass, had shown that building and maintaining a school did not require publishing lots of research. By the 1880s Weierstrass was old, ailing, and suffering from the slings and arrows that his colleague, Leopold Kronecker, hurled at the conventional foundations of analysis. Berlin was beginning to lose its once dominant position within German mathematics, and Klein stood poised to move into a new role as the leader of an internationally acclaimed school of function theory that drew heavily on Weierstrassian ideas packaged in Riemannian garb. To see this second phase of Klein's career in a broader light, it helps to consider the role and function of mathematical schools in the German universities of the nineteenth century.

MATHEMATICAL SCHOOLS AT THE GERMAN UNIVERSITIES

Since the term 'school' has traditionally been used in a very loose fashion among mathematicians and historians of mathematics, a few words should be said about how the notion might be anchored down a bit. One way to do this is by taking heed of what historians of chemistry and other disciplines have recently written about the nature of the laboratory research schools that came to play such a prominent role in the nineteenth century. One such investigator, Gerald Geison, has attempted to identify the most salient characteristics of scientific schools in general, and many of the features he describes can be readily found in the typical mathematical research schools of the nineteenth century [Geison, 1981]. Unlike seminars, whose institutional role and function were clearly spelled out in the statutes that governed their operation, research schools had no formal status as such. Nevertheless, they assumed a prominent role at many German universities, particularly during the middle decades of the nineteenth century when scientific and mathematical research was strongly influenced by the widely divergent regional conditions that prevailed throughout the German states.

Essentially, research schools have been understood as locally-based groups of researchers working under the supervision or guidance of an acknowledged leader and pursuing a shared set of research interests. Above all else, the common denominator among such schools has been taken to reside in the structure and dynamics of their core groups along with the attendant norms and inner values evinced in the research practices of their members. Determining membership in a school or even the very existence of such a setting can be quite problematic owing to the voluntary character of the enterprise. Clearly, the leader of a school had to be not only an acknowledged authority in the field but also someone capable of imparting expertise to his pupils. This carried with it, among other demands, an obligation to supervise doctoral dissertations and postdoctoral research. This supervisory function, however, could take almost any form depending on the working style adopted by the school's leader. In the final instance, this kind of arrangement depended on an implicit reciprocal agreement between the professor and his students, as well as among the students themselves, to form a symbiotic learning and working environment based on the research interests of the professor. Unlike laboratory research schools, however, the principal aim of a typical mathematical school was *not* to promote a specific research program nor was its purpose to engage in a concerted effort to solve a problem or widen a theory. These were merely potential means but not the the main end of these schools, which primarily sought to produce new mathematicians rather than new mathematical results. The success of a given school was thus judged not so much by the immediate results produced by its younger members but by their later achievements as mature creative mathematicians.

Taking this much for granted, we may safely agree to confine the term 'mathematical school'—assuming the notion is to have any force at all as an analytical category—to those situations where the presence of a cohesive working group, with the sense of belonging that membership in such a group entails, can readily be discerned. This means, on the one hand, abandoning the commonly used label of 'school' in connection with localities like Göttingen or Berlin, which might better be called research centers at which from time to time schools, like those of Klein and Weierstrass, happened to thrive. By the same token, historians should avoid the trap of identifying intellectual traditions or research styles with schools, as has often been done, for example, when contrasting Riemannian and Weierstrassian function theory. Since Riemann founded no substantial school as such—Gustav Roch died at an even younger age than his mentor and Friedrich Prym largely worked in isolation in Würzburg—the comparison ought to be made in terms of methodological issues rather than by invoking teacher-pupil relationships that only capture in a very limited way the real issues at stake. Following these precepts, we can easily identify numerous examples of mathematical research schools at the German universities during the nineteenth century: among the more important were those headed by Jacobi, Weierstrass, Clebsch, Klein, and Lie. Still, some of the era's most influential figures, notably Gauss, Dirichlet, and Kummer made their marks without cultivating a school-like atmosphere.

In Klein's case, the school he headed at both Leipzig and Göttingen during the second period in his career represents the apogee of his efforts as a master teacher. The model he adopted then had much to do with the dramatic collapse he suffered in late 1882 in an effort to match Poincaré's sensational work which broke open the general theory of automorphic functions. During Klein's earlier creative period at Erlangen and the *Technische Hochschule* in Munich, his students' research reflected the broad spectrum of interests that Klein initially pursued in rather haphazard fashion. His second period, on the other hand, was largely devoted to systematic efforts to develop various parts of a broad terrain dominated by the theory of algebraic functions. Klein's general strategy aimed to erect an imposing Riemannian-style edifice designed to compete with the program pursued by Weierstrass in his famous cycle of lecture courses on complex function theory. Thus, Klein's own lecture courses during his second period concentrated on a geometric approach to the theory of elliptic, hyperelliptic, and Abelian functions. But while these courses provided the basic foundations for his research program, the real dirty work was to be carried out by his pupils [Parshall and Rowe, 1994, 175–228].

The late 1880s and early 1890s was also the period when dedicated young mathematical talent proved particularly difficult to come by at the German universities owing to a dramatic drop in enrollments. Once again, though, Klein managed to make the best of a situation other mathematicians might have considered intolerable. The German universities had long been attracting ambitious students from abroad, and this trend continued to gain momentum during the 1880s when numerous North Americans began to find their way into mathematics lecture halls. Intellectually, mathematicians from the United States drew on a wide variety of European resources. Nevertheless, at this decisive stage in the incipient development of their research community the younger generation of Americans looked to Germany's mathematicians and mathematical institutions as their principal models, and no one in Germany could compete with Felix Klein when it came to exciting foreigners about mathematics and engaging them to produce it. Incredible as it may seem, during this second stage in his career Klein attracted more talented American youth than all the other mathematicians in Europe combined! In view of this singular achievement, it is instructive to consider more closely how Klein conducted the seminars he ran during this period. But before doing so, and by way of contrast, a few passing remarks ought to be made about another foreigner who exerted a considerable, though less lasting influence on American mathematics, namely James Joseph Sylvester, who was appointed as the first professor of mathematics at Johns Hopkins University at its founding in 1876 [Parshall and Rowe, 1994, 59–146].

Sylvester's inimitable enthusiasm for mathematics and poetry were legendary, and for his many admirers the Englishman's romantic eloquence fully compensated for the skills he otherwise lacked as a teacher. In fact, Sylvester rambled rather than taught, but he soon attracted an audience of devotees who learned how to think along with him. This group formed the nucleus of Sylvester's mathematical 'laboratory', a term he used to describe the experimental working style he and the members of his school adopted. Relying heavily on guesswork, conjecture, and argument by example and counterexample, Sylvester's students groped their way along week by week trying to match wits with each other as well as with their teacher. Their work on combinatorics eventually bore fruit, but the results were far less impressive than the 'laboratory' environment that produced them, itself an experiment that could hardly be replicated. Indeed, when Sylvester retired after seven years at Johns Hopkins, its mathematics department quickly lost the momentum his presence had provided. Perhaps no mathematician could have created a viable mathematical school in the United States so early and with so few resources, but Sylvester's idiosyncratic style meant that his students had nothing to fall back upon once their mentor had left their midst, and consequently none of them went on to successful research careers.

Following an abortive attempt to hire Felix Klein as Sylvester's successor, the Hopkins program languished in the hands of the astronomer Simon Newcomb. Like Klein, Newcomb had plenty of ambition, but he lacked any real appreciation for European research in pure mathematics and this proved disastrous for the fledgling enterprise Sylvester had launched. Graduates of leading universities in the United States thus found few options for undertaking further study in their native country. Word soon spread, however, that Klein's lectures and seminars offered the kind of advanced training many were hoping to gain. Several of the young men and women who found their way to his

Leipzig and Göttingen lecture halls returned to the United States with a burning desire to incorporate the traditional German research ethos into the fabric of the American mathematical community. In this respect, Klein served as the spiritual father figure for the generation that made higher mathematics a visible part of the mathematical culture in the United States.

One of Klein's most fervent American disciples was Frank Nelson Cole, who later taught at Columbia University. He is remembered today in connection with the American Mathematical Society's Cole Prize, named in honor of the man who served for many years as Secretary of the AMS. At Leipzig, Cole attended Klein's seminar along with Princeton's Henry Fine, and he later wrote lengthy accounts of Klein's work for the Society's Bulletin. Cole's description of the orderly manner in which Klein conducted the Leipzig seminar makes for a vivid contrast with the extemporaneous style Sylvester cultivated at Johns Hopkins:

> [His] management of the *Seminar* has always been exceptionally efficient, even among the German models. It is Klein's custom to distribute among his students certain portions of the broader field in which he himself is engaged, to be investigated thoroughly under his personal guidance and to be presented in final shape at one of the weekly meetings. An appointment to this work means the closest scientific intimacy with Klein, a daily or even more frequent conference, in which the student receives generously the benefit of the scholar's broad experience and fertility of resource, and is spurred and urged on with unrelenting energy to the full measure of his powers. When the several papers have been presented, the result is a symmetric theory to which each investigation has contributed its part. Each member of the *Seminar* profits by the others' points of view. It is a united attack from many sides of the same field. In this way a strong community of interest is maintained in the *Seminar*, in addition to the pleasure afforded by genuine creative work [Cole, 1893, p. 107].

This kind of carefully orchestrated research environment—so characteristic of Klein's second period—not only represents the extreme opposite of Sylvester's laboratory approach it also contrasts sharply with the open-ended atmosphere that emerged in Göttingen during the third phase of Klein's career. The reasons for this striking shift take us back to the theme of Prussian educational politics touched upon earlier.

KLEIN'S ASCENT TO POWER AS A WISSENSCHAFTSPOLITIKER

Klein, who was occasionally introspective, sometimes thought of himself as a Faustian figure. He longed for knowledge, but he also loved power: either to exercise it himself or to exert it through others. Most of those who acted in concert with Klein did so in subservience to his authority, but his success also required an ability to serve those in still higher positions, most notably Friedrich Althoff, the controversial ministerial official who practically single-handedly ran the entire Prussian system of higher education [vom Brocke, 1980]. As a measure of Althoff's authority, when he died in 1908 his position in the Prussian Ministry of Culture had to be broken up into four new ones in order to reduce the workload to reasonable proportions! From the beginning of his administrative career, Althoff showed little patience with the vaunted ideals spouted by German professors in defending the integrity of their specialized fields of expertise [Zassenhaus, 1973]. He thus set himself the task of breaking down the system of inbred schools by establishing special centers for integrated research at designated Prussian universities. Instituting such a reform from above was not easy, and Althoff encountered strong resistance from distinguished scholars like the sociologist, Max Weber, a spokesman for the traditional ideals of academic integrity based on the sovereignty of the individual faculties. Weber and other critics saw in Althoff's policies an attempt to undermine academic and disciplinary freedoms within the Prussian universities.

The so-called 'Althoff system' involved a complicated network of *Vertrauensleute*, mainly university professors, who dutifully served the man at the top. Such unofficial contacts enabled Althoff to obtain detailed information about the academic standing and personal qualities of practically every potential candidate for even the lowliest positions within the Prussian system of higher education. Klein entered gradually into this intimate sphere of advisors to Althoff beginning with his appointment at Göttingen in 1886. Like his *de facto* boss, Klein had the kind of vast ambition and immense appetite for work that Althoff admired. Yet it was not until the mid 1890s that he attained a strong enough position that he could—with Althoff's backing—begin to unveil major new plans for reshaping the mathematical sciences and related fields in Göttingen. Much to his chagrin, Klein found during his first six years in Göttingen that his alliance with Althoff was rather weak and ineffectual. For this reason, it will pay to reconsider in detail some of the events that signaled the transition from the second to the third phase of Klein's career.

As has long been recognized, the year 1892 marks a major break in the internal power relations within the German mathematical community.

The *Deutsche Mathematiker-Vereinigung* (German Mathematicians Union or DMV) had been founded only two years earlier, just before Kronecker's unexpected death in late 1891. This latter event enabled Weierstrass to enter his long-anticipated retirement, thereby opening the two principal chairs in mathematics at Berlin. In the course of filling these, Althoff set off a chain reaction of appointments that ultimately reconfigured the power structures within the German mathematical community while setting the scene for the ensuing conflicts and struggles of the 1890s. As is also well known, Klein and Göttingen eventually emerged as the big winners once the dust had cleared [Biermann, 1988, pp. 150–152]. Not coincidentally, after 1892 Klein suddenly lost interest in pursuing the research program that had previously occupied his attention for a good decade. No longer facing a strong rival school in Berlin—Weierstrass had retired, and his successor, Schwarz, rather unexpectedly entered a period of semi-retirement—Klein clearly hoped to turn over the burden of training young researchers to someone who was still in the prime of his career. So, after gaining Friedrich Althoff's confidence, he began to contemplate plans for launching something more ambitious than a mere extension of his established research school. His idea was to draw on his extensive network of scientific contacts in order to transform Göttingen into a new kind environment reflecting a diverse range of research activities, but with pure mathematics situated firmly in the center binding the operation together. High on his agenda, though rarely articulated, was a deep-seated belief in mathematics as a collective enterprise, and Klein's organizational innovations did much to instill a sense of community that soon gave Göttingen mathematics an atmosphere all its own.

Still, this process took considerable time and, initially at least, the outcome would undoubtedly have appeared unlikely to the actors directly involved. Indeed, contemporary views ran in precisely the opposite direction, and for good reason. In the first place, it came as a major surprise to many when Klein was passed over for the two openings in Berlin, one of which went to a favorite Weierstrass pupil, Hermann Amandus Schwarz, and the other to the algebraist, Georg Frobenius, another leading product of the Berlin milieu. In one sense, Schwarz's departure from Göttingen was a great relief to Klein, since he could now try to shape the research orientation there along lines that suited him. In particular, he wanted to have either David Hilbert or Adolf Hurwitz appointed as Schwarz's successor [Rowe, 1986, pp. 433–436]. Both had been associated with his research schools in

Munich and Leipzig earlier, and both knew one another well from their many years together in Königsberg. The problem for Klein was, first, that Hilbert had not yet attained a professorship, which meant that the Göttingen philosophical faculty would hardly be inclined to offer him a chair that traced its lineage back to Gauss, Dirichlet, and Riemann. Second, Hurwitz was a Jew, and the Göttingen faculty had an unofficial quota policy that limited the number of Jewish professors to one per discipline. Klein described these and other difficulties in a letter written to Hurwitz at the outset of the negotiations:

> Althoff was here for three days and has decided on the calls to Berlin ... [Concerning Schwarz's replacement] you will probably have guessed that I want to recommend you and Hilbert as the only two who, together with me, are in a position to assure Göttingen a place of scientific distinction ... Naturally, I will name you first and Hilbert behind you. There are, however, a series of difficulties associated with your being called ... First, there is the problem of your health ... Secondly, there is the much subtler difficulty that you are, not only personally but also in your mathematical style, much closer to me than is Hilbert. Your coming here could therefore perhaps give our Göttingen mathematics a too one-sided character. There is thirdly—I must touch on it, as repugnant as the matter is to me, and knowing full well your justified sensitivity to this—the Jewish question. Not that your call as such would present difficulties; these I would be able to overcome. The problem is that we already have [Arthur] Schönflies, for whom I would like to create a firm position as salaried *Extraordinarius*. And having you and Schönflies together is something I will not get past either the faculty or the Minister.[1]

Klein's quandary was actually even more difficult than he let on in this letter to his star pupil. For one thing, he failed to mention the problems he faced due to the jealousies aroused within his school. Klein clearly wanted to appoint a younger mathematician at Göttingen, and this meant passing over his former pupil, Ferdinand Lindemann, the senior figure at Königsberg, who expressed bitter disappointment when he learned of Klein's decision. Far more consequential than this internal strife, however, was the conflict Klein faced in the Göttingen faculty, since neither the departing Schwarz nor his ally, Ernst Schering, favored Hurwitz's appointment. After an intense debate, the faculty compromised on the following list of candidates, presented to the Ministry in rank order: 1) Heinrich Weber, 2) Adolf Hurwitz, 3) Friedrich Schottky [Rowe, 1986, p. 434]. Since the Ministry could freely choose from this list (occasionally such *Berufungsliste* were even ignored altogether),

Klein was counting on Althoff to reach over Weber and choose Hurwitz instead. He even made it clear to Althoff that, in view of the faculty's stance vis-à-vis Jewish scholars, he would be willing to drop his efforts on behalf of Schönflies in order to ensure Hurwitz's appointment.[2]

This strategy might well have worked out had not the situation then taken a surprising new turn when Frobenius, who had not yet officially accepted Kronecker's chair in Berlin, began to express an interest in the vacancy in Göttingen. Caught by surprise, Klein realized at once that he now had a chance to gain the services of one of Berlin's stellar pupils. Moreover, since Frobenius was a leading expert in algebra, his appointment would obviate the problem of overlapping research interests that Klein had raised in his letter to Hurwitz. Hardly able to contain his excitement, Klein rather tactlessly informed Hurwitz that, had he known there was any possibility of appointing Frobenius, he would have placed the Berlin-trained mathematician's name first on his list of nominees (Hurwitz still had no idea that Weber's name, and not his, headed the faculty's list).[3] Klein then invited Frobenius to visit him in Göttingen, afterward waiting in suspense to learn his decision. It came quickly, and soon after Frobenius opted for Berlin, Althoff offered the Göttingen chair to Heinrich Weber, thereby thwarting Klein's initial plans. Although a distinguished mathematician, Weber was the kind of quiet scholar typical of an era that was fast receding below the horizon. His congenial, easy-going personality contrasted sharply with Klein's energetic manner, which fit the new mood of the post-Bismarckian age so well. As it turned out, the two men got along just fine, but at the time Klein felt that he had lost considerable face and that his position within the Göttingen philosophical faculty had been badly damaged. Having argued fervently for Hurwitz's appointment, only to see the opposition's candidate, Weber, called instead, he wrote Althoff a whining letter proclaiming that he was now in an intolerable position. In Klein's view, the situation could

... only be somewhat remedied by having Schönflies named *Extraordinarius*. On the one hand, it is known that I have been working on his appointment for years; on the other, that my efforts have only met with resistance, so that I only dispensed from doing so as Hurwitz's call sttod in question. Should Schönflies now be passed over, this impression [i.e., of Klein's impotence] will become a virtual certainty. I would then be forced to advise young mathematicians not to turn to me if they hope to make further advancements in Prussia.[4]

Not long after Klein wrote this letter, Schönflies received the coveted appointment as associate professor in Göttingen. Hurwitz, on the other hand, fared less well, perhaps partly because of Althoff's own unhappiness about the course of events in 1892. Anticipating these problems, Klein raised the specter of anti-Semitism in a letter to Hurwitz, predicting that his former pupil's 'chances of succeeding Weber in Marburg or obtaining a call anywhere else in Prussia were unfavorable'.[5] Shortly thereafter Hurwitz took a position at the Zurich Polytechnical Institute, where he remained until his death in 1919, despite his expressed wish to return to a Prussian university. Whether or not anti-Semitism actually played a part in hindering Hurwitz's career remains an open question, but as to the pervasive character of this sentiment within the German universities there can be no doubt. One mathematician who knew about it from firsthand experience was Klein's older friend and ally, Paul Gordan. When Gordan learned that Klein's effort to engineer Hurwitz's appointment had backfired, he wrote to him as follows:

I am sorry to hear that you were not called to Berlin, as your all-embracing spirit would have brought order to the mathematical relationships in Germany ... It was just that you recommended Hurwitz for Göttingen; Hurwitz deserves this distinction. That your recommendation did not go through, however, is a fortune for which you cannot thank God enough. What would you have had with Hurwitz in Göttingen? You would have taken on the complete responsibility for this Jew; every real or apparent mistake by Hurwitz would have fallen on your head, and all his utterances in the faculty and senate would have been regarded as influenced by you. Hurwitz would have been considered nothing more than an appendage of Klein.[6]

It seems unlikely that Klein would have felt consoled by Gordan's remarks, as the outlook from his perspective in April of 1892 looked very bleak indeed. Althoff, after all, was a man for whom actions spoke more loudly than mere words, and his own behavior hardly suggested that he wanted to disrupt the established mathematical order, largely dominated by Berlin, in order to install a new regime in Göttingen under Klein's leadership. Thus, it appeared very much as if Klein's broader ambitions had been stymied once again. But just two months later he got a tremendous break, though not from Althoff nor even from within Prussia. It came rather in the form of a very attractive offer from the Bavarian Ministry of Culture, which hoped to induce Klein to accept the chair at Munich University vacated by Ludwig Seidel's retirement. Even before Klein had the official offer in hand, he took up negotiations with Althoff, who matched the Bavarian conditions right down the line, thereby enabling the rejuvenated Göttingen mathematician to decline graciously [Toepell, 1996, pp. 208–212, 370–378].This was just the chance Klein had been waiting for: to serve the Prussian state. It had taken him fifteen years to become a professor in Prussia, and another six had passed before he could demonstrate his loyalty by declining a lucrative offer from the outside. The next year, marked above all by his triumphal trip to the Chicago Mathematics Congress as official representative of the Prussian government, saw Klein beginning to unfold some of the grandiose plans that would dominate his attention during the third phase of his remarkable career.[7]

GÖTTINGEN IN THE ERA OF KLEIN AND HILBERT

If the year 1893 saw Klein finally realizing his long-sought ambition to become a *Wissenschaftspolitiker*, one whose clout extended well beyond the discipline of mathematics, it would be grossly misleading to suggest that he already had some kind of blueprint in mind for building a new type of mathematical center in Göttingen. Many of his subsequent actions were improvised, and it took more than a decade before all the main players and institutions in Klein's Göttingen venture had found their appropriate niches. Karl-Heinz Manegold has described the numerous obstacles Klein had to overcome in launching the *Göttinger Vereinigung zur Förderung der angewandten Physik* in 1898 [Manegold, 1970]. This organization, an unlikely mixture of scholars and industrialists, spearheaded Klein's efforts to promote applied research at Göttingen. Initially, Klein approached the Krupp firm in hopes of gaining financial backing for this venture, but the leadership of the steel giant reacted indifferently. After this setback, Klein managed to win over the politically-active general director of the Bayer chemical works, one Henry Böttinger, who also happened to be an ally of Althoff's. Still, Klein's venture faced strong opposition from quarters that evaded even Böttinger's and Althoff's extensive spheres of influence, namely powerful engineers and technologists like Alois Riedler and Adolf Slaby. The latter had a direct line to the ear of Kaiser Wilhelm II, a strong proponent of technical education. While Manegold deals with these highly visible political debates and their eventual resolution in considerable detail, his account fails to take note of the serious problems that continued to plague the Göttinger

Vereinigung up until the appointments of Carl Runge and Ludwig Prandtl in 1904. During the preceding six years, in fact, Klein was continually frustrated by a series of unhappy appointments that left the future of Göttingen's new professorship in technical physics unsettled. Thus, it took another half dozen years *after* the founding of the Göttinger Vereinigung before Klein's Göttingen venture finally began to stabilize. Indeed, the year 1904 marks the true emergence of the modernized Göttingen scientific community as depicted in Herbert Mehrtens's provocative book.

Just two years earlier Klein had managed to avert another major crisis that arose when Hilbert received a call from Berlin. No mathematician in Germany since Gauss had declined such an opportunity, and it was clear to everyone involved that this tempting offer constituted a true test that would determine which of the two German centers held dominance in pure mathematics. By this time, Hilbert had more than fulfilled Klein's prediction that he was the 'rising man' on the scene, and Minkowski had long been hinting that his friend would do well to remove himself from Klein's long shadow.[8] So Hilbert was willing to play for big stakes, but, to his delight, Althoff was more than willing to match them, creating a third *Ordinariat* in Göttingen for Minkowski, a move Hilbert later effusively called 'perhaps unique in the annals of Prussian higher education' [Hilbert, 1910, p. 355].

Two years later, with the appointments of Runge and Prandtl, even the Prussian Ministry was taken aback by the sudden convergence of manpower and resources that had sprung to life in Göttingen, and no one with eyes in their head could have failed to see that the Göttingen mathematicians represented the dominant bloc within the German mathematical community.[9]

Yet Göttingen's mathematicians, led by Klein and Hilbert, were by no means content to rest on their laurels [Pyenson, 1979]. Through influential pupils like Arnold Sommerfeld and Max Born, both Klein and Hilbert strove to make inroads into theoretical physics, a field that had only begun to establish itself as an independent discipline. Thus, unlike the Berlin mathematicians of this period (Frobenius, Schwarz, and Schottky), the Göttingen crowd wanted not only to dominate pure mathematics but also to extend that dominance beyond the traditional disciplinary boundaries established by their predecessors.

Klein's initial efforts to accomplish this found concrete expression in the *Encyklopädie der mathematischen Wissenschaften*. Although this project gave considerable visibility to recent work in analysis and geometry, its strength was especially evident in the applied volumes: mechanics (volume 4, edited by Klein and Conrad Müller), physics (volume 5, edited by Arnold Sommerfeld), and astronomy, geophysics, and geodesy (volume 6, edited by Karl Schwarzschild, Emil Wiechert, and Phillip Furtwängler, respectively). A number of the more important articles were written by foreign authorities, for example, the two classic essays by H.A. Lorentz on Maxwell's theory of electromagnetism and his own theory of the electron. Particularly noteworthy contributions came from Paul and Tatyana Ehrenfest, who surveyed the foundations of statistical mechanics, as well as the young Wolfgang Pauli, whose essay on relativity theory quickly established itself as a classic. The editors and authors of these and many other articles in the *Encyklopädie* operated within Klein's extensive network of scientific contacts. Their names and institutional affiliations thus serve as clues to the various complicated alliances he managed to build between Göttingen and other leading scientific centers, including Munich, Leiden, and Cambridge. Klein's longstanding connections with leading British mathematicians and natural philosophers deserves special mention in this regard.

Like Helmholtz, who had done so much to promote Maxwell's theory in Germany, Klein was one of the few German scientists who took a deep interest in the British mathematical—physical tradition. Moreover, he had maintained contacts with British scientists ever since his first visit to England in 1873, when he befriended Arthur Cayley, J.J. Sylvester, R.S. Ball, and W.K. Clifford. One of Klein's main goals with the *Encyklopädie* was to showcase fields of applied research that had been cultivated by leading British figures. In a whirlwind tour of the British Isles during the late 1890s he managed to gain the support of several leading physicists for his project. Thus, A.E.H. Love offered two articles on hydrodynamics, Horace Lamb contributed an essay on acoustics, Edmund Whittacker dealt with celestial orbits, perturbation theory, and dynamical systems, and the Anglo-American Ernest W. Brown wrote on lunar theory. Klein's success in this endeavor reflects the strength of his international network in both England and the United States. The two traditional European centers, Paris and Berlin, on the other hand, were conspicuously underrepresented in Klein's *Encyklopädie* project.

Seen in retrospect, one can readily identify some of the strategic goals that motivated Klein's efforts and schemes in Göttingen. With the help of Althoff, he expanded his own formidable network of contacts and used these in order to transform Göttingen into an intensely interactive scientific center [Rowe, 1989]. By taking the lead in this initiative, rather than developing such a plan in cooperation with his colleagues in the natural sciences, Klein could push forward a vision of the Göttingen community in which he and his fellow mathematicians stood at the hub of the whole enterprise. One of the cornerstones in Klein's overall strategy for building up mathematics in Göttingen was reliance on diverse intellectual and material resources with the potential for mutual reinforcement. While this might sound rather trite and obvious, such a tendency ran straight across the grain of established organizational structures within the German university system. As semi- autonomous institutions, Germany's philosophical faculties, particularly in the Prussian universities, had proliferated throughout the course of the century largely by virtue of the specialized research interests typically undertaken in various scientific schools. Such schools had, of course, no official status, but they led to informal scientific networks that clearly played a conspicuous part in setting research agendas and shaping careers.

Both Klein and Althoff saw little use for the kinds of narrowly focused research schools that adorned many German universities. Like his former teachers, Plücker and Clebsch, Klein had a long-standing suspicion of the Berlin mathematical tradition, and he opposed the kind of disciplinary purism promoted in Berlin, sensing that this tendency weakened the position of mathematics vis-à-vis other neighboring disciplines. His Göttingen experiment thus meant consciously breaking with the notion of 'mathematics for mathematics sake' which had served as the dominant ideology in Prussia since Jacobi's day. Frobenius, who saw himself as the prime protector of this legacy, loathed everything associated with Klein and his Göttingen enterprise. He and the other leading representatives of Berlin mathematics—Schwarz, Fuchs, Schottky, and Schur—thus took no part in Klein's massive *Encyklopädie der mathematischen Wissenschaften*, a project Frobenius once dubbed as 'senile science'.[10] But Klein had no interest in undermining traditional pure mathematics; he thought of Göttingen mathematics as representing in microcosm the full range of the discipline, including number theory, algebra, and higher analysis, thereby paying Gaussian rigour its rightful due.

Hilbert's vast research program, unveiled in part in his famous address of 1900 at the Second International Congress of Mathematicians in Paris, fit beautifully within Klein's overall conception.

Volker Peckhaus has described how Hilbert sought to create institutional support for research in the foundations of mathematics, even by competing with the Göttingen philosophers for control of positions in philosophy [Peckhaus, 1990].

Hilbert's penchant for power typically operated within the discipline, though with forays into epistemology and the foundations of physics in the name of his program to axiomatize all exact knowledge. Klein, on the other hand, had far more experience with the actual problems that stand in the way of successful mathematical 'foreign relations', namely convincing those outside the discipline of the importance of mathematical research. Klein's own teaching activities were part and parcel of his diplomatic efforts to promote mathematics, and he always emphasized the importance of tangible manifestations of mathematical productivity. As a particularly famous instantiation thereof, he touted the merits of building mathematical models which he placed on prominent display. The impressive collection of mathematical models and instruments found today exhibited throughout the Göttingen Mathematics Institute serves as a reminder of this Kleinian legacy, one whose purpose was twofold. While the display reflected to some extent the concerns of practical men, engineers and architects, its overwhelming message was symbolic and achieved through an intermingling of technical mathematical knowledge with aesthetically-pleasing geometrical representations. Some who contemplated the deeper meaning of the many plaster models of beautiful algebraic surfaces came away with a deeper appreciation of their mathematical properties. But whether or not studying such models furthered ones technical understanding of the mathematics involved, the larger purpose could still be served, namely to evoke in the observer a sense of wonderment. For not a few who studied in Göttingen, the sight of all these riches brought with it a sense of mystery and awe. Evidence of this can be found not only from contemporary testimonies but also the commercial success enjoyed by the firms of L. Brill and later M. Schilling in marketing copies of many of the models Klein promoted to leading departments of mathematics in Europe and the United States.

Klein's efforts to make mathematical knowledge visible, his numerous appeals to *Anschaulichkeit*, went hand in hand with policies designed to make pure mathematics an indispensible part of applied research. Those policies gained a strong foothold when the *Göttinger Vereinigung* began providing money to build a series of new institutes in applied fields. But Klein's intention all along was not just to establish a series of such instibues at Göttingen but rather to place the mathematicians physically in the middle of this new research complex. Thus, by 1910 he had persuaded the organization's leading industrialists to purchase the land required to erect his most cherished dream: a new mathematical institute in the immediate vicinity of the bustling scientific center on the south end of the town. Only then, would the importance of mathematics to the whole enterprise become as transparent to the outsider as it had long been to those familiar with life on the inside of the institutes' walls. The war and its ensuing economic hardships dashed all hopes for realizing such an institute, but in an ironic turn of events, Richard Courant managed to negotiate a deal with the Rockefeller Foundation that led to the opening of Göttingen's new mathematical quarters in the Bunsenstrasse in 1929 [Reid, 1976]. By then, Klein was dead and the heyday of the Hilbert era was long over.

If contemporary mathematicians continued long afterward to identify the massive talent that gathered around Hilbert after 1900 as the 'Hilbert school', we can now with retrospect recognize the striking differences between this phenomenon and conventional mathematical schools. Indeed, Hilbert openly criticized the kinds of schools that dominated the German universities during the 1880s and 1890s, advocating along with Klein a more integrated model for organizing mathematical research. While Hilbert continued to espouse the optimistic views of his famous Paris lecture for the next two decades, Klein voiced concern about the future of his discipline. In particular, he felt that pure mathematics, Germany's traditional pillar of strength, had been cultivated to the neglect of applications. In Klein's view, it was high time for the German mathematical community to climb down from its ivory tower and pay more attention to the needs of modern exact scientists and engineers. Throughout all of Klein's maneuvering, he strove for a broadly balanced faculty with interests ranging from Landau's expertise in number theory to Prandtl's masterful understanding of hydrodynamics. At the same time, he wanted to break with the established German tradition of disciplinary purism, a penchant tracing back to Jacobi's Königsberg school, but which reached its zenith in Berlin during the period 1855–1892, when Kummer, Weierstrass, Steiner, and Kronecker taught there. Despite the shifts Klein made during each of the four principal stages of his career, one can still discern a consistent pattern of rejection of such purism and, in particular, opposition to those German mathematical schools that promoted various special brands of research.

NOTES

1. Felix Klein to Adolf Hurwitz, 28 Feb. 1892, Mathematisches Archiv, Niedersächsische Staats- und Universitätsbibliothek Gättingen, my translation.
2. Felix Klein to Friedrich Althoff, 7 March 1892, Rep. 92 Althoff AI no. 84, Bl. 21–22, Zentrales Staatsarchiv, Berlin.
3. Felix Klein to Adolf Hurwitz, 28 Feb. 1892, Mathematisches Archiv, Niedersächsische Staats- und Universitätsbibliothek Göttingen.
4. Felix Klein to Friedrich Althoff, 10 April 1892, Rep. 92 Althoff AI no. 84, Bl. 32–34, Zentrales Staatsarchiv, Berlin.
5. Felix Klein to Adolf Hurwitz, 11 April 1892, Mathematisches Archiv, Niedersächsische Staats- und Universitätsbibliothek Göttingen.
6. Paul Gordan to Felix Klein, 16 April 1892, Cod Ms. Klein IX, Niedersächsische Staats- und Universitätsbibliothek Göttingen.
7. On Klein's trip to Chicago, see [Parshall and Rowe 1994, pp. 295–362].
8. Klein used the phrase 'rising man' in a letter to Althoff, written on 23 Oct. 1890 (Rep. 92 Althoff B No. 92, fols. 76–77, Zentrales Staatsarchiv, Berlin). Minkowski's advice was imparted to Hilbert in a letter of 20 July 1898, see [Minkowski 1973, p. 108].
9. Elster to Klein, 18 July 1904, transcribed in Klein *Nachlass* IV.
10. F. Klein to Walther von Dyck, 24 March 1893, Bayerische Staatsbibliothek, München, cited in *ibid.*, p. 209.

REFERENCES

Biermann, K.R., [1988] *Die Mathematik und ihre Dozenten an der Berliner Universität, 1810–1933*, 2nd ed., Berlin, Akademie Verlag.

vom Brocke, B., [1980] 'Hochschul-und Wissenschaftspolitik in Preussen und im Deutschen Kaiserreich, 1882–1907: Das "System Athoff"' in Baumgart, P. (ed.), *Bildungspolitik in Preussen zur Zeit des Kaiserreichs*, Stuttgart, 9–118.

Cole, F.N., [1893] 'Klein's Modular Functions', *Bulletin of the American Mathematical Society*, 2, 105–120.

Geison, G.L., [1981] 'Scientific Change, Emerging Specialties, and Research Schools', *History of Science*, 10, 20–40.

Hensel, S., [1989] 'Die Auseinandersetzungen um die mathematische Ausbildung der Ingenieure an den Technischen Hochschulen in Deutschland Ende des 19. Jahrhunderts' in *Mathematik und Technik im 19. Jahrhundert in Deutschland*, Studien zur Wissenschafts- Sozial- und Bildungsgeschichte der Mathematik, vol. 6, Göttingen, Vandenhoeck & Ruprecht.

Hilbert, D., [1910] 'Gedenkrede auf Hermann Minkowski' *Mathematische Annalen*, 68, 445–471 (in *Gesammelte Abhandlungen*, vol. 3, 339–364).

[1932–1935] *Gesammelte Abhandlungen*, 3 vols, Berlin, Springer.

Klein F., [1923] 'Göttingen Professoren: Lebensbilder von eigener Hand', *Mitteilung des Universitätsbundes Göttingen*, 5, 11–36.

Manegold, K.H., [1970] *Universität, Technische Hochschule und Industrie. Ein Beitrag zur Emanzipation der Technik im 19. Jahrhundert unter besonderer Berücksichtigung der Bestrebungen Felix Kleins*, Berlin, Duncker and Humbolt.

Mehrtens, H., [1990] *Moderne—Sprache—Mathematik. Eine Geschichte des Streits um die Grundlagen der Disziplin und des Subjekts formaler Systeme*, Frankfurt, Suhrkamp.

Minkowski, H., [1973] *Briefe an David Hilbert*, ed. L. Rüdenberg and H. Zassenhaus, New York, Springer.

Parshall, K., Rowe, D.E., [1994] *The Emergence of the American Mathematical Research Community, 1876–1900: J.J. Sylvester, Felix Klein, and E.H. Moore*, Providence, American Mathematical Society, History of Mathematics, 8.

Peckhaus, V., [1990] *Hilbertprogramm und Kritische Philosophie*, Studien zur Wissenschafts- Sozial- und Bildungsgeschichte der Mathematik, vol. 7, Göttingen, Vandenhoeck & Ruprecht.

Pyenson, L., [1979] 'Mathematics, Education, and the Göttingen Approach to Physical Reality, 1890–1914', *Europa: A Journal of Interdisciplinary Studies*, 2 (2), 91–127.

[1983] *Neohumanism and the Persistence of Pure Mathematics in Wilhelmian Germany*, Philadelphia, American Philosophical Society.

Reid, C., [1976] *Courant in Göttingen and New York*, New York, Springer.

Rowe, D.E., [1985] 'Essay review of [Manegold 1970], [Tobies 1981], and [Pyenson 1983]', *Historia Mathematica*, 12, 278–291.

[1986] "Jewish Mathematics' at Göttingen in the Era of Felix Klein', *Isis*, 77, 422–449.

[1989] 'Klein, Hilbert, and the Göttingen Mathematical Tradition', *Science in Germany: The Intersection of Institutional and Intellectual Issues*, ed. Kathryn M. Olesko, *Osiris*, 5, 189–213.

Tobies, R., [1981] *Felix Klein*, Biographien hervorragender Naturwissenschaftler, Techniker und Mediziner, Bd. 50, Leipzig, Teubner.

[1991] 'Wissenschaftliche Schwerpunktbildung: der Ausbau Göttingens zum Zentrum der Mathematik und Naturwissenschaften', in vom Brocke, B., (ed.), *Wissenschaftsgeschichte und Wissenschaftspolitik im Industriezeitalter. Das 'System Atlhoff' in historischer Perspektive*, Hildesheim, Verlag August Lax, 87–108.

Toepell, M., [1996] 'Mathematiker und Mathematik an der Universität München. 500 Jahre Lehre und Forschung', *Algorismus*, Heft 19, München, Institut für Geschichte der Naturwissenschaften, 208–212; 370–378.

Zassenhaus, H., [1973] 'Über Friedrich Althoff', in Minkowski [1973], 22–26.

Chapter 5

THE GERMAN AND FRENCH EDITIONS OF THE KLEIN-MOLK ENCYCLOPEDIA: CONTRASTED IMAGES*

Hélène Gispert

In 1898 a German-language encyclopedia of the mathematical sciences began to appear in fascicles, published under the auspices of the Göttingen Academy of Sciences. This undertaking was initiated and directed by Felix Klein, and was pursued systematically until the First World War, eventually being completed in 1935. A French edition of the work also appeared. The work has classically been considered as a striking monument to the organization of the mathematics of the period, and of the construction and reconstruction of mathematical knowledge at that time.

A project of this kind, organized into many volumes, comprising two hundred individual articles and reaching over twenty thousand pages at its completion, is necessarily shaped by editorial choices which recall both the images the original authors have of mathematics and the images held by the editors themselves. These choices and images may be studied equally well at the scale of the individual articles, concepts, and research areas, and at the macroscopic scale of the overall structure of the project. I will focus on the latter here, seeking to explain certain of these choices and the images they induce, while suggesting certain interpretations in the light of different manifestations of mathematical life at the time.[1]

The publicity appearing in the catalogues of the publishers Teubner and Gauthier-Villars is specific about organizational issues and the nature of the enterprise.[2] The title and the overall plan, announced when the fascicles began to appear, attest to the ambitious character of the project. The 'Encyclopedia of the Mathematical Sciences, Including their Applications' was conceived in seven volumes embracing, successively, arithmetic and algebra (volume 1), analysis (volume 2), geometry (volume 3), mechanics (volume 4), physics (volume 5), geodesy, topography and astronomy (volume 6), and questions of a historical, philosophical and didactic nature (volume 7).[3,4] This encyclopedia of the mathematical sciences was explicitly part of a specific genre—that of popularization—and aims at a specific readershiped: professional mathematicians as well as astronomers, physicists and engineers. As the publishers' notices indicate, this is a genre and a readership not usually associated with the traditional specialist mathematical press.

Thus the mathematical sciences on which the two editions of the encyclopedia were intended to expound upon encompass not only 'pure' mathematics, but the applications of mathematics and its history as well. But with this first characteristic, the broad and inclusive nature of the work, we must immediately note a second. The announcement by Gauthier-Villars, in referring to 'French traditions and habits,' seems to add a relative dimension, namely a national one, to this effort at the diffusion of mathematical knowledge. How, and to what extent, these announced intentions were realized is something we examine in what follows.

THE ROLE OF APPLICATIONS: PROMOTING A CERTAIN IMAGE OF MATHEMATICS

Three volumes, constituting half of the two hundred articles in the *Encyklopädie*, treat mechanics, physics, geodesy and topography, or astronomy, rubrics which are grouped in the French edition under the heading 'Applied Mathematics'. The rest, again according to the publisher's announcement for this edition, have to do with pure mathematics.[5] This presentation takes some liberties with the actual content of the different volumes, underestimating the place of applied mathematics in the encyclopedia. For example, in the first volume 'Arithmetic and Algebra' (which actually consists of four physical volumes) one of the four volumes is entitled 'Calculus of probabilities, theory of errors, various applications'. It contains one 250-page article on numerical calculations, another of about one hundred pages on actuarial techniques, and a third of about fifty pages on mathematical economics, not completed for the French edition.[6] Thus over half the volume is devoted to 'applied mathematics' or 'mathematical applications,' the terms used by the French and German editions respectively.

A Skewed Image: The Place of Applications

The results of this kind of editorial choice may be illuminated by comparing the *Encyklopädie* with other contemporary enterprises. When the publication of the original *Encyklopädie* began, less than one quarter of the articles covered by the *Jahrbuch über die Fortschritte der Mathematik* were under the headings of mechanics, physics, astronomy and geodesy. While this fraction increased in the first years of the twentieth century it still reached only one third before the war. The *Jahrbuch* took as its mandate the publication of reviews of all articles on mathematical subjects, whether in mathematical or other scientific journals. An examination of the tables of contents of specifically mathematical journals, such as the *Mathematische Annalen*, Crelle's *Journal*, the *Journal des mathématiques pures et appliquées*, or the *Bulletin de la Société mathématique de France* further underlines the originality of the *Encyklopädie* in this regard: the portion which these journals devote to applied mathematics, no matter how the term is interpreted, is incomparably smaller.

This originality is further shown in the choice and kind of 'applied' subjects included in the *Encyklopädie*. On the one hand, all the major areas of mechanics or physics from these years are covered, at least in the German edition, which is in itself already an extension of the coverage found in the traditional mathematical press. On the other hand, the publishers explicitly requested that coverage should include experimental physics and areas of mechanics associated with technology that could arise in the laboratory [von Dyck, 1908, p. 129]. Thus the *Encyklopädie* had a dual vocation. In it, the mathematician could encounter questions of pure science, as well as those posed by applications.[7] But it was also to be a place where mathematical notions useful to the physicist, engineer, or any member of a potential 'client' discipline could be found, whether for their theoretical requirements or for their applications.[8] In a period when mathematics, engineering, and theoretical physics asserted themselves as separate fields, as shown by the existence of specialized journals in each field, this editorial dictum from the mathematical quarter challenged established ideas about the disciplinary division of labour.

The image which the *Encyklopädie* presents is thus somewhat skewed with respect to the production customarily classified as mathematical at the time. This skewing, which had been deliberate from the outset of the enterprise, is in one sense the trademark of the project as conceived by Felix Klein. Klein considered the *Encyklopädie* as a unique opportunity to revitalize contemporary mathematics, which he considered dangerously specialized, dreadfully abstract, cut off from recent developments in the sciences and technology. This has less to do with establishing a new standard in the organization of knowledge than to renew contacts with the mathematical traditions of the first three quarters of the nineteenth century including the links that had been enjoyed with applications in natural science and technology—but in the context of the new scientific and technological conditions of the beginning of the twentieth century.[9] It was in reaction to the representation which Klein made for himself of the mathematics of his time that he sought to promote this new image: the *Encyklopädie* was to be, in a photographic sense, the negative of this representation.

A German Image?

The obvious stamp of Klein's personal vision may lead us to ask to what extent this 'applied' character of the *Encyklopädie* should be perceived as German. We find an answer, for example, by considering the different developments which mathematics in Germany was then experiencing, where the most brilliant mathematicians were divided between Göttingen, under Klein's rule, and the competing school at Berlin, the home of Frobenius, Schwarz and Fuchs—whose collaboration was not solicited.

At the end of the century, Göttingen was marked by the efforts of Klein, who had succeeded at the local level in developing and institutionalizing a mathematical activity in liaison with other physical sciences and technology.[10] The *Encyklopädie* profited considerably from these efforts, both in the extent and in the quality of the volumes on applied mathematics. The fourth and fifth volumes, directed by Klein and his closest associates (among them Müller and Sommerfeld), contain for example the remarkable expositions of Paul and Tatiana Ehrenfest on the foundations of statistical mechanics, of Boltzmann on the kinetic theory of matter, of Lorentz on Maxwell's electromagnetic theory, and of Pauli on relativity theory, as well as articles by many Göttingen colleagues such as Runge (on measurement) and Minkowski (on molecular physics). Similarly, it was Klein's colleagues at Göttingen, such as the astronomer Karl Schwarzschild, who directed volume 6 (geodesy and geophysics, astronomy).

On the scale of Germany as a whole, however, and even at Göttingen—where Klein had successfully attracted Hilbert—mathematics could not be reduced to the application-related image promulgated by the *Encyklopädie*. The first decade of the twentieth century is symbolized, at least retrospectively, as much by Hilbert's *Grundlagen der Geometrie* and more generally by his famous lecture on the future problems of mathematics at the Paris International Congress of Mathematics, as by Klein's *Encyklopädie*. These masterworks recall conceptions and evoke images of mathematics and its future which differ to the point of contradiction.

The *Encyklopädie* thus permits us to grasp the complexity of an important historiographical question for a period when mathematical activities and communities were structuring themselves along both national and international dimensions: how genuine were apparently national characteristics in the representation that mathematicians made of their discipline? The insistence on applied mathematics is in fact more Kleinian than German, even if the encyclopedic enterprise, pursued under the auspices of the Academies of Göttingen, Leipzig, Munich and Vienna, affected a broader base.

The existence of a French version, whose publisher emphasized the national specificity in pre-publication advertising, gives us the chance to test further the hypothesis of national representations.

The Viewpoint of the French Edition

The *Encyclopédie* presents a markedly different profile from that of the German edition. The volumes on applied mathematics contained fifty percent fewer articles than the first three volumes on pure mathematics. While it is true that any study of the *Encyclopédie* is biased by its incomplete state (Gauthier-Villars abruptly broke off publication in 1916), the extent to which the translation did advance is nonetheless significant given the difficulties which Molk encountered, particularly for the volumes on applied mathematics.[11]

Thus in 1911, among the hundred articles of the first three volumes (described as 'pure mathematics') whose translation was planned, forty-five had appeared or were under way. By contrast, for the remaining volumes, only sixteen were being worked on and none had come out.[12] Eventually, the mechanics volume was truncated by half, and no French author was announced for the articles most applied to technological questions. The physics volume only reached 340 pages, while in the original it had been one of the largest. Furthermore it contained none of the noteworthy expositions cited above in the German edition.

In fact it seems that Molk had great difficulty finding authors for the last three volumes. Gauthier-Villars announced those responsible for the volumes—Appell for volume 4, Potier, then Langevin and Perrin for volume 5, Lallemand and Andoyer for the two parts of volume 6—but none of them appears to have had a large enough community from which to recruit potential workers. The example of Langevin, who was supposed to write the French version of the articles by Boltzmann and Lorentz, and whom Molk begged for his contribution for years without success, suggests two reasons for this problem. Besides a genuine lack of qualified authors, there was a lack of motivation among those who had agreed to participate. This differs considerably from the situation of the German edition, where Klein had been able to enlist the aid of the best scientists in mechanics as well as in physics.

This relative weakness of applications in the French edition recalls the very low participation by French mathematicians in the sections of the International Congresses of Mathematicians which were reserved for applications, whether in 'mechanics or mathematical physics' or in the newer sections devoted to 'various applications of mathematics' (in 1908) or to 'economics, actuarial science and statistics' (in 1912).[13] In the same way, the comparative study of publications of the *Société mathématique de France* and of the *Deutsche Mathematiker-Vereinigung* brings out the lack of interest which the former had in applied areas.[14] Was the situation with the *Encyclopédie*, then, a manifestation of a reticence to promote an image of mathematics which was foreign to a 'less applied' French tradition? We can only attempt a reply if we take into account the diversity of the French mathematical scene and the representations of mathematics associated with different loci.

The position of the Parisian university mathematicians who, with Paul Appell at their head, refused any place at the Sorbonne for applied science, confirms the opposition between Paris and Göttingen. There was no chair of applied mathematics in Paris at the time. It should be noted however that in Paris the chairs in mechanics, mathematical physics, and even experimental mechanics were held by mathematicians.

But the image of mathematical France would be quite incomplete if we restrict ourselves to university professors at the Sorbonne. Even in Paris, other prestigious institutions gave priority to the development of applications, including those of mathematics. Outside the purely 'academic' world, for example at the *Conservatoire national des arts et métiers* or in the *Ecoles d'application* where graduates of the Ecole polytechnique completed their studies, teachers, engineers and military men worked and published mathematical theories and results in a number of applied domains.[15] Outside Paris, it was within certain universities that this type of activity and teaching developed in the last decade of the nineteenth century, for example in Besançon with the mathematician J. Andrade.

Other extra-academic networks like the *Association française pour l'avancement des sciences* (AFAS) sought to promote mathematics in connection with its applications.[16] The mathematical section of AFAS brought together 300 speakers and published twelve hundred articles between 1872 and 1914, covering mathematics, astronomy, geodesy and mechanics. In 1884, at the initiative of its president C. Laisant, the section debated the question 'On how useful it would be to place pure mathematics more completely at the service of the applied sciences, in particular mechanics,' thus showing itself to be completely in line with the views Klein was trying to promote in the *Encyklopädie*.

Escaping the French-German Dichotomy

Historians of mathematics now pursue interests beyond the narrow confines of mathematics strictly within the academy, and the personalities associated with these developments. This allows us to take into account the multiplicity of representations of mathematics in use at various levels. Different kinds of literature of the period present us with differing images which help us to understand the composition of the mathematical sciences in the late nineteenth century.

Considered globally, and as we noted earlier, the response of the *Jahrbuch* does not agree with those of the two editions of the encyclopedia. At the scale of France, once we stop looking at things through the filter of the SMF (which gives a rather narrow image of activity in the French milieu), other prospects and other equilibria are revealed—by the *Jahrbuch* as well.[17] Taking these into account creates problems for a possible uniform characterization, labeled 'French,' of mathematical activities and representations, for example concerning the lack of interest or involvement in areas of application. Taking this into account likewise disturbs the incomplete and reductionist opposition between 'French' and 'German' specificities of mathematical production, an opposition nevertheless suggested by comparing the bulletins of the two national mathematical societies. Thus a comparison of the two editions of the encyclopedia is of interest for other components of the national landscape as well as for the encyclopedias themselves.

Further consideration reveals a second reason why it would be reductionist to consider only the French/German opposition. This project, which brought together those who Klein considered at the time to be the best specialists in each of the domains to be treated, presents a much more interesting and complex image of the mathematical geography of the period, diversifying mathematical centres (and hence peripheries) according to the subjects treated. Of the 165 collaborators announced for the German edition, over twenty per cent were neither German nor Austrian: Italian mathematicians (Enriques, Fano, Loria, Segre, ...) contributed to the geometry volume; British physicists (Lamb,

Love, Whittaker) to mechanics; and to a lesser extent, French mathematicians (Painlevé, Vessiot, Brunel) to analysis. This representation of national competencies, with its deliberate choices and dead ends, is a supplementary historiographical path offered by the the Klein-Molk encyclopedia.

MODERNITY AND THE ENCYCLOPEDIA

In making explicit the way in which Klein's editorial choices about applied mathematics worked in his overall conception of the project, we can no longer have any illusions about the encyclopedia being a faithful or objective representation of the mathematics of his time. So how should one consider the mathematics of this period, a time when classical research patterns coming out of a practice of 'normal science' mixed with the avant-garde results and approaches which were to transform the scientific landscape (in mechanics and physics as well) during the early decades of the twentieth century?

Klein and his associate editors were faced with other kinds of choices. To what extent would they or could they take this exceptional research into account? How far would the image they constructed, during these critical years, belong to the classical world of the nineteenth century, and how far to the one that was just beginning to take shape ?

Different Views on Modernity

The opinions of contemporaries were diverse. Certain of Klein's colleagues, most notably Frobenius in Berlin, called the mathematics set out in the *Encyklopädie* a 'senile science' [Rowe, 1989, 209]. Though of course this rather unnuanced judgement comes from a polemical context, it is of some interest to examine to what extent it is well-founded.

It could be that Frobenius's critique aims principally at the first volume of the *Encyklopädie*, and especially that part of it devoted to algebra. This is not the place for a detailed examination of this particular domain, nor of any other in the *Encyklopädie*, whether theory of numbers, analysis, geometry, probability, or any of the areas in the last three volumes mentioned above.[18] I would like instead to examine the position of Frobenius with regard to several general characteristics of the various volumes.

It is obvious at once that much contemporary research, including the most innovative, was in fact included in the German version. Other new topics appeared in the French version, depending on their date of publication, due to the delay of several years (sometimes ten or more).[19]

These reports, it is true, do take diverse forms. Some newer research was the subject of articles specifically devoted to such topics, but most frequently, especially in the first three volumes, identification of the most recent developments on certain problems was integrated into the body of an article.

Opinions like those of Frobenius thus can not be based on the glaring absence of new results or theories, even in the volumes on pure mathematics. Instead it is the organization of the articles and volumes and the balance between different developments, their respective profiles, that come in for criticism.

Some of the editorial choices expressed in the foreword [von Dyck, 1904] play a determining role from this standpoint. The emphasis placed on the exposition of the historical development of the different branches of mathematics during the nineteenth century determined to a great extent the organization of the work, at least for the first three volumes, and compelled an 'integrated' treatment of contemporary research. The deliberate importance given to articles on applications or in applied fields, especially in the first volume, further isolated the most critical theoretical developments.

Thus, coming back to our initial question, while the most up-to-date and innovative research is not absent from the *Encyklopädie*, Klein and his collaborators did not choose to structure their exposition around this. To follow Frobenius, it remains to understand how their choice comes from a 'senile science.' The study of a particular article, that on analysis situs by Dehn and Heegaard in the geometry volume, can provide us with an entry point [Epple, 1995, 392–93].

This is one of the rare articles in the *Encyklopädie* which was written from the Hilbertian viewpoint characteristic of the period of the *Grundlagen der Geometrie*, in which the concepts, techniques, and hierarchy of problems are defined in an internal fashion, with an elimination of the original context provided by other disciplines. This strategy of elimination, which as Epple argues is one of the elements leading to mathematical modernity, appears as a counterexample with respect to the approach inspired by the 'Kleinian' values and practices which dominate the entire enterprise of the *Encyclopädie*.

The modernity which the *Encyclopädie* might seem to lack could thus be judged from the extent to which avant-garde research was not taken into account, and as a result by the degree of rupture and autonomy from the more classical mathematical environment. Faced with this criterion of modernity, it may be useful to mention another article, namely the one on set theory which appeared in the first volume of the French edition. This article was entirely rewritten by R. Baire who, motivated by a procedure quite different from that of Dehn and Heegaard, was extremely reticent to produce an exposition of set theory independent of other more classical domains to which it might be applied.[20] Note that these three authors, all profound innovators in their respective domains, may be classified in opposed camps as regards the notion of mathematical modernity provided by Mehrtens's reading of the mathematical field of this time, with Dehn and Heegaard being 'moderns' and Baire a 'counter-modern' [Mehrtens, 1990].

The encyclopedia evidently did not promote the radical view of the mathematics of its time suggested by this conception of modernity. And it does not omit contemporary research which is foreign to these new standards of mathematical activity, indeed quite the contrary. Thus assessing the modernity of the encyclopedia depends just as much on the point of view one brings to bear on the mathematical world of the time as it does on the image which Klein and his collaborators present.

For the historian, a first kind of response might be imposed by a teleological view, which would only retain from the end of the nineteenth century and the first few years of the twentieth those innovative results from which entire portions of twentieth-century mathematics emerged. In this view the modern is that which was not rendered obsolete by subsequent developments in pure mathematics. By contrast the great mass of other results and labors which fill the mathematical press of the period—even though some of this was likewise innovative—are judged as backward-looking and potentially antiquated even though contemporary. In this case the lack of modernity would apply not so much to the encyclopedia and the image it provides us, but rather to the world of mathematics as a whole.

It seems to me of interest to suggest other possible modes of response, which would consider the possible modernity of the encyclopedia less as a function of the mathematical innovations which it contains, and more on the basis of the nature of the project. It may in fact be necessary to reinterpret the meaning of this question of modernity to the extent that the encyclopedia was not conceived for the learned and academic world alone, and thus cannot be read as a kind of general survey of the mathematical (or scientific) journals of the period.

A Social and Cultural Dimension

The aim of the project—to provide a general image of the situation of mathematics in the culture of its time—led to an extension of the mathematical corpus considered of and increased the importance of other modes of interaction between mathematics and the society of the time [Tobies]. This aim, certainly one of Klein's goals, is a notable concern of the period not only in Germany (where other encyclopedic enterprises integrating mathematics flourished) but in many other countries.[21] The structure of the *Encyklopädie*, with on the one hand applications to the most varied areas and on the other a seventh volume on historical, philosophical and didactic questions, corresponds to a conception of mathematics as at once a science of the human mind *and* a science acting in the world, a conception expressed at the time by many mathematicians.

In different countries mathematical communities took part in movements for the reform of secondary teaching which were then developing across Europe, seeking to modernize the schooling of elites by giving full weight to science and mathematics. This took place on a national scale, as well as on an international level, with the creation of the international commission of mathematical instruction. The success of the international journal *L'enseignement mathématique*, which was created in 1899, and which became a leading journal of the time, shows to what extent this was a crucial question at the beginning of the century.

Confronted with the need to legitimate the importance of their discipline and its teaching both to engineers and to intellectual elites, mathematicians insisted on this double character of their discipline. One of the principal artisans of the *Encyclopédie*, Émile Borel, for example elevated mathematics to the level of the 'human sciences,' one which contributed to 'the formation of free men whose reason only yields to facts,' and insisted on 'making evident for all ... the points of contact between mathematics and modern life, the only means to prevent [them] from being suppressed one day as useless, as a financial saving' [Borel, 1904, 121].

Insofar as it participates in this defence of the discipline, the encyclopedia therefore expresses a modernity which is part of this international movement for the cultural, social and ideological renewal of the educational values of which Klein was one of the main leaders.

Thus conceived, mathematics has the vocation to interest a broad audience. The audience described in the foreword to the *Encyklopädie* includes astronomers, physicists, engineers and technicians, besides mathematicians (including mathematics teachers). The extension of the potential readership, and thus the collection of mathematical subjects to be treated, explains the presence of work ignored by the great mathematics journals of the day. This includes the vast field of what may be termed 'intermediate mathematics'; engineering mathematics is treated likewise.[22] From both of these areas, which were the objects of specialized publications in a press burgeoning in the 1890s, comes an entire series of articles, in particular in volumes one and three on numbers and geometry.

This cohabitation within the same work of different mathematical worlds ordinarily directed at different audiences, and the accompanying crossing of invisible boundaries drawn between distinct communities interested in mathematics, arose from editorial decisions which it hardly seems fair to term retrograde. For example, the new role of intermediate mathematics and its mathematical networks across Europe would seem to situate the encyclopedia within one important contemporary dynamic of mathematical activity.

Historiographical Modernity

The strong presence of history in the Klein-Molk encyclopedia, in spite of the absence of a specific volume be devoted to it (which was never undertaken) likewise inscribes the work in the mathematical reality of the day.[23] In Paris in 1900 the opening lecture of the second International Congress of Mathematicians was devoted to the historiography of mathematics, and was given by Moritz Cantor [Cantor, 1900]. He underlined the success of this branch of activity, which had come particularly into favour in the previous twenty years, a fact attested by the activities of the history of mathematics sections at these first congresses. The Heidelberg congress of 1904, for example, adopted a resolution on the teaching of the history of mathematics in plenary session. This was motivated by the fact that the discipline had become of an undeniable importance both from the point of view of pure mathematics and from the pedagogical standpoint.

In its way, the encyclopedia provides a measure of the favour which the history of mathematics then enjoyed among mathematicians. Many of the authors, finding themselves opportunistically cast in the historian's role, included memoirs and treatises from historical authors in the bibliographies which began their articles, and added many footnotes pointing to non-contemporary work or to recent publications in the history of mathematics. The encyclopedia thus presents us with the image of a mathematical milieu of an astonishing cultivation, conversant with contemporary historical research on the pre-modern period.

This historical work was amplified in the French version, thanks in part to the omnipresent interventions of Eneström, and here the use of recent historical work makes the *Encyclopédie* a showcase of historiographic modernity.[24] Even the style of Eneström's contributions stresses this modern dimension. In his notes, which appear throughout the French edition, he sought to specify exactly what was known and what was not, to correct anachronisms in the reading of older mathematical works, and present ongoing debates among specialists.

The history of mathematics scene, as it is presented in the two editions, seems to have undergone a profound renewal from the 1860s, at least according to the numerous references in the first volume. These are principally concerned with the medieval West and the Renaissance, with the Arab world, India and East Asia, areas which had, since that decade, been the object of numerous works by about fifty historians and philologists from many countries, though mostly Germany. By comparison work in the field of Greek mathematics seems to be concentrated among a small number of specialists. The world of mathematics was thus growing both in space and in time. However, this expansion of horizons, which was engendered by an impressive number of editions of sources, does not seem to have overcome the dominant historiographic model. The encyclopedia is thus the reflection of a conception which overestimates the Greek contribution, seen to be at the origin of all subsequent mathematical developments, and the mathematicians of the seventeenth century. The role of the Arabs—to whom nothing original is attributed—of India, and of the Medieval period is minimized.

However, the essential feature of Klein's project with regard to the history of mathematics does not reside in giving more or less developed expositions of the ideas and results of past centuries. The written recommendations of the editors to the authors are clear: historical developments in their greatest generality should be given from the beginning of the nineteenth century [von Dyck, 1904, XII-XIII]. At this point we go beyond footnotes, and entire paragraphs are devoted to history—even entire articles in the geometry volume.

Klein thus hoped to begin constructing the history of nineteenth-century mathematics across the different volumes, assigning a very precise role to this as-yet-unwritten history. In depicting the progress of the mathematical sciences in the course of the nineteenth century, he drew attention to modes of development which he felt were threatened but which seemed to him still desirable, for both present and future.

In some cases the versions of the history of the nineteenth century presented in the two editions do not agree.[25] The French edition modifies the German one by offering different or complementary version of developments in which certain actors do not have the same importance. The historical accounts in the two editions, by their very nature, seek to identify milestones in the development of mathematical ideas. Yet these accounts show that, for the nineteenth century—a century in which scientific communities began to be structures in a national framework—these milestones were not universally agreed upon. This history of mathematics, could therefore be read in a variety of ways according to the traditions in which the authors were enrolled. These variations, however, were not merely nationalistic in origin.

CONCLUSION

In this study of the German and French editions of Klein-Molk encyclopedia I have focused on some of the structural choices made by the editors in the definition and execution of their project. As we have seen, the interpretations that can be made call into play the multiplicity of mathematical representations which were then at work at national levels, as well as the diverse forms of mathematical activity which were developed in society at the time. The very ambition of the encyclopedia, an attempt at the diffusion of knowledge in a style and to a readership which were not those of the specialized mathematical press, allows the historian to take into account certain aspects of this diversity which have rarely been considered.

By basing my account principally on this particular characteristic of the encyclopedia I have tried to add some new images to the existing gallery representing mathematics at the dawn of the twentieth century. The settings, the actors, and the motivations which enter the scene at that time all extend our frames of reference. Mathematics is enlisted in trajectories and interactions at once scientific, cultural and social, suggesting new images and new views of the period.

Of course my choice in no way exhausts the readings which could be made of the *Encyklopädie*, of its French edition and of its involvement in this period. The richness, breadth and complexity of the work call for further studies which, following the example of some earlier ones, would extend the illumination of the relations between this major work and its epoch which I have begun here.

NOTES

* Translated by T. Archibald
1. Few studies have appeared to date about the articles of the German and French editions of the encyclopedia. This is all the more reason to point out certain recent studies which explore the vision of mathematics proposed by certain authors in specific areas. The global study of the editors' assumptions which I present here rests on these works and the discussions I had with their authors, which have been particularly helpful. I will refer to them at various places in the text in order to provide concrete illustrations of my themes.
2. The announcement for the French edition is presented in appendix 2. In the text this edition will be described as the *Encyclopédie*. The comparison with that for the German edition (described as the *Encyklopädie*) [von Dyck 1904] brings out certain nuances of difference which I will not detail here.
3. The full title of the German edition is translated literally here. This was not precisely retained for the edition in French by Molk which began in 1904. Published by Gauthier-Villars until 1916 on the private initiative of Teubner, the original publisher, the unfinished French edition was titled literally 'Encyclopedia of the pure and applied mathematical sciences'.
4. See Appendix 1, which presents the structure of the *Encyclopédie* as it was announced in the Gauthier-Villars catalogue of 1904. We will not present here the few differences which exist between this outline and the one announced in the *Encyklopädie*. Volume 7 was never published, a point to which we shall return below. Historical developments are however integrated into most of the articles of the different volumes.
5. The classification and distinction thus established between pure and applied mathematics in the title and in the presentation of the table of contents only exist in the French edition.
6. In the recent republication of the French version by J. Gabay, the publisher has reconstituted in a supplementary volume the complete and detailed table of contents of the French Edition [Gabay, *Encyclopédie des sciences pures et appliquées*, Volume 8].
7. The formulation of the advertisement by Gauthier-Villars (appendix 2) is more ambiguous as regards to the source of the questions of pure science than is the original Teubner text which explicitly mentions applications. See as well [von Dyck 1904, XI] point 3 on 'Allgemeine Grundsätze für die Bearbeitung der Artikel'.
8. I wish to thank Tom Archibald for discussions on the physics volume. As regards probability and statistics Michel Armatte notes several areas which are ignored as domains of application, such as the biological and social sciences and certain areas of engineering, while as other sectors such as insurance and actuarial science are treated to an exorbitant amount of space [Armatte].
9. The same vision of past and contemporary nineteenth-mathematics is expressed in Klein's *Vorlesungen über die Entwicklung der Mathematik im 19. Jahrhundert*, an other historical undertaking he realized later in the 20th century. See for example the introduction Klein wrote in the second part of the *Vorlesungen* [Klein, Teil 2, 1].
10. On the Göttingen mathematical tradition see [Rowe 1989].
11. In the correspondance of Emile Borel at the Institut Henri Poincaré (Paris), a certain number of letters of J. Molk attest to these difficulties.
12. See the 1911 announcement of Gauthier-Villars
13. See for these developments Chapter 6 of [Gispert 1991].
14. For the comparison between these pubications see [Tobies and Gispert 1996]. We note in regard to the more important place of applications in the DMV the particular influence which Klein exerted there.
15. For extra-academic research in these contexts see [Fontanon, Grelon 1998], [Crépel 1990], and, for the region outside Paris, [Nye 1986].
16. On the AFAS, see [Gispert 1999a].

17. A detailed comparison for the cases of AFAS and the SMF of these different balances and the conceptions which grounded them is provided in [Gispert 1999a].

18. Some articles on the German and French editions of the encyclopedia address this on certain points, though it is essentially historical work that remains to be done and brings into play historiographic questions which I mention in the following paragraphs. For the volume on number theory see [Goldstein], for probability and statistics see [Armatte], and for the foundations of geometry see [Bkouche].

19. The French version took as its mandate the updating of developments and references since the appearance of the original. This work, sometimes carried out with the agreement of the German author, was in some cases well done, in others not done at all because of the abrupt breaking off of the project, both from the point of view of translation and that of publication.

20. On Baire's motivations, which were shared by Borel, Lebesgue and others French young analysts, see [Gispert 1995]

21. On the other encyclopedic enterprises in Germany, see [Tobies].

22. See [Ortiz].

23. An international congress of mathematical philosophy was held in Paris at Easter, 1914, in order to prepare the philosophical part of this last and specific volume [*L'Enseignement mathématique*, 1914, pp. 55–56, pp. 370–378]. The work, and therefore the volume which was never published, was stopped by the war. I am not aware of such preparatory work for the historical part.

24. I will not stress here the specific character of the *Encyclopédie* where many articles became much large than in their original version because of the new importance of historical developments. See [Gispert, 1999b].

25. In the geometry volume, for example, this concerns the articles on projective geometry, or on analytic and synthetic geometry in the nineteenth century where the roles of Möbius and Chasles are treated differently; and in the volume on geodesy the French version gives much weight to the developments of the seventeenth and eighteenth centuries 'when France was at the head of the other nations.' See [Gispert, 1999b].

REFERENCES

Armatte, M., 'Les probabilités et les statistiques dans l'*Encyclopédie des sciences mathématiques pures et appliquées*: promesses et déceptions', in Gispert, H., Verley, J.L., (eds.) *L'Encyclopédie des sciences mathématiques pures et appliquées*, (1904–1916). *Traduire ou adapter l'entreprise de F. Klein*, to appear.

Bkouche, R., 'La pensée géométrique au tournant des XIXe et XXe siècles', Gispert, H., Verley, J.L., (eds.) *L'Encyclopédie des sciences mathématiques pures et appliquées*, (1904–1916). *Traduire ou adapter l'entreprise de F.Klein*, to appear.

Borel, E., [1904] 'Les exercices pratiques en mathématiques dans l'enseignement secondaire,' *Conférences du Musée pédagogique*, Paris, Imprimerie nationale, 107–131.

Cantor, M., [1900] 'Sur l'historiographie des mathématiques', *Compte rendu du 2e Congrès International des Mathématiciens*, Paris, 27–42.

Crépel, P., [1990] 'Contrôle statistique de qualité: victoire et défaîte des artilleurs français', *Matapli* (23), 20–30.

'Chronique, Congrès de philosophie mathématique', *L'Enseignement mathématique* (16), 1914, 55–56; 'Chronique, Le premier congrès de philosophie mathématique', *L'Enseignement mathématique* (16), 1914, 370–378.

Epple, M., [1995] 'Branch Points of Algebraic Functions and the Beginnings of Modern Knot Theory', *Historia Mathematica* 22, 371–401.

Fontanon, C., Grelon, A., (eds.), [1998] *Les professeurs du Conservatoire national des arts et métiers, dictionnaire biographique 1794–1955*, 2 vols., INRP-CNAM.

Gabay, J., (ed.), [1995] *Table des matières et tribune publique. L'Encyclopédie des sciences mathématiques pures et appliquées*, tome 8.

Gispert, H., [1991] 'La France mathématique. La Société Mathématique de France (1870–1914)', *Cahiers d'histoire et de philosophie des sciences*, 34.

[1995] 'La théorie des ensembles en France avant la crise de 1905: Baire, Borel, Lebesgue ... et tous les autres', *Revue d'Histoire des Mathématiques*, 1, 39–81.

[1999a] 'Réseaux mathématiques en France dans les débuts de la Troisième République', *Archives Internationales d'Histoire des Sciences*, 142, 122–149.

[1999b] "Les débuts de l'histoire des mathématiques sur les scènes internationales et le cas de l'entreprise encyclopedique de Felix Klein et Jules Molk', *Historia Mathematica* 26, 344–360.

Goldstein, C., 'La paix des étoiles: les tomes de théorie des nombres dans les deux versions de l'Encyclopédie' in Gispert, H., Verley, J.L., (eds.) *L'Encyclopédie des sciences mathématiques pures et appliquées, (1904–1916). Traduire ou adapter l'entreprise de F. Klein*, to appear.

Klein, F., [1927] *Vorlesungen über die Entwicklung der Mathematik im 19. Jahrhundert*, Teil 1, 1926, Teil 2, 1927, Berlin, Springer.

Mehrtens, H., [1990] *Moderne—Sprache—Mathematik. Eine Geschichte des Streits um die Grundlagen der Disziplin und des Subjekts formaler Systeme*, Frankfurt, Suhrkamp.

Nye, M.J., [1986] *Science in the Provinces. Scientific Comunities and Provincial Leadership in France, 1860–1930*, University of California Press.

Ortiz, E., [1996] 'The Nineteenth-century International Mathematical Community and its Connection with those on the Iberian Periphery' in Goldstein, C., Gray, J., Ritter, J., (eds.), *L'Europe mathématique*, Paris, Editions de la Maison des sciences de l'homme, 323–343.

Rowe, D., [1989] 'Klein, Hilbert and the Göttingen Mathematical Tradition', *Osiris*, 5, 186–213.

Tobies, R., 'Mathematik als Bedansteil der Kultur—Zur Geschichte des Unternehmens *Encyklopädie der Mathematischen Wissenschaften mit Einschluss ihrer Anwendungen*' in Gispert, H., Verley, J.L., (eds.) *L'Encyclopédie des sciences mathématiques pures et appliquées, (1904–1916). Traduire ou adapter l'entreprise de F. Klein*, to appear.

Tobies, R., Gispert, H., [1996] 'A Comparative Study of the French and German Mathematical Societies before 1914' in Goldstein, C., Gray, J., Ritter, J., (eds.), *L'Europe mathématique, Mathematical Europe*, Paris, Editions de la Maison des sciences de l'homme, 409–432.

von Dyck, W., [1904] 'Einleitender Bericht über das Unternehmen der Herausgabe der Encyclopädie der mathematischen Wissenschaften', *Die Encyklopädie der mathematischen Wissenschaften mit Einschluss ihrer Andwendungen*, t. 1, vol. 1, Leipzig, Teubner, I–XX.

[1908] 'Die Encyklopädie der Mathematischen Wissenschaften', *Atti del 4. Congresso Internazionale dei Matematici*, vol. 1 , Roma, 123–134.

☞ Vient de paraître tome I vol. 1 fasc. 1:

ENCYCLOPÉDIE

DES

SCIENCES MATHÉMATIQUES

PURES ET APPLIQUÉES

**PUBLIÉE SOUS LES AUSPICES DES ACADÉMIES DES SCIENCES
DE GÖTTINGUE, DE LEIPZIG, DE MUNICH ET DE VIENNE
AVEC LA COLLABORATION DE NOMBREUX SAVANTS.**

ÉDITION FRANÇAISE

RÉDIGÉE ET PUBLIÉE D'APRÈS L'ÉDITION ALLEMANDE SOUS LA DIRECTION DE

JULES MOLK,

PROFESSEUR À L'UNIVERSITÉ DE NANCY

EN SEPT TOMES.

I. ARITHMÉTIQUE ET ALGÈBRE:
Rédigé
en allemand par FR. W. MEYER à Königsberg,
en français par J. MOLK à Nancy.

II. ANALYSE:
Rédigé
en allemand par H. BURKHARDT à Zurich,
en français par J. MOLK à Nancy.

III. GÉOMÉTRIE:
Rédigé
en allemand par FR. W. MEYER à Königsberg,
en français par J. MOLK à Nancy.

IV. MÉCANIQUE:
Rédigé
en allemand par F. KLEIN et C. H. MÜLLER à Göttingue,
en français par P. APPELL, de l'Institut, à Paris.

V. PHYSIQUE:
Rédigé
en allemand par A. SOMMERFELD à Aix-la-Chapelle,
en français par A. POTIER, de l'Institut, à Paris.

VI₁. GÉODÉSIE ET TOPOGRAPHIE:
Rédigé
en allemand par PH. FURTWÄNGLER à Potsdam et
F. WIECHERT à Göttingue,
en français par CH. LALLEMAND, du Bureau des longitudes, à Paris.

VI₂. ASTRONOMIE:
Rédigé
en allemand par K. SCHWARZSCHILD à Göttingue,
en français par H. ANDOYER à Paris.

**VII. QUESTIONS D'ORDRE HISTORIQUE,
PHILOSOPHIQUE ET DIDACTIQUE.**
(Ce tome est encore à l'état de projet.)

PARIS,
GAUTHIER-VILLARS

LEIPZIG,
B. G. TEUBNER

—

1904

☞ L'Encyclopédie sera publiée en 50 livraisons environ (de 10 feuilles grand-in-8°) paraissant, autant que faire se pourra, de trois mois en trois mois. Le prix de la livraison se montera à peu près à 50 cts. (— 40 Pfennige) par feuille de 16 pages.
☞ Paru: Tome I vol. 1 fasc. 1; 160 p.; frs. 5.— (— Mk. 4.—)

Gauthier-Villars, 1904

AVIS

Cette Encyclopédie est un exposé simple, concis et, autant que possible complet, des résultats acquis dans les diverses branches de la science mathématique. On y trouvera des renseignements bibliographiques permettant de suivre le développement des méthodes propres à chacune de ces sciences; on s'est d'ailleurs particulièrement attaché à suivre ce développement depuis le commencement du 19e siècle.

Pour les Mathématiques pures, on insistera sur les définitions et sur l'enchaînement des théories, sans donner des démonstrations. Pour les applications des mathématiques, les diverses sciences techniques seront exposées avec de larges développements: de la sorte, le mathématicien pourra facilement prendre connaissance des questions de science pure qu'il aura à traiter; l'astronome, le physicien, l'ingénieur pourront, eux aussi, se reporter aux solutions des problèmes qui les intéressent.

Chacun, sans études préalables spéciales, peut aborder la lecture de l'Encyclopédie et y puiser des données générales sur l'ensemble des mathématiques pures et appliquées. Cet ouvrage rendra également service aux spécialistes, désireux de se documenter sur quelque point particulier de la science.

La rédaction est assistée d'une Commission composée de délégués des Académies des sciences de Göttingue, de Leipzig, de Munich et de Vienne. Cette commission comprend actuellement MM. Walther von Dyck à Munich, Gustave von Escherich à Vienne, Otto Hölder à Leipzig, Félix Klein à Göttingue, Louis Boltzmann à Vienne, Hugo von Seeliger à Munich et Henri Weber à Strasbourg.

L'édition française de l'Encyclopédie est divisée en sept tomes, comprenant chacun trois ou quatre volumes grand in-8, qui paraissent par livraisons. On trouvera à la fin de chaque volume un index alphabétique se rapportant aux matières contenues dans ce volume.

Dans l'édition française, on a cherché à reproduire dans leurs traits essentiels les articles de l'édition allemande; dans le mode d'exposition adopté, on a cependant largement tenu compte des traditions et habitudes françaises.

Cette édition française offrira un caractère tout particulier par la collaboration de mathématiciens allemands et français. L'auteur de chaque article de l'édition allemande a, en effet, indiqué les modifications qu'il jugeait convenable d'introduire dans son article et, d'autre part, la rédaction française de chaque article a donné lieu à un échange de vues auquel ont pris part tous les intéressés; les additions dues plus particulièrement aux collaborateurs français, seront mises entre deux astérisques. L'importance d'une telle collaboration, dont l'édition française de l'Encyclopédie offrira le premier exemple n'échappera à personne.

Les tomes I-III, consacrés aux mathématiques pures, sont rédigés dans l'édition allemande sous la direction de MM. François Meyer à Königsberg (Algèbre et Géométrie) et Henri Burkhardt à Zurich (Analyse); dans l'édition française ils seront rédigés, d'après l'édition allemande, sous la direction de M. Jules Molk à Nancy. Les articles de l'édition allemande sont dus à MM. [...]; dans l'édition française, ils seront exposés par MM. [...].

Les tomes IV-VI, consacrés aux mathématiques appliquées, sont rédigés dans l'édition allemande par MM. Félix Klein à Göttingue (Mécanique), Arnold Sommerfeld à Aix-la-Chapelle (Physique), Ph. Furtwängler à Potsdam et Emile Wiechert à Göttingue (Géodésie, Topographie et Géophysique), Charles Schwarzschild à Göttingue (Astronomie). Dans l'édition française ils seront rédigés, d'après l'édition allemande, pour les questions d'ordre général par M. Jules Molk à Nancy, et plus particulièrement pour la Mécanique par M. Paul Appell à Paris, pour la Physique par M. Alfred Potier à Paris, pour la Topographie, la Géodésie et la Géophysique par M. Charles Lallemand à Paris, pour l'Astronomie par M. Henri Andoyer à Paris. Les noms des savants et des ingénieurs auxquels sont dus les différents articles de mathématiques appliquées seront publiés ultérieurement.

Il serait superflu d'insister davantage sur l'intérêt que présente l'Encyclopédie. Cet ouvrage a sa place marquée dans toutes les bibliothèques scientifiques.

Jules Molk, Nancy, 1904.

Chapter 6

MATHEMATICS AND NATURAL SCIENCE IN THE NINETEENTH CENTURY: THE CLASSICAL APPROACHES OF POINCARÉ, VOLTERRA AND HADAMARD

Jeremy Gray

By the end of the nineteenth Century, the classical triumvirate of elliptic, parabolic, and hyperbolic partial differential equations was well established, and classical mathematics had achieved remarkable success in describing the physical world. To investigate the depth and extent of this success, this chapter examines three areas in which leading applied mathematicians tried to build upon that success in the years 1890–1914:

1) Thomson, Heaviside, and Poincaré on the telegraphist's equation;
2) Hadamard's journey from non-linear problems to ill-posed problems, via Kovalevskaya's Theorem;
3) Volterra on Green's theorem, elasticity theory, and his generalisation of complex function theory to 'functions of lines'.

The ideas of Poincaré, Hadamard, and Volterra about the role of mathematics in physics prompt some conclusions about the partial successes of these branches of applied mathematics at a time when physics was entering its era of greatest change.

INTRODUCTION

The theme of this chapter is the relationship of mathematics to natural science, seen, in the context of this book, from a perspective that embraces such issues as the rise and fall of branches of mathematics and their institutional status. Its particular focus is on the work of Poincaré, Volterra, and Hadamard.

There are a number of ways in which this is a difficult issue to grasp. For example, a fuller discussion would take up the work of Tazzioli [Tazzioli, 1993] and others on the Italian school of mathematical physics created by Betti and Beltrami, which influenced Volterra. These Italians at Pisa were geometers and mathematical physicists, the subjects being linked by the hypothesis of the elastic ether: differential geometry suggested methods in elasticity theory and thence in Maxwell's (and, later) Hertz's theories of electro-magnetism.

In this world of national schools there were rivalries and alliances, narrow views and broad. A singular role was played by Poincaré, who from his earliest contacts with Klein to his death in 1912 seems to have been a remarkably independent figure. He was happy to be involved with Italian mathematicians (his famous paper on the electron was published in the *Rendiconti* of Palermo) and seems to have set himself two institutional tasks. I would suggest that by his example and the clarity of his expositions he hoped to renew the involvement of French mathematicians with mathematical physics, and to be a channel for bringing foreign ideas, such as those of Maxwell, Hertz, and Lorentz, to France. He adopted the role of the expert, first among equals, who could adjudicate disputes and reach out, through his many popular essays, to the larger educated public. All this rested, of course, on his ability to produce significant original contributions. And all this was done without there being in any sense a school around Poincaré; in fact, the next generation of French mathematicians tended to keep away from what Poincaré was doing.

It is in this context that any assessment of status must be made, and it clear that any answer must be shifting, partial, even contradictory. There is no sense in which subjects and disciplines are prioritised and every educated mathematician has to come to the same list. The German case is the best known, thanks to the achievement of Jungnickel and MacCormmach [1986]. There, there were recognisably distinct professions of physicist and mathematician, and within physics there were traditions of experimental and theoretical kinds. As the nineteenth century came to an end the balance shifted more and more towards theory, while at the same time the most energetic leader of a Mathematics Department, Felix Klein in Göttingen, was about to spend the next twenty years trying to attach his department to technology and engineering.[1] The French were much more experimentally minded in their physics, which lends a certain piquancy to Poincaré's role. Recently, Smith and Wise have attempted, quite plausibly, to suggest that a meaningful distinction can be drawn between Lord Kelvin's Glasgow-based attraction to useful science and Maxwell's Cambridge-based theoretical science. So the view from a Mathematics Department over the domain of physics would depend on where you were nationally (and, of course, within that country too). Mathematical physics, applied mathematics, whatever it is called, is disputed territory: mathematicians responded to problems and priorities that others (physicists and even technologists) also wished to determine.

The resolution of a problem might depend on the discipline to which you belonged. Significant, even vital applied mathematics can be done in ways that strike physicists as irrelevant (I shall adduce some examples below). Good physics can be done in ways that make mathematicians despair; in this period, the name of Heaviside springs to mind. The virtual identity of complex function theory in a single variable and harmonic function theory in two dimensions that Riemann did so much to bring about, fails in three and more dimensions, thus depriving mathematicians, but not physicists, of the tools of their trade. The whole question of experiments in physics, which has occupied historians of science a great deal recently, is not one with which a mathematician can feel at ease. Questions of validity must be surrendered, even left in dispute. There may be no answers that satisfy philosophers of science (or so it has been said) which enable one to say when a theory succeeds or fails, but the practice of physicists is instructive and interesting. Likewise there are times when no amount of clever mathematical work seems to be going in the right direction, and the physicist and applied mathematician call it off, even though the mathematician can look upon a body of attractive new theorems. This is interesting, especially when these two or three figures are united in the same person.

There is a further sense in which success is a term with many meanings. There can be a body of proven results, patently valuable in this or that scientific context, and there can be research challenges. There is no reason at all to suppose that these two go together: theories can stagnate, new results can flood in. There is a fundamental psychological difference between a theory with an active past, and a theory with an active future, between theories that someone might teach and subjects that they might research.

It is a familiar observation, which I shall call the Rebirth Thesis, that in the history of physics around 1900 there was a widespread feeling that physics was nearly exhausted—the fundamental discoveries had all been made and only a few details remained to be worked out. However, that the rapid discoveries of quantum physics and general relativity soon made much of classical nineteenth century physics obsolete. This striking disparity gives grounds for caution when considering the role of mathematics in the extraordinary success of the classical Natural Science in the nineteenth century. It suggests a paradox: that much of what is undoubtedly a success also became somehow irrelevant,

Moreover, there is no analogue in mathematics to the disappearance of classical physics, nor to the feeling that, around 1900, there simply wasn't much left to do. This should make us speculate on the degree to which the successes of classical approaches in mathematical physics were extra-ordinary, and what confidence should be placed in such successes.

In this context, the historian of mathematics looks at the relationship between mathematics and the Natural Sciences around 1900 as one where the extraordinary success of mathematics continued without much change, even as physics was having its second revolution. It resembles the servants and bureaucrats who go from one monarch to the next, or even from monarchy to republic or dictatorship, and whose continuities in the end count for as much as the changes wrought by the new regime. One such mathematical shift was the creation of a theory of integral equations and of functional analysis. This has been described several times, notably by Dieudonné [1981] (whose account of Volterra was recently challenged by Fichera [1994]) and Lützen [1982]. This remarkable mathematical theory is exactly the branch of mathematics that Wigner found unreasonable: an independent creation strikingly well adapted to the new quantum mechanics. But it does not quite address the problem we are pursuing, because it concerns the new physics and not the old.

Instead, I shall try to measure the success of classical mathematical physics by looking at what its leading figures found still had to be done in the period from, say, 1900 to 1914. One transformation that is most striking is the transition from a handful of partial differential equations in mathematical physics, to an awareness that much lies beyond them, with hitherto unsuspected qualitative behaviour. From the classical examples of the wave equation, the heat equation, and Laplace's equation (potential theory), there had been distilled a classification into first-order partial differential equations, and the hyperbolic, parabolic and elliptic families of second-order linear partial differential equations. By 1910, there was a vivid awareness of equations of mixed type, of the relevance of non-linear equations, and of the subtleties of initial conditions.

Each of the first three classical equations is connected to an immense theory, and has a solid base in applied mathematics. The wave equation drives the theory of sound and the transmission of light. The heat equation is the equation for the theta function, and so lies at the base of the Jacobian theory of elliptic functions; its implications for physics are immense. Potential theory has as a special case in two variables the theory of harmonic functions, and so, thanks to Riemann, the whole of complex function theory; while its applications were described by Kellogg as the theory of Newtonian gravity (Kellogg, [1929], p. 1). For almost two hundred years it has been possible to write very large books on any one of these equations, and in so doing describe a large piece of the natural world. So the story I want to tell is the overthrow of what I shall call the classical triumvirate. In that light, I shall return to the extraordinary success of the classical approaches, which enthroned the triumvirate, the better to see what those approaches did and did not accomplish.

THE TELEGRAPHIST'S EQUATION (IN THE BACKGROUND, RIEMANN ON SHOCK WAVES)

The first example shows how robust the triumvirate was, and how astute mathematicians were at using it. Darboux, in a remarkably thorough study of this equation (*Surfaces*, 2, Chapter III) wrote what he called Euler's equation in the form

$$E_{\beta\beta'}: \quad \frac{\partial^2 z}{\partial x \partial y} - \frac{\beta'}{x-y}\frac{\partial z}{\partial x} + \frac{\beta}{x-y}\frac{\partial z}{\partial y} = 0,$$

where β and β' are constants. (It was loosely derived from the partial differential equations that Euler wrote down in his *Institutiones Calculi Integralis*, 3, Section 2, Ch 3, which are more general than the one that bears his name; the one that gets closest is a special case.) The special case $\beta = \beta'$ studied by Euler [§ 328, p. 217] is also the one Riemann studied. He introduced the adjoint of this equation, and showed that the general solution of the Euler equation can be expressed in terms of simple solutions of the adjoint equation.

Darboux pointed out that the special case reduces, via a substitution, to the equation:

$$\frac{\partial^2 f}{\partial x \partial y} = \frac{\beta(1-\beta)f}{(x-y)^2}. \tag{1}$$

He then showed that any solution $Z(\beta)$ of this equation is of the form $(x-y)^\beta Z(\beta, \beta')$, where $Z(\beta, \beta')$ denotes a solution of Euler's equation. An ingenious argument invoking the hypergeometric equation then enabled Darboux, following Appell [1882], to find the general solution of Euler's equation.

Riemann's paper is one of the major papers in applied mathematics during this period. It gives a rigorous method for solving certain types of hyperbolic partial differential equations, and describes how shock waves form. In it, Riemann predicted that in the solutions the pitch varies as the wave propagates.[2]

If in the equation (1) one substitutes $\beta - x$ for x and then lets β tend to infinity, one eventually obtains the equation $\frac{\partial^2 f}{\partial x \partial y} = f$, which is the telegraphist's equation. This equation was written down for the first time by Kirchhoff in his [1857], profoundly studied by Heaviside (see for example his [1876]) and has been well analysed historically in Yavetz [1995]. It is interesting to explain the physical interpretation of this equation, as is done in the books by Webster [1933] and Yavetz.

$$A \frac{\partial^2 U}{\partial t^2} + B \frac{\partial U}{\partial t} + CU = \frac{\partial^2 U}{\partial x^2}$$

The telegraphist's equation above describes the propagation of electricity in a long straight wire. It is satisfied by the current at any point, and by the potential. The constants that appear in the equation involve the conductance, the self-inductance , the resistance and the leakage conductance of the wire. In the ideal case of no resistance and no leakage B and C vanish, and the equation reduces to the wave equation. In the case when the inductance is negligible by comparison with the resistance, the constant A may be taken to be zero, and the equation is parabolic. It is in fact the one-dimensional heat equation, and was treated in this spirit by William Thomson (later Lord Kelvin) in 1855 when he was advising on the laying of Atlantic cable. It follows that the maximum electrical effect at a distance x is inversely proportional to x^2.

Despite the success of that enterprise, the character of the equation is very different when neither resistance nor leakage is negligible. In this case neither B nor $\dot{C}$ vanish, and the equation can be reduced to the form:

$$\frac{\partial^2 U}{\partial t^2} = \frac{\partial^2 U}{\partial x^2} + U$$

(provided there is some resistance and some leakage). The method of separation of variables combined with the theory of Bessel functions now shows that the general solution with initial conditions at $t = 0$ of $U = F(x), \frac{\partial U}{\partial t} = G(x)$ is sum of three terms. One is a wave propagating with velocity proportional to $1/\sqrt{x}$, as before, although now it is exponentially damped. But the other two terms are like the tail of a wave that never dies away. Therefore, when an attempt is made to transmit a periodic wave down the wire the velocity and wavelength depend on the frequency and the waves undergo dispersion. Unless, and this was to be Heaviside's remarkable discovery in 1887, the values of the physical constants can be so adjusted that the rate of dispersion is zero. This can be done both mathematically and physically, it merely requires that the leakage be non-zero. Far from being an inconvenience, this condition is necessary for the production of distortionless telephony. The signal becomes fainter over distances, but this can be corrected for by fitting amplifiers. Long distance telegraphy had dealt with distortion by accepting a low transmission rate, so as to separate the pulses. Telephony required much higher frequencies; with some leakage and a deliberately high self-inductance it became distortionless. Long distance communication was reborn—although the money for the first successful patents went to the American electrical engineer Michael Pupin in 1901, and not to Heaviside. The reader is referred to Yavetz's fascinating book for a detailed account of the whole story, which explains among other things why it took Heaviside from 1881 to 1887 to make the simple observation about distortion, and how his work fitted into contemporary ideas of electro-magnetic theory.

Solutions of the telegraphist's equation were given using Riemann's method by Picard [1894], the year after Poincaré had solved it by means of complex Fourier integral methods under the initial conditions that $U(x, 0) = f(x)$ and $\dfrac{dU}{dt}(x, 0) = f_1(x)$ when $t = 0$ [Poincaré, 1893]). Poincaré considered that case when the functions $f(x)$ and $f_1(x)$ are given by polynomials inside the interval $b < x < a$ and vanish outside it. He quickly found using the theory of residues that U was zero outside the interval $b - t < x < a + t$, and had four discontinuities, at $a \pm t$ and $b \pm t$, which propagate at the speed of light. Other initial conditions were investigated. For example, if $f(x) = 0$ for all x, and $f_1(x) = \pi/2$ inside $(-\varepsilon, \varepsilon)$ and 0 outside, then the solution is given by a Bessel function on the interval $(-t, t)$ and is zero outside. In any case, the head of the disturbance moves with a finite speed, as is the case with the transmission of light but not of heat, and the head, once it has passed, leaves behind a disturbance which never vanishes, which does not happen with the wave equation.

In his short paper, Poincaré did not mention the ingenious discoveries of Heaviside. There is no reason why he should have done so, but the omission in his *Cours sur les oscillations électriques* (in the pages after

183) is more significant. There he surveyed a considerable amount of mostly French experimental work, with a view to deciding between the old theory of electro-magnetism (due to Kirchhoff) and the modern theories of Maxwell and Hertz. One reason for Poincaré missing Heaviside's trick is that the simplifications Poincaré performed depend on $B^2 - 4AC \neq 0$, but the special case arises exactly when $B^2 - 4AC = 0$. So the experimental work was given a theoretical twist and technological implications were not mentioned. In the opinions of Webster and Yavetz, it is only with Poincaré's work that the equation attracted the attention of mathematicians.

FROM NON-LINEAR PROBLEMS TO ILL-POSED PROBLEMS VIA KOVALEVSKAYA'S THEOREM

The second example takes us from Poincaré to Hadamard, and involves a growing sophistication among mathematicians about non-linear problems. Problems which can be handled by the calculus of variations, such as the Hamilton-Jacobi approach to dynamics, are linear or non-linear according as the input that determines a specific problem. A remarkable indication of what can happen in this context was provided by Poincaré in his celebrated study of celestial mechanics, [1890] recently well described in Barrow-Green [1997] and Diacu and Holmes [1996]. Confronted with a non-linear problem such as the motion of a small body moving under the gravitational influence of two large ones which it does not itself influence (such as the restricted three body problem, where the small body also lies in the common plane of the two large ones) the tradition had been to examine the problem locally. Mathematicians and astronomers looked for power series solutions, which they expected to converge only locally (establishing any convergence at all may be difficult, because the coefficients are often functions of the time variable), or sought to replace the problem with a linear approximation, which would only be a good enough approximation over a small domain. Poincaré set himself the task of discovering what might happen globally, with results that continue to resonate through mathematics. Most notably, he discovered what are called homoclinic points, points in the neighbourhood of which solution curves exhibit virtually chaotic behaviour. This is a dramatic indication of how very different the global behaviour of solutions can be from what any local analysis might suggest.

It was observed by Szolem Mandelbrojt [1953, p. 600] that there is an interesting thread that runs through much of Hadamard's work and which starts here.[3] In one of the very few papers by Hadamard which relates directly to what Poincaré had done (which is striking, because their working lives overlapped for 20 years) Hadamard studied the long-term behaviour of geodesics on manifolds of negative curvature [Hadamard, 1897]. He discovered that four types of behaviour might occur:

(I)   the geodesic might be closed;
(II) the geodesic might approach a closed curve (itself a geodesic) asymptotically;
(III) the geodesic might wander off to infinity; or
(IV) the geodesic might wander between neighbourhoods of several different closed geodesics.

But the long-term behaviour is hard to determine. Hadamard showed that an infinitesimal change in the initial direction of a geodesic that does not go to infinity can produce a geodesic with any of the above behaviour.

Mandelbrojt observed that when Hadamard wrote his 'Notices sur les travaux scientifiques' in 1901, he referred to Poincaré's work and his own as falling into 'the category of ill-posed problems' because 'infinitesimal changes in the initial data ... can entail a total and absolute change in the result'. On this account, Hadamard's old friend from their time together at the École Normale and later in Bordeaux, Pierre Duhem [1906, pp. 226–230], dismissed these topics as examples of deep mathematical problems 'useless to the physicist' because they oversimplify the underlying physical model (one might wonder how a more complicated model will avoid the undesirable features of the simpler one, but it is not impossible).[4] The thread (the idea of well- and ill-posed problems) is one we shall pick up again shortly.

To pursue this thread, we must add a further element to our story: the idea of initial or boundary conditions. In principle, such data picks out a unique solution from the vast range of solutions to the given differential equation. The simple problem of motion in a plane under a gravitational field has solutions that are conic sections. Specify the starting point and the initial velocity and the motion is then completely determined. In this case, although any conic may arise if you specify the initial point, the dependence on the initial conditions is continuous: varying the initial conditions infinitesimally does not change the conic if it is an ellipse or a hyperbola. The belief that solutions always depend continuously on the initial conditions was so deep that Hadamard had to fight very hard against it.

There are other sources for the disquiet. The important thesis of Kovalevskaya, [1875], which made her name, can be well illustrated by this example, taken from Cooke [1984, pp. 33–34]. The heat equation (a linear partial differential equation) is

$$\frac{\partial u}{\partial t} = \frac{\partial^2 u}{\partial x^2}.$$

The initial conditions if you fix a value of x, say $x = x_0$ are fine: if

$$u(x_0, t) = f(t) \,, \frac{\partial u}{\partial x}(x_0, t) = g(t),$$

where f and g are analytic at a time t_0, then there is a unique solution of the differential equation analytic in a neighbourhood of (x_0, t_0). But if instead you try to fix an initial value of t, it can all go wrong. Kovalevskaya gave the example of $u(x, 0) = (x - 1)^{-1}$, but as Cooke points out, the example $u(x, 0) = (1 + x^2)^{-1}$ is equally devastating mathematically but has the advantage of being physically plausible. In this case, the power series solution of the differential equation diverges for all values of t other than $t = 0$. There is no solution which is analytic in x and t around the time $t = 0$. Kovalevskaya's thesis was taken up and extended by Poincaré in his own thesis, in work that he then reconsidered when investigating the existence of solutions to the Hamilton-Jacobi equations that formed the rigorous spine of his study of celestial mechanics.[5]

One of the deepest responses to Riemann's paper on shock waves and the investigations of Kovalevskaya is contained in the famous book by Hadamard, *Leçons sur la propagation des ondes* [1903]. In the second of its six chapters, Hadamard began to consider the propagation of waves in deformable media, a line of enquiry he had first picked up from Hugoniot. The question at issue is the motion in advance of the wave and after. Hadamard investigated what are called the conditions of compatibility.

There are kinematic and dynamic questions, just as in elasticity theory there are geometrical questions and questions about forces. The wave is defined as follows: At time $t = 0$ a particle is at a, at time t, it is at x. The wave front $F(t)$ with equation $f(a, t) = 0$ divides the medium into two regions M^+, M^-, where M^+ is $f(a, t) > 0$, and M^- is $f(a, t) < 0$. Let ϕ be a function of interest, a so-called dynamical variable, such as pressure, velocity, density, temperature. Physics implies some equations for these, in the form of differential equations when they are continuous, and in the form of constraints when discontinuous.

Wave fronts may have strong discontinuities (when by definition the dynamical variable are discontinuous) or weak (when some first order derivative jumps). In the weak case there are possibly some geometrical or kinematical conditions—they depend on the existence of the wave. In Chapter 5, Hadamard showed that acceleration waves are very different from shock waves. For example, only shock waves can cause vortices.

Hadamard now embarked on a series of studies of partial differential equations, indeed it may be said that this subject formed (for the second half of his life) his life's work, and Hadamard lived a very long time indeed. In his [1908a], he tackled the difficult question of the equilibrium of clamped plate via a Green's function argument (this is a fourth order partial differential equation, generalising Laplace's equation). In his [1905] and [1908b] he wrote on the Cauchy problem for general, linear second order partial differential equations. He based his approach on the work of Kirchhoff, Beltrami, and especially his friend Volterra, who had done the light (and acoustic) case, i.e. constant coefficients; Hadamard took up variable coefficients. This work on the Cauchy-Kovalevskaya problem led Hadamard to formulate the concept of the well-posed problem. These are problems whose solutions depend continuously on the data. The concept became familiar to Anglophone mathematicians with his lectures at Yale, given in 1920 and published as [1922] (where he cited Poincaré's address to the International Congress of Mathematicians in 1897 to the effect that these are problems which physics poses—a mathematician would never have thought of them—but to which also physics suggests something of the answer).

The famous example of the heat equation and other partial differential equations is now in view. The equation gives $\dfrac{\partial^k u}{\partial t^k}$ as an analytic function in the variables t, the x's, and partial derivatives of the unknown function of orders up to k, and it has a unique solution for Cauchy data given at $t = 0$ (in conformity with the original notations, the roles of t and x have been switched from the earlier example). Hadamard asked what would happen if the partial differential equation was not given by an analytic function. In 1917, lecturing at Zürich, he said that he had found that mathematicians often thought that because any function can be well approximated by analytic functions, a non-analytic partial differential equation could be replaced by a 'nearby' analytic one, with only slight effect on the solutions. 'But', he went on, 'in my opinion this objection would not apply, the question not being whether such an approximation would alter the data very little, but whether it would alter the solution very little'. Then followed his example.

It is the partial differential equation $\dfrac{\partial^2 u}{\partial x^2} + \dfrac{\partial^2 u}{\partial y^2} = 0$ with the Cauchy data $u(0, y) = 0$ and $\dfrac{\partial u}{\partial x}(0, y) = u_1(y) = A_n \sin(ny)$ where $n \gg 0$ and A_n becoming very small as n grows large (for example, $A_n = 1/n^p$). The data can be made as small as you wish. But the equation has as its solution $u = \dfrac{A_n}{n} \sin(ny) \sinh(nx)$, which is very large for any value of x other than $x = 0$.

However, as Hadamard went on to observe, the behaviour of the equation $\dfrac{\partial^2 u}{\partial x^2} - \dfrac{\partial^2 u}{\partial y^2} = 0$ with the Cauchy data $u(0, y) = \phi(y)$ and $\dfrac{\partial u}{\partial x}(0,y) = 0$ is quite different, and is well-posed, because of the solution formula $u(x,y) = [\phi(x + y) + \phi(y - x)]/2$.

Maz'ya and Shaposhnikova point out that at first Hadamard defined well-posedness in terms of existence and uniqueness of the solution, and suggested that continuous dependence on initial data is important only for the Cauchy problem. However when Courant and Hilbert (*Methods of Mathematical Physics*, 2, p. 227) incorporated the continuous dependence into the definition of well-posedness, he moved to agree with them.

FROM GREEN'S THEOREM TO WORK OF VOLTERRA

Mention has been made several times of Volterra. To follow his arguments, however briefly, it is best to recall two aspects of the work of Green. One concerns his reciprocity theorem. This says given a region S with boundary σ and two differentiable but not necessarily harmonic functions U and V defined in S, and where Δ denotes the Laplace operator $\sum \dfrac{\partial^2}{\partial x^2}$ that the sum of a volume integral involving V and ΔU and of a volume integral involving $\operatorname{div} U$ and $\operatorname{div} V$ equals a surface integral involving V and the normal derivative of U. The import of this theorem, which is just an exercise in the calculus (first-order Taylor series approximations) is that one can compare volume integrals with surface integrals. So, by letting U be a harmonic function or letting $V = 1$, or letting U or a normal derivative vanish on σ, one can get uniqueness theorems for harmonic functions. Another version of the reciprocity theorem says that if both U and V are (not necessarily harmonic) functions then a volume integral involving V and ΔU, U and ΔV is equal to a surface integral involving V and the normal derivative of U, U and the normal derivative of V.

The other aspect of Green's work concerns what are called Green's functions. If the origin is an interior point of S (for simplicity) a Green's function for S is one that behaves like $1/r$ near the origin, is otherwise a harmonic function inside S and vanishes on the boundary σ. We can take the reciprocity theorem with U a harmonic function and $V = G - 1/r$ (which is another harmonic function within S) and deduce that a harmonic function is determined by its values on the boundary.

In the 1870s Betti generalised the reciprocity theorem and the use of Green's functions to obtain results in elasticity theory. His work, and that of Beltrami, began a generation of Italian activity, in which the names of Cerruti, Somigliana, Lauricella, and ultimately Volterra figure prominently, and to which Boussinesq in France and others also contributed. Volterra's contribution was to extend the Green's Theorem approach to the case where the functions involved are analytic, but not single-valued, on the multiply-connected domains he specialised in studying. Indeed, the theme of many-valued analytic functions was an important one for Volterra, who had found a significant mistake in one of Kovalevskaya's papers at just this point (it is in her paper on elasticity theory).

With this background all too briefly sketched, we can now savour the interesting perspective on the changes in mathematics and physics in the early years of the twentieth Century that is provided by two extensive sets of lectures given by Vito Volterra, one in Sweden in 1906 (published as [1912]) and one in America in 1909.

He began in 1906 by acknowledging that mathematical physics was undergoing a period of crisis, and he cited accounts by Poincaré and Picard. But he said, even if some concepts will be replaced, 'a part of mathematical physics has a good chance of surviving the flood. It represents, perhaps crudely, but certainly very simply, a large part of the known facts, relates them together, and is useful beyond any discussion. The history of science offers examples of analytic theories of certain phenomena created under the influence of certain principles and which have survived the collapse of those principles'—he gave the example of optics as one of many.[6]

He went on to observe that many theories in mathematical physics may be treated under a single heading, and specifically reduced to one of three (his count) types of partial differential equation (elliptic, hyperbolic, parabolic and mixed). In his opinion, even if the analogies this suggests lapsed, along with the basis in physics, the building that remained would be so solid, and so useful, that it would continue as one of the most beautiful chapters in analysis. For example, he observed that physicists had sometimes created mathematical physics on the hypothesis of a continuous medium, sometimes by appealing to discrete molecules and action at a distance, but either approach could lead to differential equations, because a limit process effected the passage from the discrete to the continuous.

In the elliptic case, Volterra knew that there was an old but profound analogy between complex function theory and the motion of two-dimensional liquids, brought about by Laplace's equation. In these lectures he sought to extend the analogy to higher dimensions. In the hyperbolic case, he was pleased to show, for the first time, that the Weierstrass-Kovalevskaya method was related to that of Kirchhoff, and so to the other method in the field, due to Riemann and Green. The parabolic case he found less advanced (and he noted when the essay was republished in 1910 that this case had recently made great strides).

He began with the equations of elasticity, which form a system of elliptic partial differential equations. His main question concerned what he called cyclic (and we call multiply-connected) regions that have been distorted; in particular, he asked if the solutions are monodromic. To investigate this, he observed that he had been able to extend Green's theorem to all the problems in the calculus of variations. In the case of elasticity, this was the approach of Betti. But if displacements are many-valued, that approach fails. By Lecture 5 he was discussing the simpler case of harmonic function theory in 3-dimensions, and the different types of connectivity in 3-dimensions. He singled out two:

a) every closed curve can be shrunk to a point (in modern terms, the first homotopy group vanishes)
b) every closed surface can be shrunk to a point (in modern terms, the second homotopy group vanishes)

Seeking, as he did, to generalise complex function theory to the higher-dimensional setting, Volterra introduced his 'functions of lines' in this context as follows, drawing on a chain of ideas he had been following for twenty years since 1887.

Volterra's work on 'functions of lines' has been discussed by Vesentini [1992], who summarises it this way. Let V be a smooth, connected, oriented Riemannian n-manifold and M an compact embedded sub-manifold of dimension p. This means that coordinates $(x_1, \ldots, x_n)$ can be introduced locally such that points on M have coordinates of the form $(x_1, \ldots, x_p, 0, \ldots 0)$. We are interested in the set $A_p(V)$ of all such manifolds M, which Vesentini drily notes is 'very rich'. A real or complex 'function of lines' (henceforth functional in the sense of Volterra, or simply a functional, the term was introduced by his friend Hadamard) of rank p is a function from $A_p(V)$ to **R** or **C**. Vesentini discusses a number of questions that Volterra confronted. What, for example, is a good definition of continuity of a functional? How can one study the (first) variation of a functional? What restrictions have to be imposed on the elements of $A_p(V)$ and on the functionals before one can get results?

Under some natural restrictions, it is possible to define the first variation of a functional, ϕ, and to obtain an expression for it as an integral, $\delta\phi = \int_M i(x)\lambda$, where $i(x)$ is a function depending on the vector field along which the variation is taking place and λ is a $p+1$-form. Functionals of the first degree are defined to be those for which the dependence on X vanishes, and the first variation is expressed by the integral of a $p+1$-form. (There is another condition, but I suppress it here.) The expression 'first degree' arises because it is linear on disjoint sub-manifolds.

If now λ, λ' and ω are forms such that $\lambda' - \lambda = d\omega$, then $\phi'(M) = \phi(M) + \int_M \omega$ defines a functional associated to λ'. As Vesentini observed, but Dieudonné did not, we are very close to De Rham theory here, and it is apparently an open question which elements of the De Rham cohomology group $H^{p+1}(V)$ are expressible as functionals of the first degree. Another comparison, which Vesentini did not make, would be with geometric measure theory and the theory of integral currents.

To obtain the connection to complex function theory, Volterra let ϕ and ψ be two complex functionals of the first degree and λ and σ their associated $p+1$-forms. The quotient σ/λ defines a complex function f, and if $d\psi = fd\phi$ then ϕ and ψ are said by Volterra to be isogenous functionals. This is a generalisation of the Cauchy-Riemann equations. If now S is a $p+1$-dimensional sub-manifold such that $H_1(S) = 0$, then the fact that $d\sigma = 0$ implies that $\int_S f\lambda = \int_s fd\phi = 0$, which generalises the Cauchy Integral Theorem.

In Stockholm, Volterra considered a region which satisfies (a) and (b) and a surface integral taken over an 'open' surface, i.e. one with a boundary . The integral depends on the boundary—so it is a functional. In this case, it is a function of closed curves (boundaries). It is linear in these curves $V(s + s') = V(s) + V(s')$, so it is a functional of the first degree.

He then showed that one can similarly invoke differentiation, which leads to the first variation of a functional, and pointed out that this is not abstract, but real and practical—consider the potential of a (magnetic) field due to a current in a closed curve, and vary the curve.

In Lecture 6 he returned to his theme that the natural generalisation of analytic functions was to the function of lines. He observed that isogeneity was an equivalence relation, and he showed how to define when a function was isogenous to a functional, and in this way again generalised the Cauchy Integral Theorem.

I shall say no more about these lectures, except to observe that he hailed as the greatest triumph of recent mathematical physics the analogy of vibrations of elastic body and the electro-magnetic theory of light due to Hertz. This analogy was singled out for further attention in the Clark Lectures. There he began by saying that a mathematician would deny any difference between elasticity theory and electro-magnetic theory, because the form of the differential equations and the method of their solution agree in each case. On the other hand, this coincidence has led to a simple and natural transition between the two theories, extending to the motion of bodies. Great discoveries in analysis as so often have their roots in natural sciences, and every improvement of analytic methods has implications for mathematical physics. Either interpretations of the mathematics are sought, or rigorous proofs.

Volterra got down to hard work by showing how to derive Maxwell's theory using the calculus of variations. Having done so, he observed that it implies mechanical explanations or models for electro-dynamics, as in Kelvin's vortex theory and Larmor's work, or that of E. and F. Cosserat [1909]. Indeed, as Poincaré has observed, infinitely many mechanical explanations are possible. For example, following Beltrami [1880] one could regard the theory as that of optics in a space of non-zero curvature. Next he observed equations that can be derived from calculus of variations satisfy a reciprocity theorem akin to Green's. He illustrated this point with numerous examples, before turning to what he called the Minkowski world, in 2, 3, and 4 dimensions. He gave an account of the equations of optics in anisotropic media, before concluding with Minkowski's transformations of Lorentz's equations and their consequences.[7]

For the second of his three long Lectures, Volterra took up what he called old and new problems in elasticity theory, noting that they were erected on different physical bases: molecular (Navier *et al.*) or energy theory. The theory of elasticity had been related by Beltrami to curvature: if you accept Hooke's law, you can express the equations of elasticity theory as integral equations, and they are linear; but if you do not (and it is only an approximation) then the equations cease to be linear. The transformations of the equations of elasticity theory akin to assuming space of non-zero curvature—although he noted that even in Klein's opinion physical space was at most only slightly non-Euclidean.

He then turned to integration methods for partial differential equations, observing that the Minkowski trick offered a way of regarding elliptic and hyperbolic equations as essentially the same. This was interesting because problems of equilibrium in elasticity theory are elliptic, whereas as problems involving motion are hyperbolic. Volterra suggested that one could pass from one to the other by letting time be imaginary. He then updated his remarks on general methods of solution, the Kirchhoff-Green approach, and the problems of many-valued functions (with some photographs of materials under stress). Finally he concluded with some general existence Theorems, which, he said, do not interest physicists much. Perhaps the most novel part of the lectures, by comparison with the Stockholm series is the reference to Fredholm, Poincaré, Hilbert, and Schmidt, integral equations and eigenvalue methods. Mainstream linear functional analysis had arrived.

QUESTIONS OF STATUS

How do these developments bear on questions of Rise and Fall—the ups and downs of mathematics and mathematical physics, and the internal and external driving factors? The first observation to be made concerns the attitudes of physicists. The changes that Volterra alluded to, however vaguely, rapidly ran through physics, not once but twice in the space of less than a generation. The effect was dramatic because physics was taken to be, and probably still is taken to be, a hierarchical subject. This is the reductionist paradigm recently and eloquently defended by Steven Weinberg, the high road to nuclear and then particle physics [1993]. Even in the 1920s it was eclipsing the successes of Einstein's special and general relativity theories. This reductionism, which attached high status to the work of Hertz, Lorentz, and Poincaré, contributed to placing such topics as elasticity theory in the sidelines. More research is needed to establish the point, but let me claim for definiteness that in the period 1900–1914 elasticity theory was always important, but never of paramount importance. That might be because it failed to generate truly convincing physical models, or because the mathematics was so intractable, or because the ether was discredited.

A second opinion of physicists was voiced by Duhem and indirectly by Volterra. Duhem disparaged the subtleties of non-linear, almost chaotic behaviour; Volterra observed that physicists did not care much for existence Theorems. A good way to see this is with Poincaré's remarks about physics not only setting problems one would otherwise miss, but also hinting at the answers. There was (and is today) a feeling among physicists that one could have, if not too much mathematics then at least too little scientific intuition. Consider Boltzmann, or Einstein before general relativity, or some of the differences between Poincaré and Lorentz. It is easy enough to make logical sense of this disparity. There are two big gaps in mathematics at any stage in its development: between what you believe and what you can prove, and between necessary and sufficient conditions. Very often a mathematician can prove that some condition, X', is necessary for some desirable property, Y, to hold, but it is harder to establish sufficiency. On the other hand, examples are known where some property X'' is too weak, and X'' fails to imply Y. The physicist wishing to show that some natural property, X, implies another, as it might be, Y, finds the mathematician's X' too restrictive, but rejects the examples of X'' as irrelevant. Neither side can, however, quite characterise X so as to show, mathematically, that X implies Y.

In this context, it is interesting to see Volterra endorse Poincaré's opinion that interpretations come cheap, and indeed there will be infinitely many of them. I continue to think that this cannot be a physicist's view, and that in adopting it mathematicians cut themselves off from a source of insight that physicists typically call explanation (when it works). It must be said however, that they cut themselves off from a large amount of ephemeral entities as well.

The examples of bad behaviour of the solutions under slight changes in the initial conditions became almost paradigmatic in this respect. It became a dogma that problems in physics are well-posed, and ill-posed problems would give rise to behaviour so unstable as to be evanescent. This has the happy incidental effect of shutting out much behaviour that it is (perhaps one should say was) almost impossible to analyse mathematically. So I am happy to take this list from Maz'ya and Shaposhnikova of problems that are ill-posed in nature: gravimetrics, spectroscopy, radio astronomy, atmospheric soundings, and modelling of optimal systems. So here we have an example of a success—the isolation of the concept of well-posedness—that is in some sense a conspiracy of silence. What Hadamard discovered is not a profound insight into nature, but only a very good and useful idea.

But there are other groups who evaluate mathematics, and indeed produce it. The first example, the telegraphist's equation, is a success for telegraphy even in its over-simplified form, and a success for telegraphists in its properly understood form. The practical implications were immense. The external factors here were crucial in creating a context for the work and a real test of the solutions.

The story of elasticity theory is less dramatic, but not by much. The fact that such work was done with increasing importance as the nineteenth Century ended and the twentieth Century began is of course connected with the building of huge bridges, and large buildings in metal and reinforced concrete. When the ether is abandoned, and elasticity theory loses its connection to fundamental physics, it is left with its engineering applications. This is a story in which the names of Saint-Venant and Navier figure prominently, and most practitioners (Volterra among them) refer to the history by Todhunter and Pearson [1886]. The same importance attaches to aero- and hydrodynamics (not discussed here) and the same caveats must be applied. The mathematics around the Navier-Stokes equations is often too hard, and remains so to this day, because the problems are in some sense unstable. The success of these branches of mathematics is only, if you will forgive the pun, partial.

But let us notice also what is very remarkable in the work of Volterra. He saw his work on functionals (his functions of lines or curves) quite straight-forwardly as a generalisation of complex analysis, and in working it through he saw glimpses of what later became De Rham cohomology and even geometric measure theory. More precisely, he was motivated by the desire to extend complex function theory to higher dimensions, in order to study harmonic functions. His approach was a generalisation of the calculus of variations; the concept of isogeneity provides the generalisation of the Cauchy-Riemann equations. I think the reason this insight was not taken up by others is that the road to linear functional analysis was opened up, which is rich and difficult enough. Volterra was always interested in non-linear problems, and non-linear functional analysis is a much more recent development. That is something like Dieudonné's opinion too, but he used it as a reason to diminish Volterra's contribution. Perhaps by looking at what was done in the past, rather than keeping so clearly in mind what is done today, we can see some things more clearly. Whether or not Volterra's work is a central chapter in the development of modern functional analysis, it is surely a vindication of the value that thinking about problems in physics can have for mathematicians, and that remains a remarkable sort of, and source of, success.

I began with what I called the rebirth thesis, and with the political metaphor of mathematicians switching their allegiance from one regime to the next. I am not sure what the priorities and status claims within science were around 1900. They are hard to disentangle both from the record, and from the priorities that drive the researches of historians. It seems clear that it varied from country to country, place to place, person to person.[8] With that complication in mind, I claim that elasticity theory, aero- and hydro-dynamics were never regarded as truly fundamental. This was so, I believe, because the physicists approached these subjects with a range of patently ad hoc simplifications of their own, while the mathematicians could never offer a truly satisfactory theory (measured, say, by comparison with the remarkable identity that unites harmonic function theory and complex function theory in the two-dimensional setting).

Quantum theory, Einstein's special and general theories of relativity, and then quantum mechanics in the 1920s amount to a rebirth of physics, giving it a whole new set of fundamental concepts, and it is a fascinating, and by no means completely told, story to tell how mathematicians created a new mathematics alongside it.[9] But that story is one involving functional analysis, integral equations, and Hilbert space. What remains is a story of macro-physics: the large-scale behaviour of elastic solids (and fluids) however much that might have to be rooted ultimately in a truly atomic theory. The mathematicians who stayed with these topics resemble the loyal courtiers who continue to work for the old regime, even when life is more active elsewhere. Doubtless they, like the politicians they resemble, have their own rationalisations for doing as they did, and the analogy should not be insisted upon too much: there surely were important questions in both the mathematics and the physics remaining to be done (there still are). But it would seem that the disdain for physical models shown by Poincaré, Volterra, and Hadamard carries a price. It leaves such mathematicians only able to make one contribution: rigorous, profound mathematics. That proved elusive. Classical mathematics, like classical physics, was not so remarkably successful after all.

NOTES

1. See Parshall and Rowe 1994.
2. A brief history going from Riemann via Christoffel and Hugoniot to Hadamard and beyond will be found in Hölder [1981].
3. See the forthcoming study of Hadamard by Maz'ya and Shaposhnikova, *Hadamard* American and London Mathematical Societies, in press.
4. English translation [1954] pp. 139–141.
5. It was then discussed by Koenigsberger in his [1894].
6. Une partie de la physique mathématique a bien des chances de se sauver du naufrage. Elle représente en effet, peut-être d'une manière grossière, mais certainement d'une manière très-simple, une grande partie des faits naturels connus, les relie ensemble et a une utilité pratique hors de toute discussion. L'histoire des sciences nous offre l'exemple de théories analytiques de certains phénomènes qui ont été créés sous l'influence de certains principes et qui ont résisté la chute de ces principes. [1906] p. 2, in [1957] p. 64.
7. Minkowski's transformation sends t-> it, and switches Newtonian space-time to that of Lorentz.
8. For the example of two institutionalised views in Cambridge alone, see Warwick [1992, 1993].
9. It is a story that historians of physics avoid as well; try searching the existing accounts for a realistic assessment of von Neumann's work and its impact.

REFERENCES

Appell, P., [1882] 'Sur les fonctions hypergéométriques de deux variables', *Journal math. pures appl.*, (3) 8, 173–217.
Barrow-Green, J.E., [1997] *Poincaré and the Three Body Problem*, Providence, American and London Mathematical Societies, History of Mathematics 11.
Beltrami, E., [1880] 'Sulle equazioni generali dell'elasticità', *Annali matematica pura e applicata* (2) 10, 82.
Cooke, R., [1984] *The Mathematics of Sonya Kovalevskaya*, New York, Springer.
Cosserat E, Cosserat, F., [1909] *Théorie des corps déformables*, Paris.
Courant, R., Hilbert, D., [1962] *Methods of Mathematical Physics*, 2, 2nd edition, New York, Wiley Interscience.
Darboux, G., [1915] *Leçons sur la théorie générale des surfaces*, 2, 2nd ed., Paris, Gauthier-Villars.
Diacu, F., Holmes, P., [1996] *Celestial Encounters. The Origins of Chaos and Stability*, Princeton, Princeton University Press.
Dieudonné, J., [1981] *History of Functional Analysis*, Amsterdam, North-Holland.
Duhem, P., [1906] *La théorie physique: son objet, sa structure*, Paris, Rivière (English transl., *The Aim and Structure of Physical Theories*, Princeton, Princeton University Press, 1957).
Euler, [1924] *Institutiones calculi integralis*, 3, in *Opera Omnia*, (1) 13, F. Engel and L. Schlesinger (eds.), Leipzig, Berlin, Teubner.
Fichera, G., [1994] 'Vito Volterra and the birth of functional analysis' in Pier, J.P., (ed.), *Development of Mathematics, 1900-1950*, Boston and Basel, Birkhäuser, 171–184.
Hadamard, J., [1897] 'Sur certaines propriétés des trajectoires en dynamique', Mémoire couronné en 1896 par l'Académie: Prix Bordin, *Journal math. pures appl.* (5) 3 331–387 (in *Oeuvres*, 4, 1749–1805).
[1903] *Leçons sur la propagation des ondes*, Paris, Hermann.
[1908a,] 'Sur le problème d'analyse relatif à l'équilibre des plaques élastiques encastrées', Mémoire couronné en 1907 par l'Académie: Prix Vaillant, *Mémoires présentés par divers savants à l'Académie des Sciences*, 33 (in *Œuvres*, 2, 515–629).

[1905] 'Recherches sur les solutions fondamentales et l'intégration des équations linéaires aux derivées partielles', *Ann. Ecole Normale Supérieure* (3) 22 101–142 (in *Œuvres*, 3, 1195–1235).

[1908] 'Théorie des équations aux derivées partielles linéaires hyperboliques et du problème de Cauchy', *Acta Mathematica*, 31, 333–380 (in *Œuvres*, 3, 1249–1296).

[1922] *Lectures on Cauchy's Problem on Linear Partial Differential Equations*, New Haven, Yale University Press. (reprint, New York, Dover, 1954).

[1968] *Œuvres de Jacques Hadamard*, 2 vols., Paris.

Heaviside, O., [1876] 'On Duplex Telegraphy', *Philosophical Magazine*, (5) 1, 32–43 (in *Electrical papers*, 1, 53–64).

Hölder, E., [1981] 'Historischer Überblick zur mathematischen Theorie von Unstetigkeitswellen seit Riemann und Christoffel' in Butzer, P.L., (ed.), *E.B. Christoffel*, Boston and Basel, Birkhäuser, 412–434.

Jungnickel, C., McCormmach, R., [1986] *The Intellectual Mastery of Nature*, 2 vols, Chicago, University of Chicago Press.

Kellogg, O.D., [1929] *Foundations of Potential Theory*, Berlin, Springer, (reprint, New York, Dover, 1959).

Kirchhoff, G., [1857] 'Ueber die Bewegung der Elektricität in Drähten', *Annalen der Physik und Chemie* (4), 100, 193–217 (in *Gesammelte Abhandlungen*, 131–155).

Klein, C.F., *Klein's Evanston Colloquium Lectures and other works*, eds. Rowe, D.E. and Gray, J.J., to appear.

Koenigsberger, L., [1894] 'Ueber die von *Poincaré* gegebene Erweiterung des Cauchyschen Satzes von der Existenz der Integrale gewöhnlicher Differentialgleichungsysteme', *Journal reine angew. Math.*, 113, 115–127.

Kovalevskaya, S., [1875] 'Zur Theorie der partiellen Differentialgleichungen', *Journal reine angew. Math.*, 80, 1–32.

Lützen, J., [1982] *The Prehistory of the Theory of Distributions*, New York, Springer.

Mandelbrojt, S., [1953] 'The mathematical work of Jacques Hadamard', *Amer. Math. Monthly*, 60, 599–603.

Maz'ya, V., Shaposhnikova, T., [1998] *Jacques Hadamard, A Universal Mathematician*, Providence, American and London Mathematical Societies, History of Mathematics, 12.

Parshall, K., Rowe, D.E., [1994] *The Emergence of the American Mathematical Research Community; J.J. Sylvester, Felix Klein, and E.H. Moore*, Providence, American and London Mathematical Societies, History of Mathematics, 8.

Picard, É., [1894] 'Sur l'équation aux derivées partielles qui se rencontre dans la théorie de la propagation de l'électricité', *C.R. Acad. Sciences*, 118, 16–19.

Poincaré. H., [1879] *Sur les propriétés des fonctions définies par les équations aux différences partielles*, Première Thése, Paris, Gauthier-Villars (in *Œuvres*, 1, xlix–cxxix).

[1890] 'Sur le problème des trois corps et les équations de la dynamique', *Acta Mathematica*, 13, 1–270 (in *Œuvres*, 7, 262–479).

[1893] 'Sur la propagation de l'électricité', *C.R. Acad. Sciences*, 117, 1027–1032 (in *Œuvres*, 9, 278–283).

[1894] *Cours sur les oscillations électriques*, Paris, Gauthier-Villars.

[1916–1956], *Œuvres*, 11 vols., Paris, Gauthier-Villars.

Riemann, B., [1860] 'Ueber die Fortpflanzung ebener Luftwellen von endlicher Schwingunsweite', *Abhandlungen der Königlichen Gesellschaft der Wissenschaften zu Göttingen*, 8 (in *Gesammelte Mathematische Werke*, 3rd edition, ed. R. Narasimhan, New York, 1990, 188–207).

Smith, C., Wise, M.N., [1989], *Energy and Empire: A Biographical Study of Lord Kelvin*, Cambridge, Cambridge University Press.

Tazzioli, R., [1993] 'Ether and Theory of Elasticity in Beltrami's Work', *Archive for History of Exact Sciences*, 46, 1–38.

Thomson, W., (later Lord Kelvin) [1855] 'On the Theory of the Electric Telegraph' in *Mathematical and Physical Papers*, 6 vols, Cambridge, Cambridge University Press.

Todhunter, I., Pearson, K., [1886] *History of the Theory of Elasticity*, Cambridge, Cambridge University Press.

Vesentini, E., [1992] 'I funzionali isogeni di Volterra e le funzioni di variabili complesse' in *Convegno Internazionale in memoria di Vito Volterra*, Atti dei Convegni Lincei 92, Roma, Accademia Nazionale dei Lincei.

Volterra, V., [1909] 'Trois leçons sur quelques progrès récents de la physique mathematique' in *Lectures Delivered at the Celebration of the Twentieth Anniversary of the Foundation of Clark University*, 1–82, Clark University, Worcester, Mass., (in *Opere matematiche*, 3, 389–470).

[1912] *Leçons sur l'intégration des équations différentielles aux derivées partielles, professées à Stockholm 1906*, Paris, Hermann, (in *Opere matematiche*, 3, 63–141).

[1954–1962] *Opere matematiche*, 5 vols., Roma, Accademia Nazionale dei Lincei.

Warwick, A., [1992] 'Cambridge Mathematics and Cavendish Physics: Cunningham, Campbell and Einstein's Relativity 1905–1911, Part I : The Uses of Theory', *Studies in the History and Physics of Science*, 23, 625–656.

[1993] 'Cambridge Mathematics and Cavendish Physics: Cunningham, Campbell and Einstein's Relativity 1905–1911, Part II: Comparing Traditions in Cambridge Physics', *Studies in the History and Physics of Science*, 24, 1–25.

Webster, A.G., [1933] *Partial Differential Equations of Mathematical Physics*, 2nd ed, S.J. Plimpton (ed.), (reprint, New York, Dover, 1966).

Weinberg, S., [1993] *Dreams of a Final Theory*, London, Hutchinson.

Yavetz, I., [1995] *From Obscurity to Enigma: The Work of Oliver Heaviside, 1872–1889*, Basel-Boston-Berlin, Birkhäuser.

Chapter 7

DEVELOPMENTS IN STATISTICAL THINKING AND THEIR LINKS WITH MATHEMATICS*

Michel Armatte

IMAGES AND CONFIGURATIONS OF STATISTICS

The question raised by the editors of the present book concerns the image that, at given times, mathematicians had of their science, its methods and status, at the level of the organization of knowledge as well as within society. This chapter attempts to answer this question for the case of statistics during a period spanning approximately a hundred years up to the 1930s, corresponding to what has been called the 'probabilistic revolution' [Krüger & *al.*, 1987]. But in order to sketch out an answer, what is meant by an 'image of a scientific field' has first to be specified.

If by an image we mean a representation, or a realignment, of the products of the discipline, then, in our opinion, it should necessarily involve the *actors* in the discipline, the *contents* which serve as its raw material and references, and the *recipients* for which the image is designed, as much as with the socio-cognitive *system* which it aims to modify. One could therefore hardly speak of the actual image of a discipline such as statistics; one would rather speak of an image fashioned by some group of actors X (in general labeled 'the statisticians'), for the use of another group Y (the same or, more frequently, others: physicists, biologists, sociologists, political scientists, or economists), in a socio-cognitive context Z (characteristic of a locale and period, and to which the image itself must contribute somehow). For example, the image of statistics produced in the middle of the nineteenth century by the followers of the Belgian statistician Quetelet was produced, we shall emphasize, on the basis of a mechanist conception of average and variability laws with the intent of providing scientists and politicians with a redefinition of the social contract, its understanding, and its control, in terms of *social physics*. This image differed from that promoted by the Académie des Sciences in the first half of the century: its Prix Montyon had to be awarded to purely descriptive statistical works [Brian, 1994]. The Academie excluded conjectures and inferences, taking the form of general laws characterizing man and society, and valued only a strict recording of contingent facts as had already been done by Napoleon's bureaucracy.

In the case of statistics, as for other subjects, one must relinquish the simplistic idea of a single construction in perpetual progress, in favor of a series of successive—and/or competing—configurations, each with its own set of characteristics. In relation to a group of actors, its cognitive tools, and its social position, the image of a discipline is subject to important variations. Different representations linked to both objective and subjective features of the discipline, and to both ideological and pragmatic systems within which it is situated, confront each other. The image varies under the effect of an internal dynamics or by peripheral innovations upsetting the legitimacy or fruitfulness of these representations. The discipline therefore can only be seized and described within configurations in which the various elements of the image are relatively stable. In search of historical, social, and conceptual divisions, historians and philosophers of science have labeled these configurations *paradigms* (Kuhn), *research programmes* (Lakatos), *themata* (Holton) or *reasoning styles* (Hacking).

Whichever terminology is retained, the term refers back to a mixture of concepts, processes, values, and social interactions which form the heterogeneous but interdependent elements of a *socio-logical system* of interpretation and transformation of the world. For statistics, we must now specify the various elements of this system, and so pay attention to the relationships between statistics and mathematics.

STATISTICAL THINKING AND MATHEMATICS

Usually, historians of mathematics treat this question by isolating mathematical statistics as a specific object of study, while leaving to other experts the questions of the production of information, or the interpretation and use of results, which take place, respectively, before and after the formal work of data processing. After 1930, statistics became a part of mathematics easily identifiable as such for several reasons. Based on a branch of mathematics redefined by Borel and Lebesgue's measure theory and Kolmogorov's axiomatization, i.e. probability theory, statistics found an explicit theoretical core in Ronald Fisher's 1922 program ('The object of statistical methods is the reduction of data'). It moreover acquired a new institutional autonomy in terms of national and international scientific organizations, academic curricula, and research structures. Institutions such as the International Mathematical Society or the Econometric Society emerged from learned societies put in place during the second half of the nineteenth century (like the International Institute of Statistics); university chairs of statistics appeared in Europe and the United States; finally, laboratories were set up and later grew rapidly in the context of the explosion of applied mathematics following on the war work of the 1940s.

Such a viewpoint, however, has a major drawback. It isolates a so-to-speak 'pure' cognitive element from its social and cognitive settings, and therefore completely misses its global rationale. The mathematical model of statistical information processing is but a theoretical tool of this processing itself. And this processing itself is nothing more than a complex operation, linking knowledge and practice, which consists in *basing an action or a decision on controlled inductive reasoning based on multiple observations and measurements*. Statistical reasoning is 'a cognitive equivalence and comparability space constructed for practical ends' [Desrosières, 1993].

As the title and content of Jacques Bernoulli's founding work [1713] shows clearly and, as is now being rediscovered by contemporary studies of the works of Pascal, Huygens, and the like ([Coumet, 1970]; [Meusnier, 1996]), the history of statistics is rooted in the broader question of the art of conjecture or decision-making. Persisting with the *Encyclopédistes* and up to the nineteenth century, this history remains incomprehensible if reduced to a mere series of mathematical models. The complex socio-cognitive operation covered by Bernoulli's *stochastique* can be broken down into three interdependent phases that can rightly be attributed to statistics, as is witnessed by the diverse traces (journals, conferences, treatises, etc.) left by the statisticians' activity.

The first phase deals with the construction of facts and their determinations: for a long time this was the sole legitimate activity of the discipline, It raises such fundamental questions as those of categories, nomenclatures, and taxonomies through which reality is apprehended. 'Statisticians can only count what is standardized by collective agreements, conceptualized by social customs, questioned by politicians, [and] conceptualized by legal texts' [Martin, 1997]. Statistics also deals with investigative methodologies which produce what are sometimes called 'data.' Literally meaning 'something given,' this Latin term is obviously misleading, and a broad consensus today recognizes that reality never is 'given,' but is the product of a social and material theoretical production—a costly one, for that matter. Statistics' institutions (governmental offices and administrations, International Congresses and Statistical Institutes, etc.) have devoted a large part of their activity to the production and negotiation of procedures, definitions, and nomenclatures, defining categories—social, productive, nosographic—which are necessary conditions for any assessment of reality. This same phase moreover deals with measuring scales on which these categories are projected, and with numerical methods, tables, and charts through which the same information can be reduced and easily communicated. It might be tempting to describe this phase as investigative, administrative, or descriptive, statistics, but none of these terms alone is satisfactory. Each of them gives the erroneous impression that a preexistent reality is objectively recorded, while the situation is rather that of a social, technical construction of facts by a system of statistical information combining institutions, laws and contracts, definitions, conventions, accounting frameworks, and devices for the investigation, storage, exploitation, and publication of information.

The second phase consists in the production of a general theoretical discourse on the basis of contingent facts. Sometimes termed *inferential*, this phase of statistics provides thought and proof mechanisms taking their place within a discourse founded on repetitions and large numbers. This reasoning can be either of an inductive type if going from observations to regularities and laws, or of an hypothetico-deductive one if observations serve to the validate or refute of a set of *a priori* hypotheses (often in the form of a model). In the nineteenth century, the distinction between this phase and the preceding one was an important issue for the discipline. Scientific societies, like the Royal Statistical Society or the Académie des Sciences, wished both to contain statistics within the sole domain of data production and to discourage any form of speculation. It was felt that the latter was too dangerous an operation to be left to people other than mathematicians and philosophers, who alone had at their disposal both the mathematical instruments allowing scientific control over induction and the social structures allowing political control over the legitimacy of general discourses. In this phase, statistics was thus at the crossroads of problems of inferential logic (which can be structured mathematically) and questions of social legitimacy illustrated by numerous historical controversies. Let us think, for example, of the debates surrounding vaccination in the eighteenth century, pauperism in the nineteenth, or eugenics at the beginning of the twentieth, all of which provided contested contexts within which the legitimacy of some forms of statistical inference were elaborated.

Be it individual or collective, decision and action were the concern of the third phase. Right from the start, statistics was preoccupied with the problem of providing optimal decision-making rules for questions involving uncertain consequences. Thus maritime insurance, lifetime annuities, court testimonies, and decisions of justice, for example, provided occasions for the elaboration of these rules by the new calculus of expected value or utility. In the twentieth century, this problem of action surfaced again in operational research and collective-choice theory. In this phase again, statistics is a place of tension between scientific and political modes of thought. The problem of assessing things and men so as to be able to take the right political decisions embraces not only the original question of the State and its power (as the etymology of the word statistics reminds us), but also the general question of the articulation of science with policy. Thus it touches upon the very basis of forms of social organization, from the Enlightenment [Brian, 1994] to the present [Desrosières, 1993].

These three phases constitute three forms of statistical thinking. The first refers to definitions and the very existence of facts; it deals with their numerical treatment through syntactic operations of definitions and ordering. The second phase is concerned with discourse and semantics; it constructs meaningful representations of their organization and deals with their pertinence. The third and final phase refers to pragmatic actions; it provides rules for individual or collective action, and deals with their effectiveness in view of desired objectives and their value with respect to a given ethical system.

If we therefore accept that statistical objects must be envisioned according to the three viewpoints of syntax, semantics, and pragmatism, then it becomes clear that as a discipline statistics cannot be reduced to a branch of mathematics. In order to construct facts, laws, or rules for action, it may entertain privileged links with several branches of mathematics either relevant to its subject, or else created or enriched by its problems (formal logic, set theory, algebraic structures, linear algebra, number theory, analysis, probability theory, etc.). But, as a simplistic instrumentalist vision might have it, these two forms of relationship between mathematics and other areas of knowledge—which we may call *applied* and *motivated* mathematics—in no way exhaust the possibilities. Case studies conducted with the science studies framework of the 1980s, as well as work specifically dealing within mathematical statistics [Stigler, 1986]; [Armatte, 1995], have shown that multiple, complex links weave a fabric in which innovations in mathematics and other disciplines follow from simultaneous social constructions.

Statistics was born in the classical age at the intersection or two kinds of certainties. Founded on definitions, principles, and axioms, the absolute certainties of syllogistic and mathematical thought clashed with physical or moral claims. Based on observation and experience necessarily considered less trustworthy, since depending on the senses—and hence the result of an uncertain induction—the latter claims merited only some degree of confidence or probability. Hobbes, Hume, Pascal, Buffon, and Bernouilli developed this theme of the duality of knowledge in relation with the necessity/contingency opposition. At the same time, they established a new discipline— Pascal's *géométrie aléatoire* or Bernouilli's *stochastique*—which aimed to mend the gap between mathematics and natural philosophy by means of new methods largely inspired by mathematics itself. Condorcet and Laplace's program was nothing but the conception of mathematical tools allowing for rational thought based on facts—in the same way as one could think rationally on the basis of definitions—and, from there, the expression of the rules for finding rational courses of action. In a narrow sense, statistics is the set of these tools, designed for the rational manipulation of uncertain objects. In a wider sense, however, statistics embraces a philosophy of knowledge and action; it plays a role in the construction of facts as much as in the formulation of decision-making rules.

Today, statistics is still at the intersection of other disciplines. While some 'administrative' statisticians produce information in the form of public or private data bases, others who may be economists, sociologists, biologists, or pure statisticians, publish studies treating this data with the help of methods designed in mathematics laboratories or departments by yet a third category of statisticians. To dissociate these three forms of statistics (production of statistical information, treatment of this information, and the mathematical basis for this treatment) to insist on seeing statistics as a branch of mathematics is therefore to obstruct the understanding of the past forms of statistical thinking. Moreover, it is also to miss many past and present rearrangements of disciplines which, under the guise of intermediary disciplines such as biometrics, psychometrics, and econometrics, combine these three levels. Finally this attitude remains blind to enterprises such as data-collecting, somewhere in between information science and technology, between the production and the use of information.

Albeit irreducible to a branch of mathematics, statistical thinking has nevertheless had a wealth of relationships with mathematics whose troubled history is worth describing together with that of the paradigms of the discipline. Sometimes the internal dynamics of mathematical probability played an essential role—eg. for integral and differential calculus, the theory of generating functions and Fourier transforms before 1820, and measure theory in analysis or matrix theory in algebra at the beginning of the twentieth century; at other times problems stemming from physics, biology, or the social sciences nurtured—more than caused externally—numerous theoretical developments in statistics, including some of its mathematical models. In the nineteenth century— from Lapace's death in 1827 to the probabilistic revolution of the 1920s—the apparent absence of mathematical probabilities and statistics becomes a fertile period from the point of view of the construction of a particular inductive logic in the sciences of observation.

Having explored the notion of image in the particular case of statistics, we now turn to a typology of forms of statistical thinking from the point of view of the historian of this discipline. Two specific approaches will complete this picture. The first one uses a particular corpus—a significant sample of the treatises of the discipline—to attempt an objectification of typological characters. Produced by statisticians themselves, this material arguably forms the most susceptible of providing an image from the actors' viewpoint. Using other materials, the second approach specifically concerns the relationship of statistics with mathematics.

TYPOLOGY AND CHRONOLOGY OF STATISTICAL THINKING

The Triple Origin of Statistics

Whether classical ([Laplace, 1814]; [Todhunter, 1865]; [Meitzen, 1886]; [Westergaard, 1932]) or more recent ([Hacking, 1990]; [Stigler, 1986]; [Porter, 1986]; [Kruger & al., 1987–89]; [Lécuyer, 1980]; [Desrosières, 1993]; [Brian, 1994]; [Armatte, 1995]), the historiography of statistics has credited this discipline with a double origin. The first was descriptive science of the State, or society anatomy, that was taught in seventeenth century German universities by professors such as Conring, Achenwald, and Schloesser, i.e. some kind of political geography developing a discourse and a fairly non-numerical system of representation for the State's 'remarkable attributes'. Until the early nineteenth century, this science was developed everywhere in Europe through the rise of census and inquiry practices, like those, in France, of old-regime *intendants* and Napoleon's *préfets*. This intense administrative scientific activity led to the establishment of institutions (statistical bureaus, learned societies) and to publications (the first treatises of statistics by Peuchet (1805), Heuschling (1847), or Mone (1834), which, viewed from today, might be more accurately described as essays in rational geography than textbooks of statistics). This 'statistics' explicitly rejected 'the methods, which by enigmatic formulae, algebraic calculations, or geometrical figures, would seek to present or analyze what can be said much more simply, naturally, and without obscurity' [Peuchet, 1805].

The second origin of statistics lies in English political arithmetic by Petty (1623–1687), Graunt (1620–1674), and Davenant (1656–1714). Through a quantification of social phenomena, they wished to identify regular, reproducible accounting or functional equations dealing with their numerical data, and which could therefore shed light on princes' decisions. The King's law linking prices and quantities of traded wheat, the drawing up of mortality tables for calculating annuities by Graunt (1662), de Witt (1671), Halley (1692), Struyck (1740), Kersseboom (1742), Deparcieux (1746), Süssmilch (1741, 1761), Euler (1767), etc.; Lavoisier's estimates of France's wealth through the number of ploughs; Moheau's (1778), Condorcet's (1784), and Laplace's (1785) estimates of the French population via birthrates—all belong to this tradition which persisted into the first third of the nineteenth century. While the word 'statistics' scarcely figures in this tradition, the inferential and conjectural core of a new way of reasoning with numbers was put in place. This reasoning however had to face two difficulties.

A necessary condition for legitimate inference, the availability and reliability of data, requires an efficient statistical system. Until the Napoleonic period, this system was lacking—or, more accurately, restricted to local operations carried out by enlightened administrators concerned with population movements, conscription, or taxes. The systematization of censuses and regular inquiries, as well as the publication of statistical output, was the long-term output of nineteenth century institutions, whether national (in the French case, the *Statistique générale de la France* [SGF], established by Thiers in 1840 and the various ministries' statistical bureaus) or international (Quetelet's International Statistical Congress holding nine sessions between 1853 and 1876, followed by the International Institute of Statistics in 1885).

According to Ian Hacking [1990], an 'avalanche of printed numbers' was an essential condition for the emergence of statistical regularities. A famous French example, judiciary statistics, were regularly published from 1827 onwards and became an important factor in triggering important statistical studies on criminality and, more generally, moral statistics (Guerry, Quetelet, etc.).

As for inductive inference, it is associated with uncertainty (Hume) and fluctuations in sampling that could not be harnessed without a new theory invented at the end of the seventeenth century by a few 'jansénistes et hommes du monde.' At the beginning of the nineteenth century, this theory formed the third leg of a tripod on which statistics rested. Relying on a careful rereading of Pascal, Huygens, Montmort, Arbuthnot, and Bernoulli, historians such as Ernest Coumet [1970], Ian Hacking [1975], Lorraine Daston [1988], and Norbert Meusnier [1996], among others, have shown that the birthplace of this theory hardly lay in mathematical considerations later applied to concrete problems, but rather in this very collection of judiciary and commercial problems concerned with estimating the true measure of an uncertain situation and with making a decision in the face of contingency and risk. Long before it served as a degree of plausibility, probability was mostly the power which proves. It is clear from the integration of Bernouilli's theorem (the weak law of large numbers) in his *Ars conjectandi* [1713], that the framework of this new theory was rhetoric (an argumentative practice) and what he called *la stochastique*, which was an art of conjecture.

At this point, this theory met with infinitesimal calculus to end up, a century later, with Laplace's *Théorie analytique des probabilités* [1812], which finally put a mathematical foundation statistics in place. During this century, a dynamic internal to analysis combined with Enlightenment philosophy to define a program of unification for the sciences, as shown by the practice of the Académie and the undertaking of the *Encyclopédie*. Several recent studies on Condorcet [Rashed, 1974]; [Baker, 1975]; [Crépel and Gilain, 1989]; [Brian, 1994] have underscored the role he saw for analysis in the unification of the sciences. He similarly emphasized the place of analysis in probability theory, to which he assigned the task of measuring the precision of observations, averages, and frequencies, as well as the conduction of optimal inductive inferences by means of the least-square method.

Probability theory had a leading role to play in the *mathématique sociale* often envisioned by Condorcet and pursued by Laplace in his

Essai sur les probabilités. In this work, Laplace moreover described the 'applications of probability theory to natural philosophy.' In his *système du monde,* terrestrial and celestial mechanics resulted from a program of reducing, by means of probability theory, measurement errors due to 'a large number of perturbative causes.' Since error theory was in effect the core of the first form of mathematical statistics, let us examine the role of mathematics in this construction.

In the beginning there was no mathematical theory to be applied, but a highly important question: how can the physical sciences be rooted in observation, as required by the Galilean revolution, if nobody reads the 'book of nature' in the same way? Yet this was exactly the case for astronomy and geodesy, where, due to the unreliability of evidence, a disparity in observations was the greatest obstacle to the application of the Galilean method, and hindered choice among competing theoretical models. Clearly this 'book of nature' was not written directly in mathematical language, as Galileo had claimed. Instead, one had to '*défalquer les empêchements*[1]' i.e. strip the object of its physical attributes, before turning it into a mathematical entity. Similarly, it was impossible to approach measurement errors as mathematical objects, viz. random variables, as long as all known systematic sources of error (human, instrumental and procedural) had not been eliminated. Only after this had been done could errors be said to stem from chance. Several practical questions then arose: How could aberrant measured values be eliminated? How could they be combined with the mean, i.e. the best representation of a 'true' value? How should this 'mean' be defined? Which distribution law had to be assumed for errors? Which optimization or adjustment criteria had to be privileged?

Above all, was mathematics able to answer these questions? This, in any case, seems to be what mathematicians sought to do between 1750 and 1820, albeit only succeeding in proving associations and sequences of thought. If one assumes the following principles: (1) the 'mean' must be an average; (2) the likelihood of observations must be maximized (or the probability of an error nil), then the error law must be Gauss's. This was his first theory (1809). But in 1823, Gauss himself preferred to minimize the mean-square error of these results. This led him to the least-square procedure without assuming that errors followed a normal law. Finally, each solution to one of the three topics of 'mean,' law, and adjustment criterion is only valid in relation to solutions of the two others. Apparently, mathematics was unable to account for anything without a detour through metaphysics, i.e. (in the sense of the eighteenth century) the choice of a more or less arbitrary, *a priori* principle—averages, least squares, or normal errors.

Far from being at the origin of error theory, mathematics therefore represented its culmination. Only after it had been detached from its physical, concrete meaning could error become a random variable, that is, a mathematical entity. On the other hand, mathematics could serve to reinforce links and arguments. At the end of the nineteenth century, Lippmann would still claim that no one knew why errors followed the normal law, experimenters taking this fact for a mathematical theorem and mathematicians taking it as an empirical result. It would be difficult to find a preexistent branch of mathematics which could have been *applied* to error theory, but it would be inaccurate to say that this question *motivated*, or drove, the development of a new branch of mathematics, which would then need to be abstracted and extricated from this context. In fact, this was an exchange through which the mathematics of chance was constructed at the same time as the solution to very concrete problems in different fields. Error theory exemplifies the simultaneous emergence of probability in similar questions of testimony and measurement. One had to wait for more than a century before probability could take an abstract, axiomatic form (for example the notion of random variable) from which the initial problem to be solved has been expunged. A history of these notions therefore must grasp both ends of the chain linking things and words, and be both linguistic and mathematical.

1820–1845: *Hibernation of Probabilities; Golden Age of Statistics*

Condorcet did not live long enough to implement every potentiality he attributed to probability theory in his *mathématique sociale*. His closest colleague at the Académie, Laplace, never cited him even while going further along the same path. But his *Théorie analytique des probabilités* [1812] was neither understood nor even read outside a restricted circle of mathematicians (Lacroix, Bienaymé, Poisson, Cournot, de Morgan, etc.). The application of these formulae sometimes bred an incredibly optimistic confidence—Laplace (mistakenly) bet 999,308 to 1 that Jupiter's mass was within 1% of the value given by Bouvard. The identification of random events with drawals from an urn of fixed composition has often been misleading. Shaky foundations, self-referential definitions (i.e. probability defined as the proportion of equally probable favorable cases), and roughly demonstrated theorems left several conceptions of probability competing against one another and therefore discredited it in the eyes of mathematicians and logicians. Controversies and debates—such as those pitting Cauchy against Bienaymé about the least-square method, Poisson against Poinsot about the application of probability to judgment theory, and the controversy about the principle of insufficient reason—could effectively lead John Stuart Mill to pronounce in the first edition of his *Logic* [1843] that 'probability theory is the scandal of mathematics.'

The downfall of probability theory however occurred in a period notable for an extraordinary development of institutions and initiatives concerned with descriptive administrative statistics. In France for example, the publication of *Recherches statistiques sur la ville de Paris et le département de Seine* was, in 1823, the Big Bang of Hacking's 'avalanche.' As a result, statistical bureaus came back with a vengeance and increased their autonomy; statistical societies were revived; and *statistique morale* was developed on the basis of administrative data and hygienists' studies (Guerry, Quetelet, and Villermé).

Essentially based on censuses and exhaustive inquiries, administrative statistics eschewed recourse to inferences, whether probabilistic or not. As we have seen, administrators and academicians enforced a strict division between the establishment of facts assigned to enlightened administrators and scholars, on the one hand, and the exercise of inductive inference reserved for geometers, on the other. Restricted to geodesy, artillery, and the actuarial domain, this exercise was moreover performed with extreme caution.

Because of the role of institutions and that of the joint structuring of information systems and social systems, the situation may have been different in Germany or Britain. In the latter case for example, where the original motto of the Statistical Society of London—'*Aliis exterendum*' persisted for over fifty years indicated a clear mistrust of unrestrained data interpretation. Yet the development of vital statistics was soon associated with problems of pauperism and public health by local institutions created by the Poor Laws. Large-scale collection of data was far from sufficient for the General Register Office's statisticians. In charge of supervising demographic and public health surveys as well as the aid granted to workhouses, these statisticians necessarily participated in the great public health debates. They moreover took part in all major currents confronting these issues in the middle of the nineteenth century—hygienics and preventive medicine (Louis, Villermé), contagionism (Moreau de Jonnès), family curative medicine (Amador), eugenics (Galton, Pearson), and experimental medicine (Claude

Bernard, Pasteur). The English statisticians, like the French and Belgian moral statisticians, could not agree to keep to raw facts. But while hygienist and moral statistician Villermé opposed the contagionist director of the *Statistique général de France* Moreau, de Jonnès, no similar conflict pitted English hygienists against a centralized administration. Their leader William Farr was therefore in a position to develop, long before his French counterparts, a true statistical study of poverty and illness factors for prevention and reform [Desrosières, 1993, chap. 5]. Differences in political contexts (the industrial revolution), administrative structure (a decentralized state), and social programs (hygienics) led Farr to a different form of statistical innovation, both in his research program (search for environmental factors) and his tools (nomenclature, sequencing, and co-variations).

1850–1885: Average Theory

The third period is marked by Quetelet's work. An erudite Belgian astronomer, he was at the center of a genuine international network of statisticians. From 1853 onwards, by analogy with the world network of astronomical and meteorological observatories, famous statistical congresses gathered with the intent of taking the 'temperature' of the various societies spread over the world. With Quetelet, statistics became a universal discipline that organized sufficiently standardized nomenclatures, procedures, and methodological rules, so that they could be exchanged and compared. And thus could the avalanche of printed numbers be generalized. Although it only half-succeeded due to difficulties in transforming scientific norms into political decisions, Quetelet's project resulted in a kind of universal administrative science of data production, which provided statistics with the appropriate basis of a natural science. Like Linneus, Quetelet indeed thought of himself as establishing a new natural science, i.e. the *anatomy* of the social body.

Furthermore, Quetelet also sought to construct the *physiology* of the social body. By this he meant a dynamical explanation of society in terms of forces and balances, borrowed from mechanics. Quetelet therefore pretended to be sociology's Newton. A disciple of the French probabilists, he revived and popularized parts of probability theory by endowing it with a new significance as a theoretical basis for *social physics*, which he put together by transferring the notions of average, error law, and estimates from astronomy to the 'moral' sciences. The actual location of a celestial body was approximated by the average of observations whose errors were supposed to be distributed on a bell curve. The same distribution observed in human characters led Quetelet to identify the average with a fictitious theoretical being, the *average man*, who played in the social sciences the same role of reference for measurement as the celestial body in astronomy. A crystallization point for various debates involving determinism, statistical laws, and free will, the average man concept would serve as a complementary paradigm to Auguste Comte's positivism for most late nineteenth century social sciences.

Throughout this transition period, statistics increasingly grew into the physiology of society, rather than its anatomical description. The notion of law became an essential part of the expression of observed regularities, and a degree of reconciliation of statistics with rather elementary mathematics took place borrowing tools from probability theory and adjustment techniques.

A second characteristic of this period is an important change in the very conception of chance. For Bernouilli and Laplace, chance was nothing but a mask for ignorance of the true causes of phenomena. Chance was epistemic, and therefore totally compatible with absolute determinism. In 1859 Darwin's *Origin of Species* and Maxwell's kinetic theory of gases appeared, both of which, in their respective domains, represented major ruptures. The Darwinian theory of evolution posited random variations of species through hereditary transmission, variations that were controlled by natural or artificial selection. Maxwell's and Boltzmann's theories identified the state of a gas with the resultant of random molecular motion (following, for that matter, Quetelet's model of a society). In both cases, chance no longer was the results of ignorance; it formed the very essence of life and matter.

1885–1925: Autonomy of Mathematical Statistics

1885 was marked by the London Statistical Society's Jubilee, the foundation of the American Statistical Society, and the birth of the International Institute of Statistics (IIS). The IIS was a revival of the sessions of the *Congrès international* (1853–1872) devoid of its political function, and so only retained its scientific and professional dimensions. In this framework, nomenclatures, measurements, investigative techniques, and standards of presentation took their modern shape. Statistics gained in professionalism and autonomy with respect to neighboring disciplines with which it had long been intermingled. This movement prompted the development of the statistical method in established disciplines such as statistical physics, biometrics, and statistical economics. In France, while statistics had vaguely been integrated into courses of astronomy, geodesy, and artillery in military and engineering schools, and into courses of economics at several institutions (*Conservatoire national des arts et métiers* [CNAM], *École des ponts et chausées, École des sciences politiques*, etc.), the teaching of statistics as a full-fledged university discipline only emerged at the end of the nineteenth century (the first chair at the University of Paris was in 1895). In the United States, inspired by the English (Galton and Pearson) and German (Fechner, Weber, Ebbinghaus, and Lexis) schools, the first courses on statistics appeared after 1880 in psychology, anthropology, and economics departments, as well as in technical schools (Mayo-Smith, Falker, Cattell, Boas, and Thorndike).

At this moment, statistics no longer was a science defined by its object of study, whether political (the State) or social (population or society), but rather by its method, whose elements were perfectly set out in Ronald Fisher's *On the Mathematical Foundation of Theoretical Statistics* [1922]. Collected statistical data was considered to be a random sample of a hypothetical population (given by a parameteric law), and the object of statistics was to reduce this data into a small number of statistical summaries. Capturing the pertinent information in the data, these summaries allowed efficient estimates of, and tests on, the parameters contained in population laws.

This abstract conception was the endpoint of a process during which statistics distanced itself from average and homogeneity dogmas in order to develop notions of dispersion and heterogeneity. In Britain, dispersion and correlation theories were constructed for Galton's, Pearson's, and Fisher's eugenic studies. Concepts as fundamental as regression and correlation originally appeared in a context of heredity research which assigned them a strong concrete meaning before they ever received well-established mathematical properties and syntax. But on these topics, as well as in sampling theory, the contributions of economists Edgeworth, Yule, and Bowley were also very important. In Germany, promoting a break with Laplace's dogma, Lexis and Bortkiewicz's school argued for the reconstruction of mathematical statistics on the basis of variable-urn schemes reflecting the hetero-geneity and instability of behavior. In France, Laplacian dogma was also criticised (J. Bertrand). But no alternative was proposed, except for the idea of economists Lucien March, and later François Divisia, (which were to be also found in Germany) that, since it owed nothing to probability theory, mathematical statistics should be constructed on the basis of sophisticated adjustments of frequency laws. Once again in this case, mathematical tools were not developed independently from research programs and epistemological, ethical, and political choices underlying them.

After 1925

As described by Hacking [1990], the emergence of probability theory was a multiple success story, at the metaphysical, epistemological, logical, and ethical levels. The axiomatic foundation of probability theory based on set and measure theories restored its standing by leaving aside questions of interpretation (subjective approaches vs. those based on observed frequencies). Thanks to inference and sampling theories and to the notion of statistical model, a strong link between statistics and probability theory was reasserted by the English school (Fisher, Yule, and Kendall). Wald's theory of hypothesis testing, followed by J. Neyman and E. Pearson's, found important industrial applications such as manufacturing control. The mathematical theory of statistics was strengthened while the discipline acquired a new autonomy.

Inferential statistics became a fully-fledged branch of mathematics, and began to be taught as such. In the United States, a 1925 survey of the American Statistical Association recorded over 300 courses of statistics in 843 universities and colleges. Under Hotelling's stimulus in particular, many statistics departments were likewise established. In France, one saw the establishment of the *Institut de statistique* of the University of Paris (by March and Darmois in 1922), the *Institut des sciences financières et d'assurance* in Lyons (by Eyraud and Gumbel), and the chair in probability theory at the *Faculté des Sciences* of Paris in 1929 (occupied by Borel, and later Fréchet). In Italy (Gini, Galvani, Benini, Livi, Niceforo, and Pietra), Germany (R. von Mises) and Sweden (H. Westergaard), a multiplication of chairs, laboratories, and journals reflected the academic presence that mathematical statistics had become.

Stochastic modeling pervaded engineering science as well as rational economics. In 1929, the crisis would trigger a powerful convergence between these domains, leading up to the creation of the Econometric Society in 1930 and the Cowles Commission's work on structural models. With respect to the rationalization of budgetary choices and economic policies, mathematical statistics relied on the irruption of Keynesian macroeconomics and planning policies in order to contribute to the construction of macroeconometric models [Tinbergen, 1936–39];

[Klein, 1950]. In the related domains of management and operational research, it revived decision-making theory and benefited from military programs for applied mathematics developed after 1940.

From the historian's viewpoint, this picture of the successive or competing forms of statistics comprises a 'probabilistic revolution' in a Kuhnian sense, as was argued by Bielefeld scholars [Kruger, 1987]. Following Bernard Cohen, one should distinguish probability theory itself, its emergence, evolution, long hibernation period, and sudden awakening in the 1920s, from those nineteenth century observation sciences which were effectively 'revolutionized' by statistical thinking. Hacking, also insists that one should distinguish the series of small revolutions in the sciences of chance (Quetelet's average theory, Galton's regression theory, Fisher's estimate theory, Neyman's test theory, etc.) from the large-scale revolution swaying the Baconian sciences in the nineteenth century as a result of the statistical method, and which Hacking calls the 'second scientific revolution.'

This picture also underscores the central idea of this chapter, namely that a history of statistics cannot be written only by following the accumulation of its formal tools. For example, one might retrace—from Arbuthnot (1710) to Neyman and Pearson (1936), via Laplace, Lexis, and Fisher—a formal history of statistical tests that would show the progressive appearance of a modern mathematical theory. But this history would remain a fiction. Like a juxtaposition of technological objects in a museum supposed to represent a line of ancestry (automobiles, for example), this history would be blind to issues, problematics, tensions, and controversies which underlay innovation. Behind reasoning processes of the type 'it is not by chance that x has this value,' or 'one could bet n against 1 that this quantity does not differ from x by more than so much,' Arbuthot inscribed his statistical test of the masculinity ratio (i.e. the numerical ratio of male babies over the total number of births) in a debate about the existence of divine Providence that pitted him against authors like de Moivre and s'Gravesande who represented other theological sensibilities. For his part, Laplace showed that a given mechanical model was consistent with observation, so that the meaning of chance was completely different. Between Fisher's 'null hypothesis' test allowing him to assume with a certain confidence level the inverse hypothesis of a certain effect and Neyman and Pearson's which assumes two possible decisions and their respective risks, the major difference was between knowledge rule and decision-making rule. These differences made the syntactical rules of these tests impossible to compare, and they alone render incomprehensible the hybrid ahistorical theory which is most often taught [Gigerenzer and Murray, 1987].

THE IMAGES OF STATISTITICS AS SEEN FROM TREATISES

An historical study of statistical treatises provides an image of the discipline in its multiple configurations depending on time periods and national cultures. By its etymology, the word *treatise*—'a didactic written account treating a subject systematically and in detail,' according to the dictionary—recalls the word *treaty*—i.e. 'a formal agreement between two or more nations.' Both coming from the Latin *tractatus*, they are both translated as *traité* in French, a fact which may serve to justify a military metaphor. Like treaties, treatises can be considered as stabilizing battle fronts, and defining a territorial, or disciplinary, domain. Treatises enforce structures on knowledge, privilege syntactical sequences, legitimate modes of problem-setting, of translations, and of recommended procedures, establish a dictionary of categories, notions, and accredited terms, and finally authorize certain actors in a field.

As for textbooks, they ultimately are special treatises aiming, through teaching, at a large diffusion of the minimum content of a discipline. In a sense, the act of transforming a treatise into a textbook is an attempt at perpetuating a local translation of knowledge structure into a paradigm of normal science, involving a 'didactic transformation' of contents [Chevallard, 1980]. Not only concerned with stabilizing controversies, textbooks also are rational reconstructions, evacuating all historicity from the representation of their field, so as to make it suitable to be diffused and reproduced at all times and places. Textbooks must therefore be considered as vectors of the institutionalization of a discipline, and reattached to other interdependent vectors such as chairs, laboratories, and publications.

The corpus of 56 treatises selected here (Appendix 1) is but a sample of all texts in statistics that could count as treatises. It exceeds somewhat the nineteenth century (it comes the period 1805–1939) and French production, since it includes foreign treatises found in Parisian libraries (notably at the *Statistique générale de France* and the *Société statistique de Paris*).

In order to reconstruct the structure of the statistical field, with its lines of fracture, its subcultures, and its salient features, each page of each treatise has been assigned to one theme taken from a list of

41 categories (a number reduced to 21 at a second stage) (Appendix 2). A theme (theoretical topic or domain of application) characterizes the issue raised at a given page independently of the thesis which is defended. From this operation, a statistical table has been constructed, which for each treatise (in rows) gives absolute or relative occurrences pertaining to each of the 21 themes (in columns).

This table has been submitted to a correspondence factorial analysis [Escofier and Pagès, 1988], whose map for the first two axes (23% and 17% of inertia) have been included (Appendix 3). The horizontal axis clearly discriminates a 'literary' statistics (on the left) from a 'mathematical' statistics (on the right), whereas the vertical axis is rather chronological and distinguishes Laplacian classical probability theory (on top) to twentieth century mathematical statistics (bottom). One finds notable exceptions with Borel next to Condorcet and Liesse near 'administrators.' The cloud's V-shape suggests that the initial dichotomy was resolved into a larger unity among treatises of economic statistics. Groups appear that should not be reified, since borders are porous and intersections exist. Because they have similar thematic profiles, texts clustering in these groups can nevertheless be usefully studied and the identification of associated themes is suggestive. They indeed are a reflection of subcultures in the disciplinary space, which others have christened *themata*, styles, or schools.

This structure has been confirmed by an ascending hierarchical classification in which several thematic groups of treatises can be clearly recognized. Seven thematic groups emerge that correspond fairly well to agreed classifications among historians of statistics: classical probability, induction theory, 'German' statistics, moral statistics, administrative statistics, economic statistics, and mathematical statistics. The point of this method is to achieve, by means of conditional probabilities, an exact description of classes from the point of view of thematic descriptors, which can be divided in two groups for each class: those over-represented in the treatises of the group and those neglected (Appendix 4).

Tables of thematic occurrences, factorial maps, and classifications are representations of the variations in time and place of the image of the discipline produced by statisticians themselves. Unless the interpretation of treatises through our thematic typology constitutes a source of bias, the faithfulness of this representation is guaranteed by the fact that it is directly taken from their writings. Of course, as for any technique enabling one to discern a structure, this representation ignores internal and external dynamical factors underlying the passage from one paradigm to another. But, recapturing the polymorph, holistic character of each paradigm, it shows the complete redistribution of characters at moments, and along lines, of ruptures.

Statistics and Mathematics: Reciprocal Images

As already argued, the history of statistical thinking can be articulated neither internally to, nor independently from, the history of mathematics. Periods roughly isolated above reveal dominant configurations in which a specific relationship of statistical thinking to mathematics was put forward, among others, as a defining character of the discipline. In the 'territorial' phase of the discipline, mathematics was completely foreign to statisticians' concerns. In the statistics of the Prix Montyon, mathematics was explicitly rejected because it was seen to be limited to the display of facts. In average theory, it was a secondary instrument merely allowing an exhibition of regularities and laws. In Fisher's statistics, finally, mathematics was constitutive of the entire discipline by providing a method for controlling legitimate inferences. Of course, these configurations do not always succeed one another and their coexistence often led to overt controversies.

The most immediate way of making this relationship more precise would be to assemble an anthology of quotations illustrating statisticians' views on mathematics. The self-evident value of this compendium might however lead to great confusion unless care is taken to refer each quotation back to the paradigm to which its author belongs. Once more, what we are looking for are more objective, and above all more collective, indices of this relation.

The first of such indices may be found in treatises themselves. Originally an annotated geopolitics, statistics was a topic where theory occupied less than 10% of treatises produced by the 'German' school. Later, theory came to occupy 39% of administrator Moreau de Jonnès's treatise and 91% of economist Bowley's (1902). Moreover, only some of theoretical developments could be called mathematical. The classification obtained from the table of thematic frequencies reveals that probability theory was totally underrepresented in 4 out of 7 classes of treatises (or paradigms): German statistics (à la Schloesser), moral statistics [Quetelet, 1835], administrative statistics [Bertillon, 1895], and early twentieth century economic statistics (Bowley and Liesse) hardly refer to it. They can however call up various types of mathematical tools, such as computation of characteristic values, adjustments of tendencies or periodic functions, correlation measurements, etc.

A more precise citation analysis [Armatte, 1995] confirms the very marginal position assigned to mathematical theorems in treatises of statistics and probability theory. With very few exceptions, statisticians never cited mathematicians without publications in statistics and probability theory. The self-closure of this professional group thus seems nearly complete.

Mathematics in the Journal de la Société Statistique de Paris.

In nineteenth century France, statisticians' only scientific society which lasted for more than a few years was the *Société statistique de Paris* (SSP), founded in 1860 by the *Polytechnicien* Michel Chevalier. A disciple of Saint-Simon and instigator of several free-trade agreements, Chevalier initially molded SSP into a lobby, an instrument of triumphant liberalism, and a support group for the hygienist movement, more than into a society for the promotion of statistics. This is made clear by the profile of the members—civil servants, administrators, medical doctors, lawyers, and publicists, rather than teachers, professors, researchers, academics, or industrialists. Statistical methodology was found in very few articles published in its journal. Even in methodological papers, only a small number could perhaps be considered as mathematical, with a marked increase after 1900 [Kang, 1989].

At that time, tools of biometrics and economic statistics (charts, indexes, correlation, coefficients) revealed an opening of the *Société* to mathematical methodologies and a gradual withdrawal from rhetorical argument. In economics, the first few serious studies of correlation and regression were reviewed by Yule at a joint meeting of British and French statisticians held in Paris in 1909. This date marked the takeoff of mathematical statistics: the studies of Yule, Pearson, Edgeworth, Moore, March, and Julin on cycles and indexes, those of Moore, Lenoir, and Lehfeldt on demand, provided the material for courses and textbooks of statistics proper.

Yet, the explosion of mathematical statistics occurred outside the institutional environment familiar to the historiography of mathematics. Indeed, looking for the image of statistics taking shape in the mathematicians' community, one finds various signs of the significant weakening of the part of mathematics devoted to probability theory, from Laplace's death up to the 1930s. These signs bear witness to a complete disregard of both questions of statistical inference in various disciplines and their mathematical foundations.

The Jahrbuch über die Forschritte des Mathematik (1868–1942)

Reviewing annually the main articles and books, this classic of mathematical bibliography through its classification system alone reveals the absence of statistics as a separate category. Only a rubric labeled 'Probability theory and applications' might have included statistics before 1914, at which point it merged with analysis. A quick survey of tables of contents shows that probability theory never took up more than 3% of the pages of the *Jahrbuch* up to the 1930s.

A deeper analysis would exceed the scope of this study. But our impression is nonetheless confirmed by Hélène Gispert's [1991] study of the *Société mathématique de France* (SMF), which emphasized 'the small number of [SMF] members' articles devoted to statistics, probability [1 to 3%], mathematical physics, and mechanics.' She moreover confirms that 'the development of statistics and probability theory was achieved outside the university and the traditional mathematical milieu represented by the SMF.' First held in Zurich in 1897, International Congresses of Mathematicians would not introduce a section devoted to 'economics, actuarial sciences, and statistics' before 1912.

The Encyclopédie des Sciences Mathématiques pures et appliquées

A fundamental source that may serve to locate probability theory and statistics in the mathematical landscape of the very beginning of the twentieth century was the *Encyclopädie des mathematischen Wissenschaften mit Einschluss ihrer Amwendungen*, edited by Felix Klein from 1898 to 1935, as well as its ambitious French version (1904–1916) edited by Jules Molk. In a recent study [Armatte, forthcoming], it has been argued that this compendium was an interesting entry point into a better understanding of the issues and controversies raised by the various mathematical approaches of chance at the turn of the century. Devised to combine historical surveys and formalized, yet didactic, presentations of main concepts—but also concerned with the identification of points of agreement and controversees and between the German and French schools—articles in the *Encyclopedia* often captured crucial features of their respective fields at the time of their writing. Albeit extremely valuable, the articles devoted to probability theory, error theory, and statistics unfortunately missed essential issues because writers restricted themselves to a didactic, watered-down version of its two essential features: (1) the important 'metaphysical' foundations that still structured the properties of the domain's mathematical objects—before its 1930 axiomatization, it is impossible to speak of measuring probability independently of its meaning and context, and to pass over this debate in the *Encyclopedia* makes the historical account of this domain incomprehensible; (2) its historical development outside mathematics—since the constitutive elements of the discipline were constructed starting from various domains in the natural and social sciences in response to specific questions of representation and inference. But, while reasserting its claim over pure and applied mathematics, the *Encyclopedia* carefully relegated applications (in physics, geodesy, economics, and actuarial science) to articles other than those dealing with probability theory and statistics. This had the effect of erasing the complex, two-way interaction between mathematical formalism and theorization of a field of application. This tension is perceptible in the article on statistics, where the French author [Oltramare, 1906] was forced to attach a long preamble to his translation of the German text in order to add nuances to Bortkiewicz's very peculiar viewpoint and to warn his readers that 'each point of view concerning the very basis of statistics is indeed widely different from others.'

> The domain to be covered is very vast; it embraces birthrates, mortality, morbidity, criminality, etc.; most often results are represented by ratios which sometimes can be seen as providing particular value of more or less definite functions. The statistician's task mostly consists in quantifying when possible the degree of precision that was obtained and, if necessary, determining the exact nature, form, and coefficients of the functions whose few particular values he has in his possession. For this task, provided of course that its principles are actually applicable, probability theory gives simple rules to follow. His first duty [therefore] is to confirm this.

The author then described three conceptions which bring out the ambiguity of the relationships between statistics, probability theory, and mathematics. In the Laplacian tradition, the first position assumed a theoretical model with Gaussian probabilities: it was then sufficient to assess the degree of approximation by which the frequency approached the unknown probability that was being looked for. Noticing the lack of concordance between observation and the model above, Lexis and Bortkiewicz had introduced a second conception according to which residues were brought back to a normal distribution through a complication of the model by calculation tricks and the consideration of variable probabilities. Stemming from the work of German authors (Knapp for example), a third position wholly rejected a recourse to probability theory, denied that statistical laws could be formulated, and fitted data with formulae considered 'as mere expedients, having none of the essential features of the formulae of rational mechanics and mathematical physics.'

Developing an actuary's vision in which statistics was reduced to computational techniques for mortality tables, the article of the *Encyclopedia* totally ignored important innovations (correlation, regression, sampling) made by the British school of biometrics and economics. As suggested by the first few pages of the article, the fundamental debate concerning the relation between statistics and probability theory is a recurrent question in the history of mathematical statistics. Even in the twentieth century, French authors as important as Lucien March (director of SGF) in the 1930s, or François Divisa (professor at the *École polytechnique* and CNAM, ex-vice-president of the International Econometric Society) in the 1950s, tried to construct a mathematical statistics without relying on probability theory.

At this point, however, this attitude went against the general tendency, since from 1930 onward mathematical statistics on the whole was strongly linked with pure mathematics, and especially probability theory. Lebesgue's and Borel's integration and measure theory, Kolmogorov's axiomatics, Fisher's theory of estimation, Neyman's and Pearson's test theory, and Bachelier's and Markov's process theory would all shake the foundations of probability theory and statistics. While benefiting from these innovations, the simultaneous emergence of probabilities in several fields (statistical physics, quantum mechanics, econometrics, finance, genetics, etc.) can however be accounted for by respective internal dynamics of each. Their increasing sophistication notwithstanding, stochastic disciplines have shown by their recent history that their development cannot be explained simply by the progress of their mathematical core, i.e. probability theory. By way of conclusion, let us quote Cournot [1843]:

> The question that remains is whether this theory is but a mind game, a curious speculation, or whether on the contrary, it studies very important, very general laws that are actually ruling the world. In order to transform

the idea of an abstract relation into a law that is actually effective in dealing with phenomena, a mathematical reasoning based on a series of identities is obviously insufficient. One has to appeal to other notions, other principles of knowledge; in a word, one must carry out a philosophical critique.

NOTES

* Translated by Fabien Jurdant and David Aubin.
1. Galileo, *Dialogo* Galilei *dei massimi systemi; Dialogue Concerning the Two Chief World Systems*, transl. Stillman Drake (Berkeley: University of California Press, 1953), second day; quoted in Israel 1996, p. 114.

REFERENCES

Armatte, M., [1991] 'Une discipline dans tous ses états: la statistique à travers ses traités (1800–1914)', *Revue de synthèse*, IVème série, n° 2.

[1995] *Histoire du modèle linéaire. Formes et usages en statistique et économétrie*, EHESS thesis, Paris.

[1998] 'Les probabilités et les statistiques dans l'Encyclopédie *des Sciences Mathématiques Pures et Appliquées*: promesses et déceptions', to appear.

Bernoulli, J., [1713] *Ars Conjectandi*, 4ᵉᵐᵉ partie, traduction N. Meusnier, IREM, 1987.

Bertillon, J., [1895] *Cours élémentaire de statistique administrative*, Paris, Societé Editions Scientifiques.

Brian, E., [1994] *La mesure de l'Etat*, Paris, Albin Michel.

Chevallard, Y., [1980] *La transposition didactique*, Paris, La Pensée Sauvage.

Coumet, E., [1970] 'La théorie du hasard est-elle née par hasard', *Annales ESC*, n° 3, mai-juin, 574–598.

Crepel, P., Gilain, C., (eds.) [1989] *Condorcet mathématicien, économiste, philosophe, homme politique*, Paris, Minerve.

Dahan Dalmedico, A., [1996] 'L'essor des mathématiques appliquées aux Etats-Unis: l'impact de la seconde guerre mondiale', *Revue d'Histoire des Mathématiques*, 2 (2), 149–213.

Daston, L.J., [1988] *Classical Probability in the Enlightment.*, Princeton, Princeton University Press.

Desrosieres, A., [1993] *La politique des grands nombres*, Paris, La Découverte.

Droesbeke, J.J., Tassi P., [1990] *Histoire de la Statistique*, Paris, Presses Universitaires de France.

Escofier, B., Pages J., [1988] *Analyses factorielles simples et multiples*, Paris, Dunod.

Fienberg, S.E., [1992] 'A Brief History of Statistics in Three and One Half Chapter: A Review Essay', *Statistical Science*, 7, n° 2, 208–225.

Gigerenzer, G., Murray D.J., [1987] *Cognition as Intuitive Statistics*, Hillsdale, London, Lawrence Erlbaum Associates.

Gispert, H., [1991] 'La France mathématique. La Société Mathématique de France (1870–1914)', *Cahiers d'histoire et de philosophie des sciences*, n° 34.

Hacking, I., [1975] *The Emergence of Probability. Logic of Statistical Inference*, Cambridge, Cambridge University Press.

[1990] *The Taming of Chance*, Cambridge, Cambridge University Press.

Insee (ed.), [1987] (1977) *Pour une histoire de la statistique*, tome 1.

Kang, Z., [1989] *Lieu de savoir social. La Société de Statistique de Paris au XIXème siècle*, Thesis, Paris, EHESS.

Kruger, L., Daston L., Heidelberger M. (eds.), [1987] *The Probabilistic Revolution*, vol 1: Ideas in History, Cambridge, MIT Press.

Kruger, L., Gigerenzer G, Morgan M.S. (eds.), [1989] *The Probabilistic Revolution*, vol 2: Ideas in the Sciences, Cambridge, MIT Press.

Laplace, P.S., [1886] (1812) *Théorie analytique des probabilités*, Paris (in *Œuvres complètes*, vol. 7).

[1886] (1825) (1814), *Essai philosophique sur les probabilités*, Paris (in *Œuvres Complètes*, vol. 7, i–cliii; reprint, Paris, Bourgois, 1986, postface B. Bru).

Lecuyer, B.P., [1980] '(Préhistoire des) Sciences sociales', *Encyclopédie Universalis*.

Martin, O., [1997] 'Les statistiques parlent d'elles-mêmes. Regards sur la construction sociale des statistiques' in *La pensée confisquée*, Paris, La Découverte.

Meitzen, A., [1890] (1886), *History, Theory, and Technique of Statistics*, ed. by R.P. Falkner, Philadelphia, American Academy of Political and Social Science.

Meusnier, N., [1996] 'L'émergence d'une mathématique du probable', *Revue d'histoire des mathématiques*, 2, (1), 119–147.

Oltramare, [1906] 'Statistique', *Encyclopédie des sciences mathématiques pures et appliquée*, I–24, Paris, Gauthier-Villars.

Porter, T., [1986] *The Rise of Statistical Thinking, 1820–1930*, Princeton, Princeton University Press

Stigler, S.M., [1986] *The History of Statistics. The Measurement of Uncertainty Before 1900*, Cambridge et London, Belknap Press of Harvard University Press.

Todhunter, I., [1949] (1865) *A History of the Mathematical Theory of Probability*, New York, Chelsea.

Westergaard, H.L., [1932] *Contributions to the History of Statistics*, London, King.

APPENDIX 1: BIBLIOGRAPHY OF TREATISES

[AFTALION 1928] Albert AFTALION, *Cours de Statistique, professé en 1927-1928 à la faculté de Droit*, (recueilli et rédigé par Jean Lhomme et Jean Priou, Paris, PUF, 1928, 313 p.

[AITKEN 1947] A.C. AITKEN, *Statistical Mathematics*, 5th ed., New York, Oliver and Boyd.

[BACHELIER 1912] Louis BACHELIER, *Calcul des probabilités*, Paris, Gauthiers-Villars, 1912, 512 p.

[BENINI 1923], Rudolfo BENINI, *Principii di Statistica metodologica*, Turin, Unione tipografico-editrice torinese, 1913 (1906), 353 p.

[BERTILLON 1895] Jacques BERTILLON, *Cours élémentaire de Statistique Administrative*, Paris, Soc.Ed.Scientifiques, 1895, 593 p.

[BERTRAND 1971] Joseph BERTRAND, *Calcul des probabilités*, 3ème édition, New York, Chelsea 1971, XLIX + 327 p.,(1ère ed.1889, 2ème ed. 1907).

[BLOCK 1886] Maurice BLOCK, *Traité de Statistique*, Paris, Guillaumin, 562 p.,(1ère édition 1878).

[BOREL 1909] Emile BOREL, *Eléments de probabilités*, Paris, Gauthier-Villars, 1909.

[BOWLEY 1902] Arthur-Lyon BOWLEY, *Elements of Statistics*, London, King and Son, 1902, 335 p., 2ème édition; (1ère éd. 1901).

[BOWLEY 1920] Arthur-Lyon BOWLEY, *Elements of Statistics*, London, King and Son, 1902, 454 p., 4ème édition; (1ère éd. 1901).

[CARVALLO 1912] Emmanuel CARVALLO, *Le Calcul des Probabilités et ses applications*, Paris, Gauthier-Villars, 163 p.

[CONDORCET 1805] Marie Jean Antoine Nicolas Caritat marquis de CONDORCET, *Elémens du calcul des probabilités et son application aux jeux de hasard, à la loterie, et aux jugemens des hommes*, Paris, Royez; réédité par IREM, Université Paris VII, 1986.

[COURNOT 1843] Antoine Augustin COURNOT, *Exposition de la theorie des chances et des probabilités.*, Oeuvres Complètes, Paris, Vrin, 1984 (ed B.Bru), 289 p., (1ère éd. 1843).

[CRUM/PATTON 1925] William Leonard CRUM et Alson Currie PATTON, *An In troduction to the Methods of Economic Statistics*, New York, A.W. Shaw Cy, 1925, v+ 488p.

[CZUBER 1921] Emmanuel CZUBER, *Die Statistischen Forschungs Methoden*, Wien, L.W.Seidel & Sohn, 1921, 234 p.

[DARMOIS 1928] Georges DARMOIS, *Statistique Mathématique*, Paris, Librairie Douin, 1928, 307 p.+ 23.

[DAVIES 1922] Georges R.DAVIES, *Introduction to Economic Statistics*, N.Y., 1922, 160 p.

[DUFAU 1840] Pierre-Armand DUFAU, *Traité de Statistique*, Paris, Delloye, 1840, 378 p.

[DUGE DE B. 1939] Léo DUGE de BERNONVILLE, *Initiation à l'Analyse statistique*, Paris, Librairie de Droit et de Jurisprudence, 230 p. 1939.

[ELDERTON 1909] W.Palin et Ethel M. ELDERTON, *Primer of Statistics*, London, A. & C. Black, 1909, 84 p., (4ème ed. 1923).

[FAURE 1906] Fernand FAURE, *Eléments de Statistique*, Paris, Larose et Tenin, 1906, 128 p.

[FISHER 1925] Ronald A.FISHER, *Statistical Methods for Research Workers*, Edinburgh, Oliver and Boyd, 1936, 6ème édition, 336 p.;(1ère éd. 1925; 13ème édition 1958)

[FLECHEY 1872] Edmond FLECHEY, *Notions générales de statistique*, Paris, Berger-Levrault, 1872, vii+44p.

[FRECHET/HALBWACHS 1924] Maurice FRECHET et Maurice HALBWACHS, *Le Calcul des probabilités à la portée de tous*, Paris, Dunod, 1924, 294 p.

[GUILLARD 1855] Achille GUILLARD, *Eléments de statistique humaine ou démographie comparée*, Paris, Guillaumin, 1855, xxxii + 368 p.

[HEUSCHLING 1847] Xavier HEUSCHLING, *Manuel de statistique ethnographique universelle*, Bruxelles, 1847, 504 p.

[JEROME 1924] Harry JEROME, *Statistical Method*, N.Y., Harper, 1924, 395 p.

[JONES 1921] D.Caradog JONES, *A first course in statistics*, London, 1921, 283 p.

[JORDAN 1927] Charles JORDAN, *Statistique Mathématique*, Paris, Gauthier Villars, 1927, 340 p.

[JULIN 1921] Armand JULIN, *Principes de Statistique théorique et appliquée*, tome1: Statistique théorique,Préface L.March, Paris, Marcel Rivière, 1921, 712 p.

[JULIN 1928] Armand JULIN, *Principes de Statistique théorique et appliquée*, tome 2: statistique économique; fasc.I: statistique du commerce extérieur et des transports, Paris, Marcel Rivière, 1923, 151 p.; fasc.II: statistique des prix et méthode des index-numbers, Paris, Marcel Rivière, 1928, 338 p.

[KEYNES 1921] John Maynard KEYNES, *A treatise on Probability*, London, Macmillan, 1921, 458 p., (rééditions 1929, 1943, 1948, 1952, 1957)

[KING 1912] Willford I. KING, *The elements of statistical method*, N.Y., Macmillan, 1912, 235 p.

[LACROIX 1833] S.F.LACROIX, *Traité élémentaire du calcul des probabilités*, Paris, 3ème éd., 1833, 352 p.(1ère éd.1816, 2ème éd.1822)

[LAPLACE 1825] Pierre Simon de LAPLACE P.S., *Essai philosophique sur les probabilités*, Paris, Bourgois, 1986, postface B.Bru, d'après la 5ème édition(1825), (1ère éd. 1814)

[LAURENT 1908] Herman LAURENT, *Statistique mathématique*, Paris, O.Doin, Encyclopédie scientifique, 268 p.

[LIESSE 1905] André LIESSE, *La Statistique. Ses difficultés. Ses procédés. Ses résultats*, Paris, Guillaumin et Alcan, 1905, 188 p., (2ème édition 1912, 3ème éd.1919, 4ème éd 1933).

[MARCH 1930] Lucien MARCH, *Les principes de la méthode statistique*, Paris, Alcan,1930, 796 p.

[MEITZEN 1891] August MEITZEN, *History Theory and Technique of Statistics*, Philadelphia, Am.Acad. of political and social science, 243 p., trad. anglaise de R.P.Falkner; édition originale : Gechichte, *Theorie und Technik der Statistik*, Berlin, 1886.

[MONCETZ 1935] A de MONCETZ, *Initiation aux méthodes de la statistique*, Paris, Sirey, 1935, 80 p.

[MONE 1834] François J.MONE, *Théorie de la Statistique.*, Louvain, 1834, xxiii + 145 p., Trad de l'Allemand et du Latin par Emile Tandel, (1ère édition allemande 1824).

[MOREAU DE JONNES 1856] Alexandre MOREAU DE JONNES, *Eléments de Statistique*, Paris, Guillaumin, 2ème éd., 1856, 460 p.;(1ère éd. 1847) .

[NICEFORO 1925] Alfredo NICEFORO, *La méthode statistique et ses applications aux sciences naturelles, aux sciences sociales et à l'art*, Trad de l'italien de R.Jaquemin, Paris, M.Giard, 1925, 632 p.; (éd.ital. 1923).

[PEUCHET 1805] Jacques PEUCHET, *Statistique élémentaire de la France*, Paris, Gilbert, 1805, 610 p.

[QUETELET 1835] Adolphe QUETELET, *Sur l'homme et le développement de ses facultés*, ou Essai de physique sociale, Paris, Bachelier, 1835, 625 p.

[QUETELET 1846] Adolphe QUETELET, *Lettre à S.A.R. le Duc Règnant de Saxe-Cobourg et Gotha*, sur la théorie des probabilités appliquées aux sciences morales et politiques, Bruxelles, Hayez, 442 p.

[QUETELET 1854] Adolphe QUETELET, *Théorie des probabilités*, Paris, A.Jamar: Encyclopédie populaire, 1854, 102 p.

[RIETZ 1924] H.L.RIETZ H.L. (ed), *Handbook of Mathematical Statistics*, N.Y, Houghton Mifflin Cy., 1924, 221 p.

[RIETZ 1927] H.L. RIETZ, *Mathematical statistics*, Chicago, The open court Cy., 1927, 181 p.; (2ème éd.1929, 3ème éd. 1936).

[SCHLOZER 1805] Denis-François DONNANT, *Introduction à la Science de la Statistique*, d'après l'allemand de M.de Schloetzer, avec un discours préliminaire, des additions et des remarques par D.F. Donnant, Paris, Imprimerie Nationale, 1805.

[SECRIST 1917] Horace SECRIST H, *An introduction to statistical methods*, New York, Macmillan, 1917, 469 p.,(2ème éd. 1921)

[TURQUAN 1891] Victor TURQUAN, *Manuel de statistique pratique*, Paris, Berger-Levrault, 1891, 564 p.

[VENN 1888] John VENN, *The logic of chance*, London, Macmilan, 3ème éd. 1888, 503 p., (1ère éd.1866, 2ème éd. 1876).

[YULE 1922] George Udny YULE, *An introduction to the theory of statistics*, Londres, Griffin, 6ème éd., 1922, 415 p.; (1ère éd. 1911).

[YULE/KENDALL 1937] George Udny YULE et Maurice KENDALL, *An introduction to the theory of statistics*, Londres, Griffin, 11ème éd., 1937, 570 p.; (14ème éd. 1950)

[ZIZEK 1913] Franz ZIZEK, *Statistical averages*, traduction de W.M.Persons, New York, 1913, 392 p.

APPENDIX 2: GLOSSARY OF THE TOPICS IN TREATISES

No	Code	Designation
1	his	History of facts and ideas
2	org	Administrative organisation
3	def	Definition of statistics: purpose, function, method
4	inv	Investigating methodology: surveys, classification, codification, tabulation, computing, publishing
5	des	Descriptive statistics: averages, mode, median, quartiles and percentiles, index, fluctuation
6	pro	Theory of probability: chance and probability, principles, calculus, expectation, bayesian probablities, analytical theory.
7	bin	Bernoulli theorem, binomial law, normal law, error theory, Poisson and Lexis schemes
8	ind	Induction: regularities, laws, causes, sampling, estimation, tests
9	cor	correlation, association, regression, covariation, fitting, interpolation, least squares
10	div	Tables, bibliograpy, exercises, other topics
11	jeu	games, lottery
12	nat	Natural sciences: astronomy, geodesy, physics
13	dem	demograpy: population, natality, mortality tables
14	eco	Economics: production, consumption, income, business cycles.
15	ant	anthropometry, biometry
16	cri	moral statistics: criminal investigation, suicide
17	act	Actuarial sciences: annuities, assurances, microeconomics
18	jug	judgements, testimonies, elections, decisions
19	eta	stat. of states: army, education, health.
20	ter	Territorial statistics
21	aut	Labour and firm statistics, other applications

APPENDIX 3: MAPPING (FACTORS 1 AND 2)

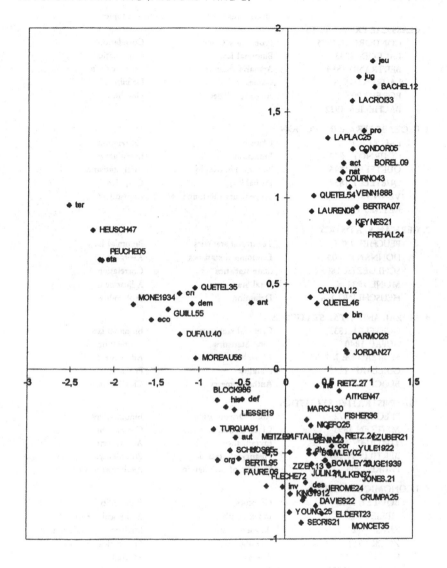

APPENDIX 4 : PROPERTIES OF CLASSES

Overrepresented / underrepresented topics

1. CLASSICAL PROBABILITY

CONDORCET 1805	Probability Calculus	Correlation
LACROIX 1833	Binomial law	Investigation
BERTRAND 1889	Actuarial Sciences	Classification
LAURENT 1908	Games	Definitions
BOREL 1909	Analytical Theory	Graphics
BACHELIER 1912		

2. PROBABILITE and INDUCTION

LAPLACE 1825	Chance	Correlation
COURNOT 1843	Induction	Investigation
QUETELET 1846	Natural Philosophy	Vital Statistics
QUETELET 1854	Probability	Graphics
VENN 1888	Judgements - testimonies	Adjustment
KEYNES 1921		

3. GERMAN STATISTICS

PEUCHET 1805	Territorial statistics	Binomial Law
DONNANT 1805	Economical statistics	Averages
SCHLOEZER 1805	State statistics	Correlation
MONE 1834	Vital Statistics	Adjustment
HEUSCHLING 1847	Definition	Probability

4. MORAL AND VITAL STATISTICS

QUETELET 1835	Criminal statistics	Binomial law
DUFAU 1840	Vital Statistics	Correlation
MOREAU DE J. 1856	Mortality	Adjustment
GUILLARD 1855	Natality	Probability
BLOCK 1886	Anthropometry	Classification

5. ADMINISTRATIVE STATISTICS

FLECHEY 1872	Adm. Organisation	Binomial law
MEITZEN 1891	Classification	Correlation
TURQUAN 1891	Investigation	Adjustment
BERTILLON 1895	History of Statistics.	Probability.
FAURE 1906	Statistic of population	Analytical theory.

6. ECONOMIC STATISTICS

BOWLEY 1902	Graphics	Probability.
LIESSE 1905	Index numbers	Analytical theory
KING 1912	Business Cycles	Statistic of population
ZIZEK 1913	Investigation	Economical statistics
SECRIST 1917	Averages	Mortality
JULIN 1910/1921	Time series	Territorial Stat.
JEROME 1924, AFTALION 1928, MARCH 1930...		

7. MATHEMATICAL STATISTICS

CARVALLO 1912	Adjustment	Vital Statistics
BOWLEY 1920	Sampling	Economical statistics.
CZUBER 1921	Correlation	Investigation
YULE 1922/1937	Association	Organisation
RIETZ 1924	Estimation/tests	Territorial and state statistics
FISHER 1925		
JORDAN 1927, DARMOIS 1928		

Chapter 8

MATHEMATICAL STRUCTURES FROM HILBERT TO BOURBAKI: THE EVOLUTION OF AN IMAGE OF MATHEMATICS

Leo Corry

The notion of a mathematical structure is a most pervasive and central one in twentieth century mathematics. Over several decades of this century, many mathematical disciplines, particularly those belonging to the 'pure' realm, developed under the conception that the aim of research is the elucidation of certain structures associated with them. At certain periods of time, especially from the 1950s to the 1970s, it was not uncommon in universities around the world to conceive the ideal professional formation of young mathematicians in terms of a gradual mastering of the various structures that were considered to embody the hard-core of mathematics.

Two important points of reference for understanding the development of twentieth century mathematics in general pertain to the ideas of David Hilbert, on the one hand, and of the Bourbaki group, on the other hand. This is also true, in particular, for the development of the conception of mathematics as a science of structures. In this chapter I explain how, and to what extent, the idea of a mathematical structure appears in the works of Hilbert and of Bourbaki.

My presentation will be based on defining the idea of a mathematical structure as a classical example of an image of mathematics. This image appeared for the first time in its full-fledged conception in 1930, in van der Waerden's classical book *Moderne Algebra*. In the work of David Hilbert we will find many of the basic building blocks from which van der Waerden's presentation came to be built. In the work of Bourbaki we will find, later on, an attempt to extend to all of mathematics what van der Waerden had accomplished for the relatively limited domain of algebra.

But before entering into all these works in more detail, it seems necessary to begin with some preliminary clarifications. The notion of an 'image of mathematics', parallel to that of the 'body of mathematics', is basic to my discussion here, and it is therefore necessary to explain how I use these terms in what follows.

Claims advanced as answers to questions directly related to the subject matter of any given discipline build the body of knowledge of that discipline. Claims which express knowledge *about* that discipline build its images of knowledge. The images of knowledge help us to discuss questions which arise from the body of knowledge, but which are in general not part of, and cannot be settled within, the body of knowledge itself, such as the following: Which of the open problems of the discipline most urgently demands attention? What is to be considered a relevant experiment, or a relevant argument? What procedures, individuals or institutions have authority to adjudicate disagreements within the discipline? What is to be taken as the legitimate methodology of the discipline? What is the most efficient and illuminating technique that should be used to solve a certain kind of problem in the discipline? What is the appropriate university curriculum for educating the next generation of scientists in a given discipline? Thus the images of knowledge cover both cognitive and normative views of scientists concerning their own discipline.

The distinction between body and images of mathematical knowledge should not be confused with a second distinction that has sometimes been adopted, either explicitly or implicitly, by historians of mathematics, namely, the distinction between 'mathematical content' and 'mathematical form.' Inasmuch as it tacitly assumes the existence of an immutable core of mathematical ideas that manifests itself differently at different times, this distinction would seem to enhance the everywhere present, potential pitfall of anachronistic and historically misleading accounts of the development of mathematical ideas. The distinction between body and images of knowledge, on the other hand, is of a different kind. The borderline between these two domains is, by its very definition, somewhat blurred and always historically conditioned. Moreover, one should not perceive the difference between the body and the images of knowledge in terms of two layers, one more important, the other less so. Rather than differing in their importance, these two domains differ in the range of the questions they address: whereas the former answers questions dealing with the subject matter of the discipline, the latter answers questions about the discipline itself *qua* discipline. They appear as organically interconnected domains in the actual history of the discipline, while their distinction may be worked out by historians for analytical purposes and usually in hindsight.[1]

Analyzing the history of mathematics in terms of the distinction between body and images of knowledge helps to attain some insights that might otherwise have passed unnoticed to historians. This is particularly the case concerning one specific, characteristic trait that we find in this area of science, but typically not in others. What I have in mind is the possibility offered by mathematics to formulate and prove, as part of the hard-core body of mathematical knowledge, certain statements *about* the discipline of mathematics, that is to say, the ability to absorb certain images of knowledge directly into the body of knowledge. In disciplinary terms, this ability is manifest in those branches usually grouped under the rubric of meta mathematics, but by no means it is strictly circumscribed to them. The peculiar capacity of mathematics to become part of its own subject-matter is what I call here 'the reflexive character of mathematics.'

The terminology introduced here will help us formulating the development of ideas covered by the present account. I will describe the introduction of the notion of structure as the adoption of a new image of mathematical knowledge, that essentially changed the conception of a specific mathematical discipline, namely, algebra. I will thus discuss the roots of this adoption and its early development. In particular, Bourbaki's contribution will be described as an attempt to reflexively elucidate this image and make it an organic part of the body of knowledge.

'PRE-STRUCTURAL' ALGEBRA

I start with a very brief account of the classical, nineteenth-century image of algebra, which the structural one eventually came to substitute. Throughout the eighteenth century, algebra was the branch of mathematics dealing with the theory of polynomial equations, including all the various kinds of techniques used to find exact or approximate roots, and to analyze the relationships between roots and coefficients of a polynomial. Over the course of the nineteenth century these problems remained a main focus of interest of the discipline, and some new ones were added, including algebraic invariants, determinants, hypercomplex systems, etc. Parallel to this, algebraic number theory underwent a vigorous development, especially with the works of Gauss, Kummer, Dedekind and Kronecker, on problems of divisibility, congruence and factorization.

Between 1860 and 1930 an intense research activity was conducted in these mathematical domains, particularly in Germany, thus giving rise to certain, central trends that can be specifically noticed in it. Among them we can mention the following: [1] the penetration of methods derived from Galois's works into the study of polynomial equations; [2] the gradual introduction and elucidation of concepts like group and field, with the concomitant adoption of abstract mathematical definitions; [3] the improvement of methods for dealing with invariants of systems of polynomial forms, advanced by the German algorithmic school of Gordan and Max Noether; [4] the slow but consistent adoption of set-theoretical methods and of the modern axiomatic approach implied by the works of Cantor and of Hilbert; [5] the systematic study of factorization in general domains, conducted by Emmy Noether, Emil Artin and their students.

In 1895, the first edition of Heinrich Weber's *Lehrbuch der Algebra* [1895] was published. This book summarized the achievements attained in the body of knowledge of this discipline, and at the same time it embodied more than any other one the spirit of the classical, nineteenth century image of algebra. If one compares it with its most noteworthy predecessors—Serret's *Cours d'algèbre supérieure* [1849] and Jordan's *Traité des substitutions et des équations algébriques* [1870]—one discovers many important additions and innovations at the level of the body of knowledge, but the image of algebra it embodies remains essentially the same one: algebra as the discipline of polynomial equations and polynomial forms. Abstract concepts such as groups, in so far as they appear, are subordinate to the main classical tasks of algebra. And, most important, all the results are based on the assumption of a thorough knowledge of the basic properties of the systems of rational and real numbers: these systems are conceived as conceptually prior to algebra.

The first two decades of the twentieth century were ripe with new ideas, and towards the end of the 1920s, one finds a growing number of works that can be identified with recently consolidated algebraic theories. These works were usually aimed at investigating the properties of abstractly defined mathematical entities, as the focus of interest in algebraic research: groups, fields, ideals, rings, and others. Van der Waerden's *Moderne Algebra* appeared—like many other important textbooks—at a time when the need was felt for a comprehensive synthesis of what had been achieved since the publication of its predecessor, in this case Weber's book. *Moderne Algebra* presented ideas that had been developed earlier by Emmy Noether and Artin— whose courses van der Waerden had recently attended in Göttingen and Hamburg, respectively—and also by other algebraists, such as Ernst Steinitz, whose works van der Waerden had also studied under their guidance. Van der Waerden masterly incorporated a great deal of the important innovations accumulated over the last years at the level of the body of algebraic knowledge. But the originality and importance of his contribution is best recognized by focusing on the totally new way of conceiving the discipline it put forward. Van der Waerden presented systematically those mathematical branches then related with algebra, deriving all the relevant results from a single, unified perspective, and using similar concepts and methods for all those branches. This original perspective, which turned out to be enormously fruitful over the next decades of research—and not only in algebra, but in mathematics at large—is what I will call here the structural image of algebra.

THE STRUCTURAL IMAGE OF ALGEBRA

The approach put forward in van der Waerden's textbook is based on the recognition that a certain set of notions (i.e., groups, ideals, rings, fields, etc.) are in fact individual realizations of one and the same underlying idea, namely, the general idea of an algebraic structure, and that the aim of research in algebra is the full elucidation of the those notions. None of these notions, to be sure, appeared as such for the first time in this book. Groups, for instance, had been introduced since the mid-nineteenth century as a main tool in the study of the theory of polynomial equations, and they could already be found in mainstream textbooks on algebra as early as 1866 (i.e., in the third edition of Serret's *Cours*). Ideals and fields, in their turn, had been introduced in 1871 by Dedekind in his elaboration of Kummer's factorization theory of algebraic numbers. But the unified treatment they were accorded in *Moderne Algebra*, the single methodological approach adopted to define and study each and all of them, and the compelling, new picture it provided of a variety of domains that were formerly seen as only vaguely related, all these implied a striking and original innovation.

The aim and contents of van der Waerden's book may be described roughly as an attempt to define the diverse algebraic domains and to elucidate their structure in detail. Of course, this characterization may appear banal and obvious to any mathematician nowadays, but by comparison to the earlier conception of algebra it is certainly not. At any rate, it raises two basic questions: [1] how is any algebraic domain defined? and [2], what does the complete elucidation of the structure of an algebraic domain entail? Algebraic domains are defined in *Moderne Algebra* in two different ways: either by endowing a non-empty set with one or more abstractly defined operations, or by taking an existing algebraic domain and constructing a new one over it, by means of a well-specified procedure (field of quotients of an integral domain, rings of polynomials of a given ring, extension of a field, etc.). In order to answer the second question, it is necessary to examine what van der Waerden actually did in his book for the various domains investigated. This description yields, in fact, an account of the essence of the structural image of algebra.

Basic to van der Waerden's analysis is the recurrent use of several fundamental concepts and questions. Among the most salient recurring concepts, one finds isomorphisms, homomorphisms, residue classes, composition series and direct products. None of these concepts, however, is defined in a general fashion so as to be *a-priori* available for each of the particular algebraic systems. Isomorphisms, for instance, are defined separately for groups and for rings and fields, and van der Waerden showed in each case that the relation 'being isomorphic to' is reflexive, transitive and symmetric. Thus the important notion, constitutive of the structural image of algebra, that two isomorphic constructs actually represent one and the same mathematical entity, appears prominently, yet only implicitly, in this textbook.

Perhaps one of the most fundamental innovations implied by van der Waerden's textbook is a redefinition of the conceptual hierarchy underlying the discipline which, in its classical image, generally granted priority to the foundations of the various number systems over algebra. In the opening chapters van der Waerden defined the natural numbers through a cursory review of Peano's axioms. He then extended them to the integers through the (informally presented) construction of pairs of natural numbers. But here he stopped: the rational and the real numbers he defined only in later chapters, and then with strict reference to their algebraic properties alone. Rational and real numbers have no conceptual priority in this book over, say, polynomials. Rather, they are defined as particular cases of abstract algebraic constructs. Thus, in Chapter III, van der Waerden introduced the concept of a field of fractions for integral domains in general, and he obtained the rational numbers as a particular case of this kind of construction, namely, as the field of quotients of the ring of integers.

Van der Waerden's definition of the system of real numbers in purely algebraic terms was somewhat more complicate. It was based on the concept of a 'real field,' recently elaborated by Artin and Schreier, whose seminars van der Waerden had attended in Hamburg. It is significant, however, that while van der Waerden was able to use this purely algebraic characterization, he did not give up completely the nineteenth-century, classical, image of algebra, and added a standard 'analytic' definition of the real numbers using Cauchy sequences (and adopting the term originally introduced by Cantor to designate them: 'fundamental sequences'). We thus find a single section in his book where $\varepsilon-\delta$ arguments appear, whereas in earlier textbooks it was standard to devote several chapters to discuss analytic questions concerning polynomials, including interpolation and approximation techniques. After van der Waerden, textbooks of algebra tended to exclude arguments of this sort. But in 1930, van der Waerden's attitude concerning the question whether a book on algebra can completely do without such considerations, even though they are not directly needed for the theories developed in it, still reflected his ties to the existing algebraic traditions.

This possibility of concentrating on determined features of a classical mathematical entity, while ignoring those that are considered momentarily irrelevant, is, no doubt, an important aspect of the study of algebraic 'structures' and one of the most striking innovations reflected in the book. Naturally, this possibility was afforded by the abstract formulation of concepts, but the mere availability of such a formulation was not in itself a compelling reason to bring about a real change in the images of algebra. Weber, for instance, had been fully aware of the abstract formulation of groups and fields when he published his textbook; in fact, it was Weber himself who had published, back in 1893, the first article where the two concepts appeared as closely related from the point of view of their axiomatic definitions [Weber, 1893]. And yet, Weber's conception of algebra was absolutely subordinated to the system of real numbers, and thus he considered it necessary to begin any discussion on algebra by elucidating all of the properties of that system.

The task of finding the real and complex roots of an algebraic equation—the classical main core of algebra—was relegated in van der Waerden's book for the first time to a subsidiary role. Three short sections in his chapter on Galois theory deal with this specific application of the theory and they assume no previous knowledge of the properties of real numbers. In this way, two central concepts of classical algebra (rational and real numbers) are presented here merely as final products of a series of successive algebraic constructs, the 'structure' of which was gradually elucidated. On the other hand, additional, non-algebraic properties such as continuity or density, were not considered at all by van der Waerden.

Within this conceptual setting, and together with the concepts which are repeatedly used in the book, one finds several problems that van der

Waerden discussed when studying each algebraic system. Thus, for instance, factorization: for every algebraic domain, some specific kind of sub-domain is usually considered, which may be taken as playing a role similar to the role that prime numbers play for the system of integers. Accordingly, the elucidation of the structure of a domain involves the elucidation of the relationship between a given element of the domain and its 'prime elements.' This question is, in fact, a specific aspect of the broader question, central to the structural concerns of the book, of the relation between given domains and the systems of their sub-domains. Galois theory, for instance, is also part of this question inasmuch as it 'is concerned with the finite separable extensions of a field K... (and it) establishes a relationship between the extension fields of K, which are contained in a given normal field, and the subgroups of a certain finite group' [Van der Waerden, 1930, p. 153].

An additional typical question that appears throughout the book concerns those new algebraic systems which are obtained from existing ones by performing certain standard constructs on them (e.g., fields of fractions of an integral domain, etc.). The question naturally arises to what extent specific properties of the original domains are reflected in the new ones. For example, if a given ring is an integral domain, the same is true for its ring of polynomials; if a given ring satisfies the base condition (namely, that every ideal in it has a finite generating set) then any quotient ring of that ring and also the polynomial ring associated with it satisfy the same condition; and so on. One may also ask which properties are passed over from a given algebraic domain to its subsystems or to its quotient systems. These kinds of questions are exemplified in van der Waerden's study of the structure of fields, which was modeled directly after the seminal work of Steinitz, who in 1910 had advanced a completely new approach for the treatment of abstract fields as an issue of intrinsic interest [Steinitz, 1910]. Van der Waerden stressed the basic role played by 'prime fields' as building blocks of the theory, and he claimed that after all the properties inherited from the prime field by its extension are known, then the structure of all fields is also known. Similar to Steinitz's had been the pioneering, but much less known, work of Abraham Fraenkel on abstract rings, a work that van der Waerden certainly knew from Emmy Noether.

Now, all these problems and all the concepts to which they apply are formulated in *Moderne Algebra* in terms of the 'modern axiomatic method.' It has been very common to consider this method as the most essential feature of the structural approach. And it is indeed the case that van der Waerden presented all algebraic theories in a completely abstract, axiomatic, way, like no other comprehensive exposition of algebra had ever done before. However, identifying the innovative character of the book with its use of the axiomatic method alone would be quite misleading. The real innovation implied by the book can only be understood by considering the particular way in which van der Waerden exploited the advantages of the axiomatic method in conjunction with all other components of the structural image of algebra which I mentioned above. The crucial issue concerns the clear recognition that all those concepts that deserve being axiomatically defined and studied in the framework of algebra are in fact different varieties of a same species ('varieties' and 'species' understood here in a 'biological', not mathematical term), namely, different kinds of algebraic structures. The central disciplinary concern of algebra becomes, in this conception, the systematic study of those different varieties through a common approach. In fact, this fundamental recognition appears in *Moderne Algebra* not only implicitly, but rather explicitly and even didactically epitomized in the *Leitfaden* that appears in the introduction to the book, and that pictures the hierarchical, structural interrelation between the various concepts investigated in the book.

Obviously, the new image of algebra presented by van der Waerden was enabled by the current state of development of the body of algebraic knowledge. However, the important point is that the former was *not* a *necessary* outcome of the latter, but rather an independent development of intrinsic value. This becomes clear when we notice that parallel to van der Waerden's, several other textbooks on algebra were published which also contained most of the latest developments in the body of knowledge, but which essentially preserved the classical image of algebra. Examples of these are Dickson's *Modern Algebraic Theories* [1926], Hasse's *Höhere Algebra* [1926] and Haupt's *Einführung in die Algebra* [1929]. But perhaps the most interesting example in this direction is provided by Fricke's *Lehrbuch der Algebra*, published in 1924, with the revealing sub-title: '*Verfasst mit Benutzung vom Heinrich Webers gleichnamigem Buche*'. All these books were by no means of secondary importance. Dickson's book, for instance, became after publication the most advanced algebra text available in the USA and it was not until 1941 that a new one, better adapted to recent developments of algebra and closer to the spirit of *Moderne Algebra*, was published in the USA: *A Survey of Modern Algebra* by Garrett Birkhoff and Saunders Mac Lane [1941].

HILBERT AND STRUCTURES

Having defined more precisely what the structural image of mathematics is, we can now more easily understand what was Hilbert's specific contribution to its rise and development. Hilbert's works in three domains are relevant to this account: invariant theory, algebraic number theory and axiomatics.

Hilbert's first important achievement in invariant theory concerns the proof of the generalized finite basis theorem [Hilbert, 1889]. Hilbert's proof that any given system of invariants in n variables has a finite basis came to solve a main open problem in the discipline, after twenty years of unsuccessful efforts by its leading practitioners. But a more general significance of his achievement was that it implicitly asserted the legitimacy of a new kind of proof which was initially rejected by many, namely, a proof of existence by contradiction. Hilbert also developed arguments that amounted to an implicit use of the ascending chain condition, an idea which Emmy Noether explicitly formulated in her own later, seminal work on factorization in abstract rings [Hilbert, 1890].

Hilbert's treatment of invariants was based on a conceptual analogy between the basic problems and the conceptual tools currently available in research on number theory and on the theory of polynomials, yet its setting remained the 'concrete' one of the theory of polynomials over the field of complex numbers. The various systems of numbers (complex, rationals, etc.) appear in Hilbert's work on invariants as the basic mathematical entities, while systems of polynomial forms are subsidiary constructs. The properties of the latter are always deduced from those of the former. Hilbert added with his work in this domain important ingredients to the future elaboration of the structural image of algebra, but he himself applied those ingredients without essentially changing the accepted images of the discipline.

The same can be said of Hilbert's work on the theory of algebraic number fields, his main domain of research between 1892 and 1899. In 1897 Hilbert published the influential *Zahlbericht* [1897], in which he presented and further developed the many recent achievements of this discipline since the times of Kummer. Dedekind and Kronecker were responsible for the bulk of these achievements, but they had addressed the whole issue from diverging points of view, whose basic orientations can be roughly characterized, respectively, as conceptual vs. algorithmic. Hilbert's adoption of Dedekind's point of view as his own leading one, together with the deep influence that his work exerted over the next generations of researchers, explains to a large extent the relative dominance of this approach over Kronecker's during the following decades.

Although this 'conceptual' approach is closely connected to the basic spirit of the structural approach, it is important not to confound the two. In his own work on algebraic number fields, and in his joint work with Weber on algebraic functions, Dedekind had treated groups, fields and ideals within a single, common formulation, yet these concepts played completely different roles in the theories where they appeared. Hilbert followed Dedekind in this distinction as well. For Dedekind, fields (not abstract fields, but rather fields of numbers) were the main *subject-matter* of research in both Galois theory and algebraic number theory. Groups, on the contrary, were only a *tool*—an effective and powerful one, it is true, yet subsidiary to the main concern of that research. In the *Zahlbericht*, while fields were ascribed a role similar to that conceded by Dedekind in his works, groups were barely mentioned at all, since results of group theory are much less needed here than, say, in Galois theory.

Hilbert also defined a ring as a system of algebraic integers in a given field, which is closed under addition, subtraction and product. An ideal of a ring is any system of algebraic integers belonging to the ring, such that any linear combination of them (with coefficients in the ring) belongs itself to the ideal. But Hilbert never described a ring as a group endowed with an additional operation, or as a field whose division fails to satisfy a certain property. Neither did he present ideals of fields as a distinguished kind of ring. Hilbert's ideals are always ideals in fields of numbers. In spite of his interest in the theory of polynomials and his acquaintance with the main problems of this discipline, Hilbert never attempted—in the *Zahlbericht* or elsewhere—to use ideals as an abstract tool allowing a unified analysis of factorization in fields of numbers and in systems of polynomials. This step, crucial for the later unification of the two branches under the abstract theory of rings, was only to be taken more than twenty years later by Emmy Noether. Obviously, the absence of such a step in Hilbert's work needs not be explained in terms of a lack of technical capabilities, but rather in terms of motivations, i.e., as an indication of the nature of his images of algebra, to which the idea of algebraic structures as an organizing principle (in the sense explained above for the case of van der Waerden) was foreign.

Finally, I mention Hilbert's axiomatic approach and its connection with the structural image of algebra. Hilbert conceived the axiomatic analysis of theories as a necessary stage in their latest stages of development, and *never as a starting point*. This conception was fully manifest for the first time in 1899 in *Die Grundlagen der Geometrie*. Geometry was the foremost example of a fully-fledged scientific discipline, amenable to a comprehensive and thorough axiomatic analysis. Until the 1920, when he dedicated much effort to study the foundations of arithmetic using axiomatic analysis, Hilbert occasionally discussed in his lectures arithmetic and logic in axiomatic terms, but whenever he dedicated some real effort to seriously discuss the application of this method beyond geometry and its deeper significance, it was always in connection with *physical* theories. What he never did, at any rate, was to discuss any of the notions later associated to algebraic structures—groups, fields, rings, ideals—in axiomatic terms.[2]

Standard axiomatic definitions of such algebraic concepts, and their analysis following the guidelines stipulated by Hilbert, were advanced over the first decade of the century in the USA, especially in the circle associated with Eliakim Hastings Moore at the University of Chicago. The activity of these American mathematicians, which came to be known as 'postulational analysis',[3] was triggered by their attentive study of the *Grundlagen*, but it took the methods developed in this book in a direction slightly different from that originally envisaged by Hilbert. Whereas for Hilbert it was the 'concrete', elaborate mathematical and physical theories which mattered, and the axioms were only a tool to enhance our understanding of them, for Moore and his collaborators it was *the system of postulates as such* that became an issue of inherent mathematical interest. Moore asked how these systems can be formulated in the most convenient and succinct way from the deductive point of view, without caring much whether this axioms actually convey any intuitive, geometrical meaning [Moore, 1902, 1902a]. His questions could equally be applied to a system of postulates defining Euclidean geometry, groups or fields, or, in fact, any arbitrarily defined system. No doubt, the point of view and the techniques introduced by Hilbert could in principle have been applied in this more general context as well, but Hilbert's actual motivations never did contemplate the realization of that possibility. Neither he nor his students in Göttingen published any work in that direction. Thus, Moore's perspective implied a subtle shifting of the focus of interest away from geometry and the other 'concrete' entities of classical, nineteenth-century mathematics, and towards the study of a new kind of autonomous mathematical domain: the analysis of systems of postulates.

The accumulated experience of research on postulational analysis eventually brought about an increased understanding of the essence of postulational systems as an object of intrinsic mathematical interest. In the long run, it also had a great influence on the development of mathematical logic in America, since it led to the creation of model theory. But as a more direct by-product of its very activity, postulational analysis also provided a collection of standard axiomatic systems that were to become universally adopted in each of the disciplines considered. The works of Steinitz, Emmy Noether and Artin, and whatever van der Waerden took from them, relied heavily on the use of these systems. They are only indirect products, however, of Hilbert's conceptions of the axiomatic method in mathematics.

A further perspective from which to analyze the relation of Hilbert with the structural image of algebra is provided by the works of his many students and collaborators. Of the sixty-eight (!) doctoral dissertations that Hilbert supervised, only four deal with issues directly or indirectly related to Hilbert's first domain of research: invariant theory. Not one of them deals with problems connected with the theory of factorization of polynomials, although at that same time important works were being published by other mathematicians—such as Lasker [1905] and Macaulay [1913]—which elaborated on Hilbert's own. Nor is there any dissertation dealing with topics that later came to be connected with modern algebra—such as abstract fields, or the theory of groups in any of its manifestations—and that knew at the time intense activity throughout the mathematical world. Moreover, a close look at the contents of the dissertations shows that Hilbert's students did not depart from the master's images of algebra. Thus, for instance, in the work of the Swiss mathematician Karl Rudolf Fueter [1905; 1907] on algebraic number fields, one finds the same functional separation between fields, on the one hand, and groups, on the other hand, that characterized the earlier works of Dedekind or Hilbert.

Likewise worthy of mention here is that, although five among the twenty-three problems that Hilbert included in his 1900 list can be considered in some sense as belonging to algebra in the nineteenth-century sense of the word, none of them deals with problems connected with more modern algebraic concerns, in particular the theory of groups [Hilbert, 1901].

But on the other hand, if one looks at the works of Emmy Noether, for instance, one understands that a direct elaboration of many ideas that had been laid down by Hilbert led very soon to develop the basic elements of the structural image of algebra. For lack of space, we cannot discuss here the works of Emmy Noether (or of Artin, for that matter) to see how she elaborated those ideas and finally crafted the new image of algebra that van der Waerden then so masterfully cast into his textbook, applying it systematically and uniformly to so many domains of algebra.

What might have been, then, Hilbert's actual attitude towards the structural image of algebra as it consolidated around 1930, after he himself had already gone into retirement? In particular, what was Hilbert's attitude towards the treatment accorded to the discipline in van der Waerden's *Moderne Algebra*? What did Hilbert think, for instance, of van der Waerden's definition of the fields of rational and real numbers—those mathematical entities laying at the heart of Hilbert's approach to invariants, to algebraic number theory and to geometry—as particular cases of more general, abstractly defined algebraic constructs? Could in his view the conceptual order be turned around so that the system of real numbers be dependent on the results of algebra rather than being the basis for it? Unfortunately, we have no direct evidence to answer these questions, yet from what I have said here, it seems fair to conjecture that such answers would be, at best, not really straightforward and at least ambiguous.

We can thus summarize Hilbert's contribution to the rise of the structural image of algebra as follows: although Hilbert's works contain most of the materials needed for elaborating the structural image of algebra, Hilbert himself neither put forward this image in its completed form in his works, nor suggested that it, or something similar to it, should be adopted in algebra.

BOURBAKI AND STRUCTURES

No name has been more widely associated with the notion of structure in modern mathematics than the name of Nicolas Bourbaki. Under this pseudonym, a group of leading French mathematicians began in 1939 the publication of an influential, multi-volume treatise that eventually covered the main areas of pure mathematics. The title of the treatise was *Eléments de mathématique* [1939] and its subtitle, 'The Fundamental Structures of Analysis'.

The initial aim of the group was to publish an up-to-date treatise of mathematical analysis, suitable both as a textbook for students and as reference for researchers, and adapted to the latest advances and the current needs of the discipline. The applicability of the topics discussed and their usefulness for physicists and engineers was a main concern during the early meetings of Bourbaki. However, given the more abstract inclinations of certain members and the way in which the writing of the chapters evolved over the first years of activities, something completely different from the classical French textbooks on analysis began to emerge. In fact, the titles of the various volumes of the treatise reflect one of Bourbaki's most immediate innovations, namely the departure from the classical view according to which the main branches of mathematics are geometry, arithmetic and algebra, and analysis. In Bourbaki's presentation, the 'fundamental structures of analysis' were: set theory, algebra, general topology, functions of a real variable, topological vector spaces; integration, Lie groups and Lie algebras; commutative algebra, spectral theories, differential and analytic manifolds, and (later) homological algebra.

A main source of inspiration for the work of the group was the model put forward by van der Waerden in his textbook on algebra. As a matter of fact, the whole Bourbaki project in its mature form may be defined as an attempt to extend to the whole of mathematics the disciplinary image put forward in *Moderne Algebra*, namely, as a recognition that different mathematical branches such as algebra, topology, functional analysis, etc., are all individual materializations of one and the same underlying, general idea, i.e., the idea of a mathematical structure. Bourbaki attempted to present a unified an comprehensive picture of what they saw as the main core of mathematics, using a standard system of notation, addressing similar questions in the various fields investigated, and using similar conceptual tools and methods across apparently distant mathematical domains.

Bourbaki gathered an exceptional collection of leading mathematicians with perhaps relatively homogenous, but nevertheless distinctive and certainly well-consolidated, opinions about what mathematics is and should be. Thus, one has to exercise special care when analyzing the work of Bourbaki, and especially when advancing any claim about the images of mathematics associated with their work. In the space available here, this problem is only more pressing. Still, I will attempt to describe in general lines what can be taken to the group's collective conceptions, although to a large extent I will stress those articulated by Jean Dieudonné, the group's most outspoken member.

In 1948 Dieudonné published, signing with the name of Bourbaki, an article that came to be identified as the group's manifesto, 'The Architecture of Mathematics'. Dieudonné raised the question of the unity of mathematics, given the unprecedented growth and diversification of knowledge in this discipline over the preceding decades. Mathematics is a strongly unified branch of knowledge in spite of appearances, he claimed, and the basis of this unity is the use of the axiomatic method. Mathematics should be seen, he added, as a hierarchy of structures at the heart of which lie the so-called 'mother structures':

'At the center of our universe are found the great types of structures, ... they might be called the mother structures ... Beyond this first nucleus, appear the structures which might be called multiple structures. They involve two or more of the great mother-structures not in simple juxtaposition (which would not produce anything new) but combined organically by one or more axioms which set up a connection between them ... Farther along we come finally to the theories properly called particular. In these the elements of the sets under consideration, which in the general structures have remained entirely indeterminate, obtain a more definitely characterized individuality.' [Bourbaki, 1950, pp. 228–29]

This characterization of mathematics was inspired, as I said, by van der Waerden's presentation of algebra, but it went much further than the latter in many respects. For one thing, what van der Waerden had left at the implicit level, the centrality of the hierarchy of structures, became explicit and constitutive for Bourbaki. For another, Bourbaki did not limit itself to promote a conception of mathematics based on a notion that, fruitful and suggestive as it might be, could not be elucidated in strictly mathematical terms. Thus, the fourth chapter of Bourbaki's book on the theory of sets [1968], defines the concept of *structure* (I use italics to denote this particular, technical use of the term), which is assumed to provide the conceptual foundation on which the whole edifice of mathematics as presented in the *Eléments* is supposedly built.

The central notion of structure, then, has a double meaning in Bourbaki's mathematical discourse. On the one hand, it relates to a general organizational scheme, that conceives the whole discipline in hierarchical terms. Like in van der Waerden's book, one does not need to have a definite definition of this notion in order to be able to grasp its meaning and, more importantly, to put it to use. Rather, one has to master the various mathematical branches presented in the treatise, and one can thus see the force of the whole structural approach and how it can be applied to additional topics not considered in the various volumes of the treatise. On the other hand, one has a formal mathematical concept, *structures*, about which Bourbaki proved some results in the relevant chapter of the treatise. Surprisingly, however, this concept plays no role at all in any of the other books of the treatise, and those few places where it is invoked, only help us understanding how little relevant it is for the issues covered by Bourbaki.[4]

But the truly astonishing point concerning this duality is the fact that many pronouncements by Bourbaki (and especially by Dieudonné) concerning the centrality of structures have been often accepted *as if* they referred to a strictly mathematical notion (such as *structures*), even when they referred, in fact, to the more general sense of the term. Sometimes the blurring of the two meanings is as explicit as it can be, and some other times it is subtler. But the confusion it has brought about can be traced in many places. The mother structures mentioned in the above quotation, taken from the 'Architecture' manifesto, is an interesting case in point. The description of the mother structures and their central role in the whole edifice of mathematics is not an integral part of the formal, axiomatic, theory of *structures* developed by Bourbaki. The classification of structures according to this scheme is mentioned several times in the book on set theory, but only as an illustration appearing in scattered examples. Many assertions that were suggested either explicitly or implicitly by Bourbaki or by its individual members—i.e., that all of mathematical research can be understood as research on structures, that there are mother structures bearing a special significance for mathematics, that there are exactly three, and that these three mother structures are precisely the algebraic-, order- and topological-structures (or *structures*)—all this is by no means a logical consequence of the axioms defining a *structure*. The notion of mother structures and the picture of mathematics as a hierarchy of structures are not results obtained within a mathematical theory of any kind. Rather, they belong strictly to Bourbaki's non-formal images of mathematics; they appear in non-technical, popular, articles, such as in the above quoted passage, or in the myth that arose around Bourbaki. Still, it is not uncommon that pronouncements about the mother structures are accepted, implicitly or explicitly, as results obtained in the framework of a standard mathematical discipline.

A second remarkable manifestation of the ambiguity inherent in Bourbaki's use of the term 'structure' concerns its historical dimensions, and therefore it is particularly interesting for the present account. Bourbaki, and in particular Dieudonné, dedicated considerable efforts to historical writing, that produced an influential historiographical body of marked Whiggish spirit [Bourbaki, 1969]; [Dieudonné, 1978, 1985, 1987]. The idea of a mathematical structure appears in it as the culminating, comprehensive and definitive stage of historical development of the whole discipline: the structural presentation of mathematics as embodied in Bourbaki's treatise was here to stay. And in saying so, reference was made not only to the image of mathematical knowledge put forward in the book; implicitly, and often also explicitly, Dieudonné asserted that mathematics had come now to be the discipline dealing not only with structures, but more specifically with Bourbaki's *structures* [Dieudonné, 1979]. As a matter of fact, Bourbaki's work did imply many important contributions to twentieth century mathematics, but the concept of *structure* is certainly not among them. Bourbaki's structural image of mathematics, on the contrary, was a main force in shaping mathematical activity all over the world for several decades after its emergence.

NOTES

1. I have discussed this scheme in detail in Corry [1989].
2. On Hilbert see Corry [1996, Chap. 3; 1996a; 1997].
3. On the American School of Postulational Analysis see Corry [1996, pp. 173–183].
4. For a detailed discussion of this issue, see Corry [1996, Chap. 7].

REFERENCES

Birkhoff, G., Mac Lane, S., [1941], *A Survey of Modern Algebra*, New York, MacMillan.
Bourbaki, N., [1939] *Eléments de mathématique*, 10 vols., Paris, Hermann.
    [1950] 'The Architecture of Mathematics', *Amer. Math. Monthly*, 67, 221–232.
    [1968] *Theory of Sets*, Paris, Hermann.
    [1969] (1960) *Eléments d'histoire des mathématiques* (Deuxième édition revuée, corrigée, augmentée), Paris, Hermann.
Corry, L., [1989] 'Linearity and Reflexivity in the Growth of Mathematical Knowledge', *Science in Contest*, 3, 409–440.
    [1996] *Modern Algebra and the Rise of Mathematical Structures*, Basel-Boston-Berlin, Birkhäuser.
    [1996a] 'Axiomática Moderna y Algebra Estructural', *Mathesis*, 12, 1–56.
    [1997] 'David Hilbert and the Axiomatization of Physics (1894–1905)', *Arch. Hist. Ex. Sci.*, 51, 89–198.
Dickson, L.E., [1926] *Modern Algebraic Theories*, Chicago, Benjamin H. Sanborn.
Dieudonné, J., [1978] *Abrégé d'histoire des mathématiques, 1700–1900*, 2 vols., Paris, Hermann.
    [1979] 'The Difficult Birth of Mathematical Structures (1840–1940)' in Mathieu, U., Rossi, P., (eds.) *Scientific Culture in Contemporary World*, Milano, Scientia, 7–23.
    [1985] *History of Algebraic Geometry. An Outline of the Historical Development of Algebraic Geometry*, Monterrey, Wadsworth (English transl. by J.D. Sally of *Cours de géometrie algébrique I* (1974), Paris, PUF).
    [1987] *Pour l'honneur de l'esprit humain. Les mathématiques ajourd'hui*, Paris, Hachette.
Fricke, R., [1924] *Lehrbuch der Algebra*, Braunschweig, Vieweg.
Fueter, R., [1905] 'Die Theorie der Zahlstrahlen', *Journal reine angew. Math.*, 130, 197–237.
    [1907] 'Die Theorie der Zahlstrahlen. II', *Journal reine angew. Math.*, 132, 255–269.
Hasse, H., [1926] *Höhere Algebra*, 2 vols., Berlin, Sammlung Göschen.

Haupt, O., [1929] *Einführung in die Algebra*, Leipzig Teubner.

Hilbert, D., [1889] 'Über die Endlichkeit des Invariantensystems für binäre Grundformen', *Math. Ann.*, 33, 223–226 (in *Ges. Abh.*, 2, 162–164).

[1890] 'Über die Theorie der algebraischen Formen', *Math. Ann.*, 36, 473–534 (in *Ges. Abh.*, 2, 199–257).

[1897] 'Die Theorie der algebraischen Zahlkörper' *Jahresbericht DMV*, 4, 175–546 (in *Ges. Abh.*, 1, 63–363).

[1899] *Grundlagen der Geometrie*, Leipzig, Teubner.

[1901] 'Mathematische Probleme', *Archiv f. Math. u. Phys.*, 1, 213–237 (in *Ges. Abh.*, 3, 290–329).

[1932–1935] *Gesammelte Abhandlungen*, 3 vols., 2nd ed. 1970, Berlin, Springer.

Jordan, C., [1870] *Traité des substitutions et de équations algébriques*, Paris, Gauthier-Villars.

Lasker, E., [1905] 'Zur Theorie der Moduln und Ideale', *Math. Ann.*, 60, 20–116.

Macaulay, F.S., [1913] 'On the Resolution of a given Modular System into Primary Systems including some Properties of Hilbert Numbers', *Math. Ann.*, 74, 66–121.

Moore, E.H., [1902] 'Projective Axioms of Geometry', *Trans. AMS*, 3, 142–158.

[1902a] 'A Definition of Abstract Groups', *Trans. AMS*, 3, 485–492.

Serret, J., [1849] *Cours d'algébre supérieure* (2d. ed. 1854, 3d. ed. 1866), Paris, Bachelier.

Steinitz, E., [1910] 'Algebraische Theorie der Körper', *Journal reine angew. Math.*, 137, 167–309. (2d. ed. (1930), ed. H. Hasse, R. Baer, Leipzig, Berlin, de Gruyter.

van der Waerden, B.L., [1930] *Moderne Algebra*, 2 vols., Berlin, Springer (English trans. of the 2d. ed.—vol. I by Fred Blum (1949), vol. II by T.J. Benac (1950), New York, Frederic Ungar Publishing).

Weber, H., [1893] 'Die allgemeinen Grundlagen der Galois'schen Gleichungstheorie', *Math. Ann.*, 43, 521–549.

[1895] *Lehrbuch der Algebra*, Vol. 1, (2d. ed. 1898) [vol. 2, 1896 (2d. ed. 1899); vol 3, 1908], Braunschweig (reprint New York, Chelsea).

Chapter 9

THE CREATION AND PERSISTENCE OF NATIONAL SCHOOLS: THE CASE OF ITALIAN ALGEBRAIC GEOMETRY*

Aldo Brigaglia

Algebraic geometry, in spite of its beauty and importance, has long been held in disrepute by many mathematicians as lacking of proper foundations. The mathematician who first explores a promising new field is privileged to take a good deal for granted that a critical investigator would feel bound to justify step by step; at times when vast territories are being opened up, nothing could be more harmful to the progress of mathematics than a literal observance of strict standards of rigor. Nor should one forget, when discussing such subjects as algebraic geometry, and in particular the work of the Italian school, that the so-called 'intuition' of earlier mathematicians, reckless as their use of it may sometimes appear to us, often rested on a painstaking study of numerous special examples, from which they gained an insight not always found among modern exponents of the axiomatic creed ... [But] In this field the work of consolidation has so long been overdue that the delay is now seriously hampering progress in this and other branches of mathematics ... Our chief object here must be to conserve and complete the edifice bequeathed to us by our predecessors. 'From the Paradise created for us by Cantor, no one shall drive us forth' was the motto of Hilbert's work on foundations of mathematics. Similarly, however grateful we algebraic geometers should be to the modern algebraic school for lending us temporary accommodation, makeshift constructions full of rings, ideals and valuations, in which some of us feel in constant danger of getting lost, our wish and aim must be to return at the earliest possible moment to the palaces which are ours by birthright, to consolidate shaky foundations, to provide proofs where they are missing, to finish, in harmony with the portions already existing, what has been left undone [Weil, 1946, p. vii].

André Weil's discussion of his predecessors in algebraic geometry raises many questions. In what sense was 'the work of the Italian school' actually rooted in a national tradition? Was 'intuition' a specific attribute of the 'Italian' school of algebraic geometry? How did this Italian tradition interact with the developing *structural* style in mathematics which was spreading out mostly from Germany? The following will briefly sketch a few tentative answers to these questions.

Luigi Cremona (1830–1903) is generally considered as the founder of the Italian school of algebraic geometry. Highly projective, his methods were heavily influenced by the French tradition (Chasles, Poncelet) and the German one (Möbius, Plücker, Staudt, Clebsch). Cremona has been credited with the first extensive use, by a geometer, of the idea of *birational transformations* (also called Cremona transformations). Among his students, one may cite E. Bertini (1846–1933), G. Veronese (1854–1917), and C. Segre (1863–1924). In Pisa, Padua, and Turin, respectively, they were the true architects of a flourishing Italian school. Graduating from Padua, Castelnuovo and Severi—as well as Enriques who received a degree from Pisa—all spent some time in Turin with Segre, a mathematician who exerted a strong influence on all of them.

In the following, the word 'school' will be used in a free way, with no specific sociological meaning. In a sense, one of the most important features of the Italian 'school' of algebraic geometry was that everyone, in the scientific milieu, could understand perfectly the meaning of being part of this school. In this context, it is important to underscore that, beginning with Cremona's students, Italian algebraic geometers were conscious of belonging to a definite scientific 'school.' In his obituary of Cremona, Castelnuovo, for example, wrote [1930, p. 614]:

> In order to give life to a school, neither the founder's qualifications nor his ability of outlining a research program that exceeds his own capacity suffices. It is moreover necessary that he succeeds in communicating his passion and his faith to his disciples, knows their demands, and directs their collaboration. Luigi Cremona was eminently endowed with these qualities.

Another important feature of the Italian school of algebraic geometry was that its growth was strictly tied to a major problem in the development of the discipline. At the end of the last century, algebraic geometry was advancing in a chaotic way—recall that Dieudonné [1974, I, ch. VI] labeled this period 'développement et chaos'—and was completely split up into mutually non-interacting tendencies. There was an arithmetic school (Dedekind, Weber), a geometrical one (the Italian school), and a 'transcendental' one (the French school of Picard, Humbert, and Poincaré). Communication between these different schools was made difficult by the lack of a common language; only with the creation of new and powerful technical tools in the 1960s (Grothendieck's theory of schemes) this was eventually achieved. In his influential *History of Mathematics in the 19th Century*, Felix Klein was therefore entitled to write [1926, I, p. 315]:

Algebraic geometry had developed like a tower of Babel: quite quickly it became clear that the various languages were not being mutually understood.

In this context, the very fact that the appendix to Picard and Simart's book on surfaces (1906) was eventually written by Enriques and Castelnuovo may be an indication of the French scholars' difficulty of truly grasping the Italians' results.[1]

Moreover, the Italian school was not strictly a *national* 'school,' but rather a working style and a methodology, principally based in Italy, but with representatives to be found elsewhere in the world. In this regard, the words used by Mumford are typical: 'The Italian school, and notably Severi, Todd, Eger and B. Segre developed a general theory of Chern classes in the algebraic case.' In other words, Mumford used (rightly from my point of view) the phrase 'Italian school' as a label for, not a strictly national community, but rather a more general concept. Similarly, algebraic geometers coming from Belgium (e.g. Godeaux), or the Unites States (e.g. Coolidge) could also be counted as members of the 'Italian school.'

Lastly, we may observe that it is especially appropriate to speak of a 'school,' here, in view of its long persistence (at least four generations spanning a whole century) and the large number of its members. An appendix lists the school's main members who had important (and sometimes decisive) contributions to the discipline.

SEGRE AND PEANO

It was during the years 1885–1891, in Turin, that many important features of an Italian style in geometry were established through the work of Segre, Castelnuovo, and, later, Enriques (who, although he only stayed in Turin for a few months, always remained in a close touch with the former two scholars). Obviously an intuitive, not very rigorous approach to algebraic geometry was already used by Cremona, Veronese, and nearly all scholars during the 1860s and 1870s. But this was, in some sense, a *naive* use of intuition; while, on the contrary, in the 1880s, the Turin geometers stressed their right to use intuition in a theoretical fashion. In 1891, G. Peano (1858–1932) clashed with Segre in a fierce debate about rigor versus intuition, which made this distinction in the use of intuition explicit.

At that moment, the two young scholars (Segre was then 28, and Peano 33) were at the peak of their creative strength. Segre had just completed his program of using hyperspatial geometry in order to achieve a sound basis for the so-called 'geometry on a curve,' and he had begun his collaboration with Castelnuovo (two years his junior). On his part, Peano had just published his most influential foundational work (*Calcolo geometrico*, 1888; *Arithmetices principia*, 1889; *Principii di geometria*, 1889). Both were thus eager to defend their respective practice in geometry: intuitive and creative for Segre, rigorous and formally unexceptionable for Peano. Before we examine briefly their important debate, let us turn to young Segre's first steps in his mathematical career.

SEGRE'S METHODS

On May 29, 1883, Corrado Segre took his degree in Turin under Enrico D'Ovidio's guidance. At this time, the very beginning of Italian geometers' methods, Segre was only 20 years old and Peano 25. The latter was deeply engaged in the rigorization of analysis, and his *Aggiunte* to his teacher Genocchi's book *Calcolo differenziale ed integrale* were going to be published in the same year—1884—as Segre's dissertation. Moreover both were engaged in the study of Grassmann's *Ausdehnungslehre*, and in 1888 Peano would publish his fundamental *Calcolo geometrico secondo l'Ausdehnungslehre di H. Grassmann preceduto dalle operazioni della logica deduttiva*.

Segre's thesis [1883a,b] was published a year after his defence in the journal of the local Academy of Science, and soon became a fundamental starting point for the development of Italian projective *n*-dimensional geometry. This paper was devoted to the general study, by geometrical tools, of *n*-dimensional vector spaces (project-ive spaces) and in particular bilinear forms defined on them (quadrics).

At the beginning of the 1880s, as Segre [1891] later pointed out, different ways to give meaning to the idea of *n*-dimensional spaces existed. A purely analytical, 'naive' interpretation considered the space R^n in an analytical way, simply covering it up with a geometrical language; in Plücker's style, *n*-dimensional objects were objects in usual space defined through *n* complex parameters (in our language, it amounts to saying that this space was not defined to be R^n, but isomorphic to it); finally, following a Euclidean style, Veronese had used an axiomatic approach to try and define multidimensional spaces, in a partially unsuccessful way in 1882, and more completely in his later influential treatise [Veronese, 1891].

Many foundational problems were left completely unresolved. First of all, an 'abstract,' general, framework in which it would have been possible to consider any n-dimensional space, remained to be found. One needed a precise definition of the idea of moving from one space to another, while giving different meanings to the names (point, line, etc.) of the objects of the abstract space (isomorphism). Beside these problems, there was that of connecting the various points of view, and finding a set of independent axioms from which it would have been possible to deduce the Cartesian representation (coordinatization).

On this matter, Segre wrote in his dissertation:

The geometry of spaces with an arbitrary number n of dimensions has already taken a well-deserved place among mathematical disciplines. Even when we consider it outside the important applications to ordinary geometry that it may give us, i.e. when the element or point of such a space is not considered as a geometrical object of the ordinary 3-dimensional space ... but as an object whose intimate essence has not been determined, we cannot refuse to consider it as a science in which every proposition is rigorous, because it can be proved by essentially mathematical reasonings; the lack of a sense representation of the objects we study has not much meaning for a pure mathematician.

Any continuous set of objects, whose number is m times infinity (i.e. such that there generally are a finite number among them that satisfy m simple arbitrary conditions) is said to be an m-dimensional space, of which these objects are elements.

Any m-dimensional space is said to be linear when it is possible to attach to each element m numerical values (real or complex) in such a way that, with no exception, to any set of values there corresponds one and only one element. The values of these quantities corresponding to the element are said to be its coordinates. If we characterize them by the ratios of m quantities with an $(m + 1)$th, they will constitute the $m + 1$ homogeneous co-ordinates of the element of the space, so that any element, with no exception, will be characterized by the mutual ratios of these homogeneous co-ordinates and, vice versa, will characterize their ratios.

Besides being confusing, the definition is clearly wrong (e.g., the correspondence is not one to one and so on because the element $(0, 0, 0, ..., 0)$ does not have any correspondent and the word 'ratio' has no meaning when the second number is 0), as Peano was to reproach Segre during the controversy.

But, notwithstanding the obvious lack of rigor, Segre's definition was an important working tool, which could effectively allow him to achieve a great number of new, more advanced, algebraic results that he was always able to interpret in an arbitrary linear space. This capacity required an idea, albeit a still confused one, of isomorphism between two linear spaces, an idea he expressed as follows:

> Any two linear spaces with the same number of dimensions, independently from the nature of their elements, can be considered as identical to one another, because, as we already noted, when we study them we do not pay attention to the nature of these elements, but only consider the property of linearity and the dimension of the space formed by the elements themselves. It follows that, being already known, the theory of linear forms of first, second, and third kind, for instance, of the line, of the plane, and of the space considered as dotted, may be used for any linear space of dimension 1, 2, or 3 contained in the linear space of dimension n-1 that one wants to study in general. Therefore one can use, for example, the theory of projectivity, of harmonic groups, of involution, etc., in the forms of the first kind. [Segre 1883a, p. 46].

Segre's method was fundamentally based on the so-called hyperspatial projective geometry. He fully applied his method to the complete and exhaustive study of quadrics (bilinear forms) in an n-dimensional projective space, and used his results to classify the quartic intersections in 3-dimensional space. In two years, he achieved many important results on second-degree complexes (1883–1884), line geometry (1883), Lie's sphere geometry (1884), the classification of collineations and correlations in a linear space of finite dimensions (1884), the surfaces of 4th degree with a double conic (1884), the sheaves of cones in n-linear spaces (1884), and Grassmann's varieties (1886).

Segre's methods depended on a massive use of the most recent results of linear algebra, particularly those obtained by Weierstrass (1868), Frobenius (1878), and Kronecker (1878) on matrices and bilinear forms. In particular, they required a general definition of linear spaces which was, in 1883, completely lacking (Grassmann's *Ausdehnungslehre* contained many ideas on the subject, but was almost completely unknown in these years).

It is worth noting that in his book on *Calcolo Geometrico*, Peano, who worked in the same mathematical department as Segre, was able to give the first complete definition of a linear (real) vector space [Brigaglia, 1996], a definition which is completely analogous to the contemporary one. We cannot understand why the two young scholars, who were working on similar problems, did not have any mutual influence, without looking more carefully at their different mathematical attitudes.

Segre's and Peano's respective approaches might be described using Hilbert's words:

> The building of science is not raised like a dwelling, in which foundations are first firmly laid and only then one proceeds to construct and enlarge the rooms. Science prefers to secure as soon as possible comfortable spaces to wander around and only subsequently, when signs appear here and there and there are that the loose foundations are not able to sustain the expansion of the rooms, it sets to support and fortify them. (quoted in [Corry, 1996, p. 162]).

Segre was working precisely on the building of 'comfortable spaces' for the ever growing needs of algebraic geometry; only as far as to 'sustain the expansion of the rooms' was he ever interested in foundations. Peano instead held the idea that 'the axiomatic approach may be used only to organize well established theories, as classical mathematics' [Lolli, 1985] and not to build new ones (as in Hilbertian mathematics).

In any case, in spite of its lack of rigor, Segre's point of view opened up new vistas towards further developments, even ones of a formal algebraic kind :

> It has been noted ... that for instance the general projective geometry that we build in this way, is not anything other than the algebra of linear transformations. This is only a difference not a fault. Provided that we do mathematics! [Segre, 1891, p. 405].

A closer look at Segre's arguments will clearly underscore the usefulness of 'hyperspatial' geometry, as illustrated by a statement made many years later by an American adept of the Italian school, C. Coolidge:

> Higher spaces have two claims on our attention. One is that of giving to our formulae the greatest possible extension to afford our theorems the widest generality, the other is to provide a suggestive and powerful technique for studying the properties of figures in lower space, which depend simply on a number of explicit parameters [Coolidge, 1940, p. x]

This was exactly the direction in which Segre moved. His ideas were based on a powerful synthesis of several mathematical theories: Plücker's idea of hyperspaces generated by an element of ordinary space, Klein's Erlangen Programm, Clebsch's and Noether's results on birational geometry, and Weierstrass's, Frobenius's, and Kronecker's algebraic results. He succeeded in making a complete and orderly theory which would provide a sound foundation for Italian algebraic geometers. His ideas completely pervaded the famous book [Bertini, 1907]. His methods and results have now become so commonly used and natural that it is difficult for us to understand the novelty of his work.

At the core of Segre's ideas, was the concept that a translation of new algebraic results into geometrical language could be used to build (using the same methods) a coherent, organic theory of homographies in a projective space, conics and their pencils, correlations and polarities with respect to a quadric in any dimension and with any concrete interpretation of the word 'point'. In a word, his aim was: 'to build a general theory of projective n-spaces'. The idea of 'translation' was very deeply rooted in his way of looking at mathematics. For example, when citing Weierstrass's [1868] important theorem, on elementary divisors, he wrote:

> This theorem was proved by Weierstrass in an analytical form, and it gives a complete answer, from a modern algebraic point of view, to any question on the different situations which may arise from a system of two quadratic forms and its invariants, giving a complete classification, from this algebraic point of view, of these systems of forms. We will try to translate them into a geometrical setting. [Segre, 1883a, p. 94].

In a framework in which the main algebraic ideas were far from clear, and modern abstract linear algebra not yet well established, Segre and his students immediately began to make extensive use of mathematical tools then only crudely developed. More and more, geometrical intuition was requested to overcome algebraic and analytical difficulties, which lay largely outside the possibilities of their times, as was expressed by the young Castelnuovo [1889b, p. 65]:

> We must acknowledge that, in establishing this result we base ourselves more on intuition (and on many verifications), than on a true mathematical reasoning ... But we let ourselves use a *result not yet proven* in order to solve a difficult problem, because we think that it is possible even with such attempts to be useful to the development of science, provided that we explicitly declare what we assume and what we prove.

Castelnuovo's results in this work were almost immediately picked up by Segre [1889] with these words: 'The ingenious demonstration that this geometer gave of this important formula might cast doubt on its absolute validity, which would be reflected on the present number and below ... ; however, confirmations found of these results lead me to think that they are absolutely true.'[2] The necessity of having recourse to intuition was being turned into a *method*. Due to the many great results obtained in the next few years through the use of this method, it would soon become the seal of an Italian style in mathematics.

PEANO'S CRITICISMS

The systematization of Segre and his students' recourse to intuition had Peano really worried. In his 'Observations' appended to Segre's article [Peano, 1891, p. 67], he ignited a fiery debate by quoting Segre's words—just cited above—adding the following comments:

> We believe that it is very strange that in authoritative journals, occupied with the purest kind of mathematics, one may write [such words] ... Be it only for an instant, we would like to have a voice influential enough among our colleagues so as to convince them of this truth: works lacking in rigor cannot make mathematics progress even by one step.

Later on in the same text, Peano explained:

> We maintain that when it is possible to find a single exception, a proposition is false; and that we cannot consider a result as being truly achieved, even if no exception to it is known, if it has not been rigorously proved. [p. 68].

This drastic attitude went hand in hand with distrust for unrestrained abstraction:

> Any writer may assume those experimental laws he likes, and he can make the hypotheses he prefers. The correct choice of these hypotheses has a great importance for the theory one wants to develop; but this choice is made by induction and does not belong to mathematics ... If an author assumes hypotheses opposite to experience, or hypotheses which are not verifiable by experience, nor are their consequences; he will be able, I grant, to prove some wonderful theories, such wonderful ones as to have us cry: what a profit [for science this would have been] had he used his capacity to the service of useful hypotheses! [p. 68].

Young mathematicians, like Castelnuovo, Enriques, and Severi, entered the debate with the goal of loosening what they held as the unduly strict requirements of rigor in mathematics. They shared an attitude according to which, hampering the younger scholars' creative strengths, formalism and excessive rigor were *old mathematics*. New mathematics was abstract but *geometrically intuitive*.

THE ACTORS' VIEWPOINTS

Based on geometrical intuition, the 'new' method was to be increasingly discussed by leading Italian scholars, and soon held as the principal source of many of their important results. Here only a few examples will be provided, which the reader may compare with Weil's words cited earlier. The crucial point to stress is that the leading scholars of the Italian school clearly thought of themselves as explorers of a new land and in no way, as 'mathematical architects' in Bourbaki's sense. This attitude becomes particularly striking when looking at their treatises (see Enriques' citation below): much more than to provide students with the 'logical structure' of mathematical results, these were written with the goal of giving a 'perspective of their coming into being'—an aim very different from, say, van der Waerden's in his almost contemporary *Moderne Algebra*.

In his address to the International Congress, Castelnuovo was thus quite explicit about his method [1928, p. 194]:

> We built, in an abstract sense of course, a large number of models of surfaces in our space or in higher spaces; and we arranged these models into two display cases. One contained the regular surfaces, for which everything proceeded as in the best of possible worlds; by analogy, the most salient properties of plane curves were conveyed to these surfaces. But when we tried to verify these properties for the surfaces in the other display case, the one containing irregular surfaces, then, the trouble began, and exceptions of every sorts turned out. At the end, an assiduous study of our models led us to *guess* some properties which *had to be true*, with appropriate modifications, for surfaces in both cases; we then put these properties to the test by constructing new models. If they stood up to the test, we looked—this was the ultimate phase—for a logical justification.

Similarly, according to Severi, analogy and experimentation fully belonged to the mathematician's practice [1937, p. 58]:

> Analogies are often precious; but in many other cases they are prisons where for lack of courage the spirit remains fettered. In any branch of science, however, the most useful ideas are those that born orphans later find adoptive parents.
>
> New mathematical constructions do not actually proceed exclusively from logical deduction, but rather from a sort of experimental procedure making room for tests and inductions—these constitute the real ability necessary for exploration and construction. And thus is it natural that theory development entails some modifications of, and adjustments to, initial conceptions and the language expressing these conceptions; and that one can reach a definite systematization only by successive approximations.

As for the exposition of results, Enriques [1949, p. x][3] observed:

> Not only to give to the exposited theories a logical structure, but also ... to provide an historical perspective of their coming into being. In this way one wants to offer the reader, not just the gift of something perfect at which one is allowed to look from the outside, but rather the vision of an acquisition and an advancement, the reasons for which one must understand and which the reader is invited to re-earn by, and for, himself, finding in the book a working tool ... To the model of a rational science logically ordered as a deductive theory, which must appear everywhere complete and perfect, and which, going down from more general concepts to particular applications, drives off uncertain, changeable appearances of reality—all of which recalling the dark past of the search or uncovering new difficulties, breaking the harmony of the system, ... we prefer the general philosophy of science resulting from modern critiques ... which goes beyond the opposition between deductive and inductive methods, to reach the consideration of deduction itself as a phase in a unique process that rises from the particular to the general in order to go down again to the particular.

Even more dramatic was the following of Enriques's claims: 'We aristocrats do not need proofs. Proofs are for commoners (quoted by Zariski in [Parikh, 1990]).'

Eventually, it is also worth quoting a passage from Castelnuovo's preface to this same book of Enriques's [1949, p. VII]:

> Will someone to continue the work of the Italian and French schools come soon, who will finally succeed in giving the theory of surfaces the same perfection as the theory of algebraic curves has reached? I hope so, but I doubt it. The observation that mathematics has in this century taken quite a different direction from the predominating one in the last century, feeds my doubts. Imagination and intuition, which, then, used to guide research, are today regarded with suspicion because of the terror of the errors to which they can lead. Theories, then, arose in response to the need felt by the mathematician for making precise various objects of his thought that already in a vague form were present in his mind. It was the exploration of a vast territory seen from a faraway peak ... Today, more than the land to be explored, the road leading there is of interest, and this road now is strewn with artificial obstacles, now makes its way free among the clouds.

In contrast to the ideas quoted above, it is striking to note that the formal achievements in algebraic geometry of the van der Waerdens, Zariskis, and Weils formed one of the pillars of a rising tide of structural ideas sweeping twentieth century mathematics as a whole. One may wonder how the above ideas coexisted with structuralist ones. In the following, I will limit myself to a few quotations and brief comments.[4]

A good starting point for the understanding of a structuralist appreciation of the work and method of the Italian school might be Fulton's following remark [1984, p. 26]:

> It would be unfortunate if Severi's pioneering work in this area were forgotten; and if incompleteness or the presence of errors are grounds for ignoring Severi's work, few of subsequent papers on rational equivalence would survive.

Well aware of recent results of the German school in algebraic geometry, Severi was sharply ironic [1940, p. 224 of *Memorie Scelte*]:

> That algebraic geometry can also be seen from the viewpoint of modern algebra—as van der Waerden for example does in his interesting works—is undoubtedly a good thing for algebra and for geometry. One also hopes that the penetrating methods of *Moderne Algebra* are soon to be used in an attack of essentially new problems, rather than only in the reconstruction of results already discovered by geometric means.

With barely hidden satisfaction, Severi added that van der Waerden's [1939, 202] treatise was more easily reconciled with 'the methods of the Italian school than with those of *Moderne Algebra* [Severi 1940, p. 225 of *Memorie Scelte*].'

As one may see, Severi's criticism attacked the main objectives of the new algebraic methods. In fact, these methods were precisely intended more for 'the reconstruction of results already discovered by geometric means' than for 'an attack on essentially new problems.' In Severi's mind, the former goal was a minor one as opposed to the latter, which was obviously the purpose of 'aristocrats' in mathematics. It must be stressed however that fundamentally new results in algebraic geometry (and not, using Severi's words, mere 'reconstructions') were only obtained by the 'structural' methods at the end of the 1950s with Hironaka's results on singularities.

Appearing almost in the same years (1932–1934), the following two quotations from books written on the same argument (resolution of singularities) exemplify (in my opinion) the strong contrast between the two different styles of writing about algebraic geometry. A result that Enriques commented on only in a few confident lines—'That in fact the indicated transformation process ends up by completely resolving all singularities was rigorously proved by B. Levi and O. Chisini [Enriques

1932, p. 10]'—received, in Zariski's hands, a rather more complex appreciation [1934, p. 18]:

> The proofs of these theorems are very elaborate and involve a mass of details which it would be impossible to reproduce in a condensed form. It is important, however, to bear in mind that in the theory of singularities the details of the proofs acquire a special importance and make all the difference between theorems which are rigorously proved and those which are only rendered highly plausible.

And Zariski then went on with many pages of detailed reasoning. In this context, young Italian scholars were completely discouraged from following new 'structural' trends of German school—a discouragement recalled by Zariski, who studied in Italy until the end of the twenties:

> It was a pity that my Italian teachers never told me that there was such a tremendous development of the algebra which is connected with algebraic geometry. I only discovered this much later, when I came to the United States. [Quoted in Parikh, 1990, p. 36]

Although the conflict between the German structural, and the Italian intuitive, practices in algebraic geometry has been stressed, we must also recognize that the two different styles were, in many a sense, strictly interrelated. In the thirties, van der Waerden wrote his many papers on algebraic geometry and developed his powerful results while always bearing in mind the great results of the Italian school (and mainly Severi's), even if he considered them as often incomplete and sometimes also wrong. While he developed his methods of commutative algebra and topology, Zariski never forgot the problems that aroused his interests when he was a student in Rome, a context which shed light on an interesting testimony of his [Parikh, 1990, p. 76]:

> I wouldn't underestimate the influence of algebra, but I wouldn't exaggerate the influence of Emmy Nöther. I'm a very faithful man ... also in my mathematical tastes. I was always interested in the algebra which throws light on geometry and I never did develop a sense for pure algebra. Never. I'm not mentally made for purely formal algebra, formal mathematics. I have too much contact with real life, and that's geometry. Geometry is the real life.

This kind of statement might well have been uttered by Zariski's Italian teachers. Similarly Solomon Lefschetz, the first to succeed in putting 'the harpoon of algebraic topology into the body of the whale of algebraic geometry' [to use his own words, 1968, p. 854], never hid his admiration for the Italian scholars' results nor his debts to them.

CONCLUSION

To conclude, I would like to stress that (counterfactually, of course) it might indeed have been possible, starting from inside the Italian tradition, to develop a structural view of mathematics. One can hint at this simply by providing a very short account of Gaetano Scorza's work, which clearly moved in this direction (Scorza was the author of *Corpi numerici e algebre* [1921], one of the first books written in a modern algebraic style).

> The theory of Abelian functions ... and some of the more elevated theories of algebraic geometry present such frequent analogies and such remarkable affinities that whoever examines them a bit more closely is led almost spontaneously to presume that they must all be embedded in a single general theory, and that this must provide the best explanation for the numerous and close ties that link them. And since in every research on this subject there is always ... a certain 'period table' [matrix], ... one is immediately led to think that the properties of exactly this matrix must play an essential role in those theories ... Through more or less recent works, ... these properties have become so numerous by now and have acquired such prominence that it seemed to me that ... the necessity arose of composing an organic and ordered treatment of them in a manner completely independent of the various concrete interpretations they are susceptible of. [Scorza, 1916, p. 127 of *Opere Scelte*].

This kind of attitude prompted Scorza to explore the deep links between 'structural' abstract algebra (particularly linear algebra) and the theory of Abelian matrices. I would like to provide here one example of his use of structural algebraic properties, while 'translating' them into geometrical ones. After defining a *Riemannian matrix* (R. M.) as the matrix of periods of an Abelian function, he attached to any R. M. an algebra A (the algebra of multiplications) and finally introduced this vocabulary:

The algebraic *dimension* of A corresponds (diminished by 1) to its *multiplicability index*.

A is isomorphic to A' if and only if their corresponding matrices are isomorphic.

A is reducible if and only if its corresponding matrix is *impure with complementary axes*.

A is a division algebra if and only if its corresponding matrix is pure.

A is simple if and only if its corresponding matrix is has no complementary axes.

This usage of an algebraic language was very close to the contemporary works of A. Albert in the United States. It moreover deeply influenced Lefschetz (who was in Italy in the years 1920–21 and got in touch with Scorza) who acknowledged:

We propose now to recall some concepts and definitions incipient in the works and only fully developed in recent writings of Scorza and Rosati. The nomenclature which we shall use is Scorza's [Lefschetz, 1921, p. 78 of *Selected Papers*].

A few years later, Lefschetz was even more explicit:

The theory of Abelian functions no doubt is one of the most important to have occupied mathematicians. It is the more so striking that our knowledge of matrices of period *per se* has until very recently remained rather fragmentary. A few fundamental theorems for an arbitrary genus *p* were well established and, more than anyone else, the late George Humbert studied the case $p = 2$ in depth. But, for this case, albeit not easy, direct calculations at least are possible; so, his methods scarcely are easily extendible. Above all, we owe to M. Scorza to have largely lifted the veil. No doubt his works would have spurred a considerable attention had they not appeared during the war, had the method he used [be different]. Belonging to projective geometry, and for the occasion applied to a question important for analysts, [this method] requires from the latter a special training which one can hardly expect them to possess. It would therefore seem highly desirable that M. Scorza's results be one day achieved analytically. I wish to testify that in the course of some of my researches his method whose elegance is assured was of considerable help to me. I thus believe to be doing something useful in summarizing here the main lines of his work [Lefschetz, 1923, p. 120].[5]

In this way, we may see that *inside* the Italian geometrical language and tradition, tendencies existed that could, in an original way, be merged with rising international trends. In conclusion, we may ask why these efforts were not successful. I have no definitive answer and may only suggest two tentative ones.

Firstly, there is a general answer. National tradition may evolve from a *motive* power in the development of science to an *academic* power thwarting the development of new ideas and methods. While useful to understand some trends in mathematical development, phrases like 'national school' or 'national tradition' are not without danger. The academic milieu often acts as individuals who, while looking for ancestors, *choose* among the many branches of their family tree the ones they prefer, while completely neglecting others. In our case, with their unmistakable stress on their intuitive and geometrical capability of overcoming technical difficulties (and, I may add, with the brilliant results they were able to achieve), Italian algebraic geometers had a clear awareness of partaking in a mathematical community deeply rooted in the cultural national terrain (which we may call *a national school*). But they (and above all Severi) often completely forgot that this school had emerged while maintaining both a deep understanding of the most recent algebraic results and firm links with the leading mathematical schools in foreign countries. Eventually, this national arrogance (perhaps encouraged for political reasons by the fascist regime) seriously hampered the further development of Italian algebraic geometry, until a point was reached when it was 'held in disrepute' (to put it with Weil's words) by the international community.

Secondly, I may suggest a more internal reason. As successful as they may have been in the previous period, Italian geometers' methods were steadily reaching the limits of their creative strengths in the face of the ever growing complexity of the problems involved. Moreover, while (as Enriques claimed) aristocrats can dispense with proofs, their methods were hardly adapted to the demands of a period when exciting victories were to be followed by a weary trench warfare. Hampering new advances, the retention of traditional methods would throw a new generation of Italian algebraic geometers into deep crisis.

In this situation, any young Italian mathematician (for example beginning his studies at the end of the 1960s) who wanted to study modern algebraic geometry was compelled to sever any tie with the national tradition in order to merge completely in the international milieu. From this situation, a very unfair opinion has resulted, even from an historical point of view, about the relevance of the contributions to twentieth-century geometry by Italian geometers. Only in recent years (starting in the 1980s) have young scholars regained a consciousness of the debts owed by new mathematics to these pioneering works; and the debate on intuition and rigor in mathematics has picked up momentum once again [Jaffe and Quinn, 1993]. The example of the Italian 'school' of algebraic geometry may become an important test case for the evaluation of different methodological assumptions when looking concretely, for how, mathematical research develops and new ideas grow.

NOTES

* English version revised by David Aubin.
1. see the correspondence Enriques-Castelnuovo in [Bottazzini *et al.* ii, 1996].
2. La démonstration ingénieuse, que ce géomètre y donne de cette importante formule, pourrait laisser sur sa validité absolue des doutes, qui se réfléchiraient sur le n° présent et plus loin ... ; cependant les confirmations qu'on trouve de ces résultats me portent à penser qu'ils sont absolument vrais.
3. It is possible to find many analogous concepts in [Enriques-Chisini, 1915].
4. For a more complete description of this issue see [Brigaglia-Ciliberto, 1995].
5. 'La théorie des fonctions abéliennes est sans contredit une des plus importantes qui aient occupé les géomètres. Il est d'autant plus remarquable que notre connaissance des matrices aux périodes *per se* soit restée jusqu'à tout récente date plutôt fragmentaire. On possédait bien quelques théorèmes fondamentaux pour le genre p quelconque et le regretté Georges Humbert, surtout, avait étudié à fond le cas $p = 2$. Mais pour ce cas le calcul direct, sans être facile, est tout au moins abordable; aussi ses méthodes ne se prêtent-elles guère à une extension facile. C'est surtout à M. Scorza que l'on doit d'avoir largement levé le voile. Ses travaux auraient sans nul doute suscité une attention plus considérable, n'était d'abord leur apparition en pleine guerre, ensuite la méthode dont il s'est servi. Relevant de la géométrie projective, appliquée en l'occasion à une question très importante pour les analystes, elle exige de ces derniers un entraînement spécial que l'on ne peut guère s'attendre à le voir posséder. Il semblerait donc fort désirable qu'on arrive un jour aux résultats de M. Scorza par voie analytique. Je tiens à rappeler qu'au cours de certaines recherches, sa méthode, dont l'élégance est hors de doute, me fut d'une aide considérable. Je crois donc faire oeuvre utile en résumant ici ses travaux dans leurs grands traits.'

APPENDIX

I give a tentative list of the main contributors to Italian school of algebraic geometry. For every name I give the name of his teachers and the places where he taught.

First generation:
   Luigi Cremona (1830–1903); taught in Bologna, Milano, Roma;
   Giuseppe Battaglini (1826–1894); taught in Napoli and Roma;

Second generation:
   Riccardo De Paolis (1854–1892); student of Cremona; taught in Pisa;
   Eugenio Bertini (1846–1933); student of Cremona; taught in Pavia and Pisa;
   Enrico D'Ovidio (1843–1933); student of Battaglini; taught in Torino;
   Giuseppe Veronese (1857–1917); student of Cremona; taught in Padova;
   Corrado Segre (1863–1924); student of D'Ovidio; taught in Torino;
   Mario Pieri (1860–1913); student of Peano and Segre; taught in Catania;

Third generation:
   Guido Castelnuovo (1865–1952); student of Veronese and Segre; taught in Roma;
   Federigo Enriques (1871–1946); student of De Paolis; taught in Bologna and Roma;
   Francesco Severi (1879–1961); student of Veronese and Segre; taught in Padova and Roma;
   Gino Fano (1871–1952); student of Segre; taught in Messina and Torino;
   Michele De Franchis (1875–1946); taught in Catania and Palermo;
   Gaetano Scorza (1876–1939); student of Bertini; taught in Catania, Palermo, Napoli and Roma;
   Carlo Rosati (1876–1929); student of Bertini; taught in Pisa;
   Beppo Levi (1875–1961); student of Segre and Peano; taught in Bologna and Torino;
   Guido Fubini (1879–1945); student of Bertini and Bianchi; taught in Catania and Torino;
   Giovanni Giambelli (1879–1953); student of Segre; taught in Messina;
   Luigi Brusotti (1877–1959); student of Berzolari; taught in Pavia;

Fourth generation:
Annibale Comessatti (1886–1945); student of Severi; taught in Padova;
Alessandro Terracini (1889–1968); student of Segre and Fubini; taught in Torino;
Eugenio Togliatti (1890–1977); student of Segre and Fubini; taught in Genova;
Oscar Chisini (1889–1967); student of Enriques; taught in Milano;
Giacomo Albanese (1890–1947); student of Bertini; taught in Pisa and Palermo;
Ruggiero Torelli (1884–1915); student of Bertini;
Beniamino Segre (1903–1977); student of Segre and Severi; taught in Roma;
Luigi Campedelli (1903–1978); student of Enriques; taught in Firenze;
Fabio Conforto (1909–1954); student of Enriques and Severi; taught in Roma;

REFERENCES

Bertini, E., [1923] (1907) *Introduzione alla geometria proiettiva degli iperspazi*, Principato, Messina; *Einfürung in die proiektive Geometrie mehrdimensionaler Raume, mit einem Anhang über algebrische Kurven und ihre Singularitäten*, German transl. by A. Duschek, Wien, 1924.

Boffi, G., [1986] 'On Some Trends in the Italian Geometric School in Second Half of the 19th Century', *Riv. Storia Scienze*, 3, 5–22.

Bottazzini, U., Conte A., Gario P., (eds.) [1996] *Riposte armonie: lettere di F. Enriques a G. Castelnuovo*, Bollati Boringhieri, Torino.

Brigaglia, A., [1982] 'La geometria algebrica italiana di fronte al XV problema di Hilbert', Atti del Convegno, *La storia delle matematiche in Italia*, Cagliari.

'[1992] 'The harpoon of algebraic topology: varietà abeliane e matrici di Riemann nella geometria algebrica italiana', *Arch. Int. Hist. Sciences*, 42, 290–316.

[1994] 'Giuseppe Veronese e la geometria iperspaziale in Italia', in *Le Scienze matematiche nel Veneto dell'Ottocento*, Venezia, Istituto Veneto di Scienze e Lettere, 231–261.

[1996] 'The influence of H. Grassmann on Italian projective n-dimensional geometry' in Schubring, G. (ed.), *H.G. Grassmann: Visionary Mathematician, Scientist and Neohumanist Scholar*, Boston, Kluwer, 155–163.

Brigaglia, A., Ciliberto C. (eds.), [1994] *Algebra e geometria: il contributo italiano, 1860–1940*, Suppl. Rend. Circolo Mat. Palermo, (2), 36.

[1995] *Italian Algebraic Geometry Between the Two World Wars*, Kingston, Queen's papers, 100.

Castelnuovo, G., [1889a] 'Una applicazione della geometria enumerativa alle curve algebriche', *Rend. Circolo Mat. Palermo*, 3, 27–37.

[1889b] 'Numero delle involuzioni razionali giacenti sopra una curva di dato genere', *Rend. R. Accad. Nazionale Lincei*, (4), 5, 130–133, in Castelnuovo, G., *Memorie Scelte*, Bologna, Zanichelli, 65–70.

[1928] 'La geometria algebrica e la scuola italiana', *Atti del Congresso Internazionale dei Matematici*, vol. 1, Zanichelli, Bologna, 191–201.

[1930] 'Luigi Cremona nel centenario della nascita', *Rend. R. Accad. Nazionale Lincei*, (6), 12, 612–615.

Castelnuovo, G., Enriques, F., [1906] 'Sur quelques résultats nouveaux dans la thèorie des surfaces algébriques' in Picard-Simart [1906], 485–522.

Ciliberto, C., [1991] 'A Few Comments on Some Aspects of the Mathematical Work of Federigo Enriques' in *Geometry and the Complex Variable*, Lecture Notes in Pure and Applied Math., 132, 89–110.

Ciliberto, C., Sernesi, E., [1991] 'Some aspects of the scientific activity of Michele De Franchis' in M. de Franchis, *Opere*, Suppl. Rend. Circolo Mat. Palermo, (2), 27, 3–36.

Coen, S., [1994] 'Beppo Levi: la vita', *Seminari di Geometria del Dipartimento di Matematica*, Bologna, 193–232.

Conte, A., [1987] 'La geometria algebrica italiana fra le due guerre' in *La matematica italiana tra le due guerre mondiali*, Pitagora, Bologna, 107–112.

Coolidge, J., [1940] *Treatise on Algebraic Plane Curves*, Oxford, Oxford University Press.

Corry, L., [1996] *Modern Algebra and the Rise of Mathematical Structures*, Basel-Boston-Berlin, Birkhäuser.

Dieudonné, J., [1974] *Cours de Géométrie Algébrique*, Paris, Presses Universitaires de France.

[1986] 'The Beginnings of Italian Algebraic Geometry', *Symposia math.*, 27, 245–263.

Enriques, F., Chisini O., [1915] *Lezioni sulla teoria geometrica delle equazioni e delle funzioni algebriche*, vol. 1, Zanichelli, Bologna.

Enriques, F., [1932] *Lezioni sulla teoria delle superficie algebriche*, ed. by L. Campedelli, Padova, CEDAM.

[1949] *Le superficie algebriche*, ed. by G. Castelnuovo, A. Franchetta, G. Pompilj, Bologna, Zanichelli.

Fulton, W., [1984] *Intersection Theory*, New York, Springer.

Galuzzi, M., [1980] 'Geometria algebrica e logica tra Otto e Novecento' in Micheli, G., (ed.), *Scienza e tecnica nella cultura e nella società dal Rinascimento a oggi*, Torino, Einaudi, 1001–1105.

Gario, P., [1989] 'Resolution of singularities by P. del Pezzo. A Mathematical Controversy with C. Segre', *Arch. Hist. Exact Sciences*, 40, 247–274.

[1990] 'La teoria classica delle equisingolarità per le curve algebriche piane', *Boll. Storia Scienze Mat.*, 10, 77–97.

[1991] 'Singolarità e geometria sopra una superficie nella corrispondenza di C. Segre e di G. Castelnuovo', *Arch. Hist. Exact Sciences*, 43, 145–188.

Harris, J., [1986] 'On the Severi Problem' *Inv. math.*, 84, 445–461.

Jaffe, A., Quinn, F., [1993] 'Theoretical Mathematics: Toward a Cultural Synthesis of Mathematics and Theoretical Physics', *Bull. Am. Math. Soc.*, 29, 1–13.

Klein, F., [1926] *Vorlesungen über die Entwicklung der Mathematik im 19. Jahrhundert*, Berlin, Springer.

Lefschetz, S., [1921] 'On certain numerical invariants of algebraic varieties with applications to abelian varieties', *Trans. Am. Math. Soc.*, 22, 327–428, (in *Selected Papers*, 41–196).

[1923] 'Progrès récents dans la théorie des fonctions abéliennes', *Bull. Sc. Math.*, 120–128.

[1968] 'A Page of Mathematical Biography', *Bull. Amer. Math. Soc.*, 74, 854–879.

Lolli, G., [1985] *Le ragioni fisiche e le dimostrazioni matematiche*, Bologna, Il Mulino.

Menghini, M., [1986] 'Sul ruolo di Segre nello sviluppo della geometria algebrica italiana', *Riv. Storia Scienze*, 3, 303–322.

Parikh, C., [1990] *The Unreal Life of O. Zariski*, S. Diego, Academic Press.

Peano, G., [1891] 'Osservazioni del direttore', appendix to [Segre, 1891], *Rivista di Matematica*, 1, 66–68.

Picard, E., Simart G., [1906] *Théorie des fonctions algébriques de deux variables indépendants*, vol. II, Paris, Gauthier-Villars.

Scorza, G., [1916] 'Intorno alla teoria generale delle matrici di Riemann e ad alcune sue applicazioni', *Rend. Circolo Mat. Palermo*, 41, 263–380, (in *Opere Scelte*, vol. II).

Segre, C., [1883a] 'Studio sulle quadriche in uno spazio lineare ad un numero qualunque di dimensioni', *Mem. Acc. Scienze, Torino*, (2), 36, 3–86, in *Opere*, a cura dell'Unione Matematica Italiana, Roma, Cremonese, 1961, vol. 3, 25–126.

[1883b] 'Sulla geometria della retta e delle sue serie quadratiche', *Mem. Acc. Scienze Torino*, (2), 36, 87–157

[1889] 'Recherches générales sur les courbes et les surfaces réglées algébriques', *Math. Ann.*, 34, 1–25.

[1891] 'Su alcuni indirizzi nelle investigazioni geometriche', *Rivista di Matematica*, 1, 42–66, in *Opere*, a cura dell'Unione Matematica Italiana, Roma, Cremonese, vol. 4, 1963, 381–412.

[1904] 'La geometria d'oggidì e i suoi legami con l'analisi', *Verhandlungen des Dritten Internationellen Mathematiker-Kongresses in Heidelberg*, Leipzig, Teubner, 1905, 109–120, in *Opere*, a cura dell'Unione Matematica Italiana, Roma, Cremonese, vol. 4, 1963, 456–459.

Severi, F., [1937] 'I sistemi di equivalenza sulle varietà algebriche e le loro applicazioni', *Atti I Congresso U.M.I.*, Bologna, Zanichelli, 58–68.

[1940] 'I fondamenti della Geometria numerativa', *Annali matematica pura e applicata* (4), 19, 1, 153–242, (in *Memorie Scelte*).

van der Waerden, B.L., [1939] *Ein führung in die algebraische Geometrie*, Berlin, Springer.

[1971], 'The Foundations of Algebraic Geometry from Severi to André Weil', *Arch. Hist. Exact Sciences*, 7, 171–180.

Veronese, G., [1882] 'Behandlung der projektivischen Verhältnisse der Räume von verschiedenen Dimensionen durch das Princip des Projicirens und Schneidens', *Math. Ann.*, 19, 161–234.

[1891] *Fondamenti di Geometria a più dimensioni e a più unità rettilinee*, Padova, Tipografia del Seminario.

Weierstrass, K., [1868] 'Zur Theorie der bilinearen und quadratischen Formen', *Monatsberiche Akad. Berlin*, 310–338, in Weierstrass, K., *Werke*, vol. 2, Berlin, Mayer & Mueller, 19–44.

Weil, A., [1946] *Foundations of Algebraic Geometry*, Providence, American Mathematical Society.

Zariski, O., [1971] (1934) *Algebraic Surfaces*, Berlin, Springer.

Chapter 10

DEFINABILITY AS A MATHEMATICAL CONCEPT BEFORE AND AFTER GÖDEL

Gabriele Lolli

In the opinion of most commentators, either logicians or mathematicians, the importance and influence of Gödel's incompleteness results were nil. According to Kreisel [1988], Gödel's theorems have had absolutely no use, and properly so, since they were tailored to refute— or confirm, it does not matter—a specific thesis about the completeness of mathematical systems. Bourbaki [1960] dutifully recalled Gödel's two incompleteness theorems, but only in the historical notes of the section 'Metamathematics' and nowhere else in the treatise. The theorems were likewise seen as a negative answer to Hilbert's program. Bourbaki hinted at their proofs and the technique used, saying that it was based on a one-to-one correspondence between metamathematical sentences and certain propositions of formalized arithmetic.

The most favorable appreciation is due to Myhill [1950], but he restricted himself to a philosophical point of view. He stressed the important conceptual distinction between the decidable and the semi-decidable which emerged from the combined effect of the completeness, incompleteness, and semidecidability theorems, and from the development of recursion theory. Semidecidability is a notion that could not even have been conceived of beforehand, and *a fortiori* no one could ever have imagined to give to it an abstract definition, nor that it could be exemplified by such important domains as logic itself.

In philosophy, not even the Gödel-Tarski theorem on the undefinability of truth has had any influence, although it is easy to imagine that it could have. Ironically, the notion of truth itself still is often used in the sloppy characterization of the undecidable propositions as true but unprovable propositions.[1]

In a sense the disparaging attitude towards Gödel's theorems is fair but in another sense, it is not. Mathematics has not changed because of (the awareness of) incompleteness. It would be possible to list names of people expecting and desiring completeness and decidability (from Leibniz to Peirce[2]) as well as people who remained doubtful, like von Neumann, or fearful of it, like Weyl:

The firm belief to be able to derive all general true judgments of elementary geometry ... from the geometrical axioms by means of a logical reasoning, is a scientific *faith declaration*: we are not able to have a true *intuition* that it is so, and still less to 'prove' it by an examination of the logical laws. If this should happen some day, this intuition would open to us the way to decide the truth or falsity of any geometrical judgment ... by methodically applying a certain deductive technique ('in a finite number of step'): mathematics would thus become *trivial*, at least in principle.[3]

Hilbert's hopes for a firm foundation for mathematics, however, depended entirely on the possibility of a finite consistency proof.

When in 1930–1931 negative answers came out and such an astonishing and definitive piece of knowledge was added to human intellectual patrimony, some welcomed it while others seemed frustrated or surprised. But, afterwards, mathematical research went on as before—or it almost did. Actually, the large portion of mathematics that was not concerned with logical issues went on its own way, but in other areas Gödel's theorems had established a new paradigm. In Kuhn's sense [1962], paradigms are scientific results, important solved problems which stand as models for future research, problem choices, available or allowed solving techniques, and accepted solutions.

In order to assess the value of Gödel's contribution, we must look at not so much the theorem itself (by 'Gödel's theorem' here, we mean his first incompleteness theorem) as its proof—or proofs. Gödel opened new pathways, not by his result, but rather through his proof. If, by solving the dramatic open problem of completeness, the theorem closed an era, it certainly opened a new one by introducing ideas and methods which would be widely picked up. Working in its wake, logicians (including Gödel himself) in the following years tackled other important problems such as decidability; they studied recursive functions; they invented the computer; they developed the mathematical theory of languages; they became computer scientists.

THE TECHNOLOGY OF PARADOXES AND DEFINABILITY

In Gödel's proof lay the vindication of the long-derided notion of definability, widely disputed by the fault of its paradoxes and conceptual difficulties. To start with, his techniques bestowed mathematical status to the notion of definability and hence to the mathematical theory of languages. What used to be called paradoxes were not at all seen as such anymore, but rather as arguments for proofs and evidence for features of incompleteness or undecidability in languages. And as a whole, mathematics acquired a deeper consciousness of its linguistic nature.

For his proof, Gödel used a very simple technology—a time-honored one, perhaps the first shrewd manifestation of pure thought later confined for centuries in a realm of strange, disturbing humor, namely the paradoxes of self-reflexivity. In the introduction of [Gödel, 1930], he underscored that the analogy of his argument with Richard's antinomy, as well as with 'the Liar', leaped to the eye, adding in a footnote that any epistemological antinomy could also be used for the proof. And actually, as we will see, at least one of them had already been used (in Finsler's argument based on Richard's antinomy).[4]

Gödel would later say that it was Wittgenstein's mistake and delusion to think that the theorem merely was a paradox, and not a true theorem. In order for the paradox to work as a proof, however, Gödel needed to endow it with a sophisticated technology which was original (when disregarding Leibniz's anticipation even if it was significant for Gödel himself)—that is, the arithmetization of languages.

In those days, antinomies were bountiful, and lately especially those connected with definability. The epistemological antinomies mentioned by Gödel had been named by Frank Ramsey [1925], who distinguished two groups of antinomies (contrary to those who attributed them all to a single phenomenon, e.g. Russell to impredicativity). In Ramsey's classification, group A encompassed antinomies relating to ordinals and cardinals (also called 'ultrafinite'), while group B contained the Liar, Berry, Richard, Weyl's heterological adjective, and the smallest undefinable ordinal. Already, Peano had suggested that, since pertaining to linguistics, Richard's antinomy was destined to remain of no concern whatsoever to the mathematician. To Ramsey, it seemed that contradictions of group B, which he preferred to call 'epistemological', were not purely logical, because all referred to thought, language, and symbolism, which were not formal, but rather empirical, matters. According to him, they were probably due to, not a mistaken logic, but mistaken ideas about logic and language; he probably meant either their use or what could be expected of them.

In the 1920s, after Zermelo's axiomatization of set theory, no one worried anymore about ultrafinite antinomies, but epistemological ones kept crossing the path of pragmatic mathematicians disturbed by nonsensical paradoxes. Just in those years, Zermelo was losing the debate against Skolem about the essential role of language in axiomatization for the characterization of definable (and existing) sets.

But these issues remained far from clear, and on the contrary, the official version was that all had been done, as far as foundations were concerned; Fraenkel for example was insensible to the issue.

The history of definability, including the way it was turned into a mathematical object, is important both in itself and because it epitomizes a trend of modern mathematics, characterized by jumps in the level of abstraction. Over the last two centuries—from equations to be solved to the theory of solvability, from computation to the theory of computability, from proofs to mathematical logic, from definitions to the study of definability—general concepts involved in the characterization of mathematical activities and objects have been turned themselves into mathematical objects. Definability moreover encompasses both logic and computability, and only its acceptance as a mathematical concept paved the way for the similar acceptance of computability. Thanks to its more obvious ties with mathematical practice, the latter thus hardly had to suffer the resistance met by the former.

DEFINABILITY AT THE BEGINNING OF THE CENTURY

The reception of the concept of definability was controversial at the time of its first appearance, not least because of its connection with puzzling and disturbing phenomena. It began to be discussed even before it was used for paradoxes, namely as documented by Moore [1982], in the context of the new theory of infinite sets and the debated principle of choice. It was first discussed in the famous *Cinq lettres*, an exchange among French analysts Borel [1905], Baire, Lebesgue, and Hadamard about new trends in analysis in Cantor's wake. The uncountable lurked as a nightmare to be avoided, and the only way it could be done seemed to stick to the definable. It was clear to them that to restrict oneself to definable sets meant to restrict their work to the countable. The question, which was posed at this moment (and later discussed by Richard), was whether the axiom of choice would hold for definable sets. The French mathematicians tried to distinguish between naming and existence (as Bernstein also independently did with a different terminology) and introduced the distinction between constructive and existential proofs. Lebesgue explicitly asked whether it was possible to give a well-ordering of the continuum (a problem later proved impossible, even when rather complex, second-order, linguistic means are allowed[5]).

The French mathematicians only discussed single definitions and specific definable objects. But as soon as 'definability', in the sense of the totality of definable objects of a certain type, entered definitions on a par with other, usual mathematical concepts, it produced contradictions. Therefore, the definability concept was, at the very beginning, denounced, in particular by Peano and Zermelo, as non-mathematical.

Peano [1906] remarked that 'exemplo de Richard non pertine ad Mathematica, sed ad Linguistica', although the claim is strange, coming from someone who among other things was to lay the basis for the algebraic theory of grammars [Peano, 1930]. In 1908, Zermelo played the irony card: he pointed to the contradictions the notion of definability led to, and to the fact that by appealing to it people could allegedly prove everything and its opposite. To confirm his charges, he picked up examples from works on the continuum problem notably by König, Bernstein, and Richard, who had arrived at different conclusions by applying definability arguments.

In his two papers of 1908, introducing, respectively, the new proof of the possibility of a well-ordering and his axiom system, Zermelo explicitly dealt with the question of definability. In his paper on the axiom of choice [1908a], he discussed some criticisms of his first 1904 proof. In particular, he noticed that König's attempt at solving the continuum problem by showing that its cardinality, not being well-orderable, was not an aleph, was reminiscent of an antinomy of Richard's type.

König's notion of finite definability was not, Zermelo argued, an absolute notion; it depended on the language and notation allowed, an observation anticipated by Hessenberg [1906]. Since to any object one could assign any denotation whatsoever, if necessary in an arbitrary manner, whether a single individual could have a finite designation was an ill-posed problem. To have the continuum problem depend on such issues made it trivial, and in a sense all questions could thus be reduced and decided, both ways, since the notion was contradictory. As a witness, he referred to Felix Bernstein who in 1905 believed that he had proved, by an appeal to finite definability, that the continuum was equivalent to the second number class, hence well-orderable. Starting from the same notion as König, he reached the opposite conclusion.

Since this short period forms a crucial chapter of the history of set theory, it is important to know what had happened in the preceding years. Without a clarification of the issue of definability, the replacement axiom could not have been formulated. Let us review the main episodes in precise order.[6]

At the 1904 Heidelberg International Congress of Mathematicians, Julius König announced that he had proved, on the basis of a mistaken interpretation of Bernstein's inequality, that the cardinality of the continuum could not be an aleph.[7] For a short while, Cantor had been very worried, but the issue was elucidated during the meeting itself [König, 1904; 1905a]; [Bernstein 1905b]. A different proof based on the concept of definability was afterwards proposed by König [1905b]. Albeit contemporary, this proof was independent, and perhaps slightly anticipatory, of Richard's antinomy [van Heijenoort, 1967, 142].

König considered the definable elements of the continuum—it is not clear whether for him they constituted a segment of the ordinals or a union of segments. In either case, being countable, they possessed a sup, which was thus also definable. Now, the first next element should also be definable, a fact that was a contradiction. Neglecting to mention that, for the soundness of the argument, the well-ordering should also be definable, König at most proved that there was no definable well-ordering of the continuum. As a variant, one may consider countable ordinals associated to definable countable well-orderings, namely subsets of natural numbers, or sequences of 0's and 1's, and their first nondefinable ordinal to get a contradiction. For König, his argument was not a paradox, but a legitimate proof by contradiction of the impossibility of well-ordering the continuum.

Believing that one needed a general language suitable for scientific discourse and that such a language for set theory should be applicable to logical and gnoseological questions, König [1907] tried to be more precise about the notion of definability (*endlich definierbar*). He introduced the wider concept of pseudofinite definitions (and by iteration pseudo-pseudofinite definitions, and so on), where a reference to infinity was implicit, such as in Cantor's diagonalization argument. With reference to this argument, Richard [1903], as well as Jules Tannery, had acknowledged that one should accept real numbers with a formation law not expressible in a finite number of words.

At the same 1904 Congress, Felix Bernstein promised a communication he never delivered, merely publishing in its stead a short note in the *Jahresbericht der DMV* [Bernstein, 1905c]—where he announced a paper in the *Mathematische Annalen* never to be published! In these years, therefore, he appeared to be very uncertain about the continuum, and raised both the possibility of its being not well-orderable owing to the equivocal class W of all ordinals [Bernstein, 1905a], and that of a neat determination of its position in the series of alephs. In his 1901 dissertation, he had proved that the cardinality of the continuum was the same as that of the set of all types of countable order.

In his *Jahresbericht* note, Bernstein introduced a distinction between existential and computable statements (*existierende* vs. *berechenbare Ausdrücke*). He described (or rather alluded to) a hierarchy of levels of computable functions, indexed by countable ordinals, where each new level was obtained by applying an operator to the set of functions of the previous levels. From these scanty hints, he drew an astonishing conclusion: although the notion of an arbitrary function without *Bildungsgesetz* was not justified, the cardinality of the continuum nonetheless was $\aleph_1$, because the totality of *endliche Gesetze* was not countable but had the power of $\aleph_1$. His idea is somewhat suggestive of Hilbert's [1925] in his later attempt at proving the continuum hypothesis by an iteration of recursive operators.

Working alone, with no connection to the German mathematical community, Richard [1905] used for his argument the antidiagonal of an infinite matrix of 0's and 1's, following Cantor's idea: $a(n,m)$ being such a matrix, its antidiagonal was defined as $1 - a(n,n)$. The matrix considered by Richard represented the sequence of all definable real numbers in the interval [0, 1]. The definition of its antidiagonal seemed legitimate, as it was obtained from arithmetical operations, but the matrix itself depended on the notion of definability. The matrix was some kind of Golem: its rows were the numbers which could be defined by means of 'all the permutations of the 26 letters of the French alphabet which are definitions of numbers'. In 1903, dealing with Fermat's theorem, Richard had already considered the enumeration of all possible phrases of the French language, with cancellation of those which did not have the required property of being an expression of its proof: this enumeration, which Richard postulated, listed every possible proof.

If Cantor's antidiagonal proved the uncountability of the continuum, what did Richard's prove? He himself seemed slightly confused. He refused to consider it a true contradiction, since he indicated a possible way out, that is, that the definition of the antidiagonal had to be rejected as impossible. Indeed, if the antidiagonal were legitimate, it would, as a sequence of 0's and 1's, be a line of the matrix. Suppose it were the q-th line, then at the moment at which this line had been inserted in the matrix (because definable), one would already have had to know the whole matrix. This remark of Richard's would later be accepted and developed by Poincaré and Russell as a refutation of impredicative definitions. Probably because he was reluctant to embrace the predicativity position, Richard [1907] later changed his explanation and claimed that the definition had to be refused because it immediately turned out to be inconsistent.

Like König and Bernstein, Richard also somehow hinted, without going into details, at a possible stratification of definability levels (a first one, a second one at which first-level definability was used and so on).

ZERMELO'S AXIOMATIZATION OF SET THEORY

Not only did Zermelo [1908b] criticize these arguments, he also tried to make them impossible in the setting of his axiomatization. With his axiom system, he wanted to settle the confusion over definability, which was raised by some constructions of sets. For him, this confusion caused plain mathematical errors, and not just antinomies. Definability was the object of the Zermelian notion of the definite (*definit*) condition: the *definit* concept went in the subsets axiom (*Aussorderung Axiom*) which said that only *definit* notions could be used. *Definit* was not further specified than in the following way: 'A question or statement is called *definit* if the fundamental relations of the domain, by means of the axioms and the universally logical laws, determine without ambiguity whether it holds or not'. The aim of this formulation certainly was to prevent the formation of such non-mathematical, or nonsensical, sets as the 'set of red objects'. To Zermelo, however, the condition also seemed sufficient to ensure that characterizations like 'definable in a finite number of words' similarly could not be used in the subsets axiom. As a result, Richard's antinomy and the paradoxes of finite denotation would fade away.[8]

For the moment, Zermelo had won, but this was a Pyrrhic victory. When people such as Fraenkel and Skolem began to build and study models of set theory, they easily realized that in order to control the iteration of set theoretical operations, they needed to have a preliminary specification of the totality of definite operations. A nice way towards this goal was offered by the precise clauses of first-order languages obtained by Weyl [1910] and Skolem [1913], while Fraenkel always resorted to a more mathematical formulation. But then, as a corroboration of the evils of definability, Skolem came up with his paradox. In 1930, when Gödel was writing his paper, Skolem [1930] and Zermelo [1929] were still debating about Zermelo's attempt at avoiding the relativism of set-theoretical notions with a sort of second-order logic. As he would to Gödel's result, Zermelo objected to the use of *restricted* means of proofs.

But in the 1920s, the linguistic treatment of mathematical questions grew well beyond the strange case of Skolem's paradox. The concept of formal systems and logical languages were perfected, commonly used, and investigated, by logicians. Following Hilbert's work, some truths had become largely accepted truisms, for example that proofs could be enumerated. These were used by Finsler in his discovery of a possibly devastating antinomy.

FINSLER'S ANTICIPATION

Building an undecidable statement with the help of Richard's technique, Finsler [1926] adopted (for contraction's sake) Hilbert's view on formalization. His intent was to prove that formal consistency did not ensure freedom from contradiction. On the contrary, formalism led to inconsistencies which were not such in the conceptual realm.

Finsler started by repeating Richard's antinomy. Imagining a finite system of symbols with an alphabetical order and grammatical rules B allowing to select out the set of admissible words, each carrying a unique, unambiguous meaning, he hardly seemed troubled by the fact that disambiguation presupposed that meaning was decidable. An object therefore was finitely definable if it was determined by one of these words allowed by B. Considering Richard's matrix and its antidiagonal, he reached a statement which he held as a formally insoluble antinomy.

> The given definition, however, becomes unobjectable as soon as we transfer it from the formal to the purely conceptual realm and leave the formal out of consideration. Then it unambiguously defines a certain binary sequence that is not finitely definable by means of B.

This was the weak part of Finsler's paper. Next, introducing the notion of 'formal proof', he defined a formally undecidable proposition as one for which no formal proof was possible for either the proposition itself or its contradiction.

> Now consider all the combinations of signs of the system B, that constitute a formal proof of the fact that in a certain binary sequence the number 0 occurs infinitely many times, or, alternatively, that it does not occur infinitely many times. Then with every such proof there is associated an unambiguously determined binary sequence, namely, precisely the one for which the proof holds.

A denumerable sequence of these proofs could be established and the associated binary sequences extracted from the proofs also formed a denumerable sequence.

Now take the antidiagonal sequence associated with this sequence and construct the proposition : *In the antidiagonal sequence just defined the number 0 does not occur infinitely many times.* This proposition is formally undecidable since the associated binary sequence cannot belong to the sequence established above. We can therefore say that the proposition is *formally consistent.* But for all that, we can see that this proposition is false.

Repeating the striking sentence quoted above, '*the proof is unobjectionable as soon as we transfer it from the formal to the purely conceptual realm and leave the formal out of consideration.*'

Drawing attention to the vagueness of both the formal system and the conceptual realm invoked by Finsler, the usual criticism directed against this argument echoes Gödel's. When in 1933 Finsler directed Gödel to his 1926 work, the latter reacted with little fairness as if Finsler would have tried to show some sort of absolute unprovability. He inaccurately contended not only that Finsler's system was not clearly defined, but also that the antidiagonal could anyway not be represented in a correct formal system. Later, in 1970, with a more peaceful mind, he acknowledged that if Finsler had specified a formal system, then his proof could have been amended and made acceptable to him [Wang, 1974; 1987]. It is true that Finsler's notion of the conceptual realm seems quite obscure (and he later claimed, for example, that Gödel's informal proof of the undecidable proposition was actually formal). But this is not where the problem lay. Finsler's proof was just as correct as Gödel's introduction to his own paper, where he sketched his proof, without giving all the details about the representation of the recursive functions (and in so doing, fell victim to some pitfalls as shown by Lolli [1992]).

Finsler also omitted details, far too many details—and, clearly, not simple ones that could be redeemed by the bulk of the paper, as in Gödel's case. Given an enumeration of all formal proofs, one should first single out those concerning infinite binary sequences. Then, from each of them, one should extract the sequence concerning the proof and of which the proof says that 0 occurs infinitely many times, or does not, in order to define the following matrix

$a(n,m) = 1$ if the sequence which the n-th proof is about has a 1 at the m-th place.

It is one thing to notice that a proof says something about a sequence, but it is quite another to pick such sequence and its properties, in a purely syntactical way, from the formal array of symbols. Had Finsler considered the kinds of word operations he needed to perform in order to provide a mathematical definition for his matrix, he would have started from extracting from a list its i-th element, substituting in a list an element in a given position and so on with all the combinatorial syntactical operations. Then, he would have had to prove the representability of these operations in a formal system, as did Gödel with primitive recursive functions.

Apart from this essential step, Finsler's matrix is not so different from Gödel's, which could be informally defined in the following way, which is hardly more difficult to read and interpret:

$$a(n, m) \begin{cases} = 1 \text{ if the sentence obtained by substituting the Gödel number} \\ \quad \text{of the numeral } m \text{ in the } n\text{-th formula is provable,} \\ = 0 \text{ if the sentence obtained is refutable,} \\ = \uparrow \text{ otherwise.}^9 \end{cases}$$

GÖDEL'S LEGACY

We have thus seen that the paradox technology was already available and indeed in use before Gödel. What was lacking, however, is clear from an examination of Finsler's proof: this was the mathematical representation of syntactical operations. This success of Gödel's occurred in three stages: (1) arithmetization of language; (2) recognition of the recursive character of operations involved through associated arithmetical functions; and going back to language, (3) representability of recursive functions in the formal system of arithmetic.

Gödel's arithmetization was a clever operation of coding and decoding. It consisted in assigning numbers to alphabetical signs, and numerical operations to syntactical ones in such a way that the arithmetical counterpart of syntax could effectively be handled numerically. Arithmetization made it possible to apply mathematical operations to language. Gödel thus was in a position systematically to use primitive recursion on natural numbers, whose justification and range, dating back to Dedekind only a few decades before, were not well established and known. Nowadays, recursion procedures are applied to linguistic structures and objects through what is known as recursion on (the length of) lists, but this technique was not available to Gödel. Such leaps into the abstract never actually occur. It was difficult to visualize the mathematical structure involved in such a new field, as opposed to the clear insight allowed by well-known objects like classes of arithmetical functions. New (mathematizable) linguistic structures actually emerged from a contemplation of their arithmetic code. Then, in due time, the correspondence ended up by being dropped, and linguistic structures themselves started to stand as proper mathematical structures. As a side product, Gödel defined an interesting class of purely arithmetical functions, which were to be studied and generalized in the following years, leading to general recursive functions.

Gödel's arithmetization of syntax provides a nice illustration of the fact that the first stage in the mathematization of a subject matter goes hand in hand, at least for psychological reasons, with the application of numbers to a new domain (or so it used to be, as numerous examples of direct mathematization without numbers now exist). What was already implicit in the conviction that words and proofs were enumerable now seemed obvious, namely that objects were just finite sequences over a finite set (or alphabet). Abstract set-theoretical language later became helpful, but in Gödel's time set theory was not already widely considered as a foundation for mathematics. It was just a mathematical theory concerned with infinite cardinals (and for his logical framework, Gödel relied on *Principia Mathematica* rather than set theory).

In the domain of syntax, numbers attached to words did not measure anything. They were just familiar labels. But as a Trojan horse, they introduced and made recognizable the appropriate kind of computable operations. With Turing machines, however, it became clear that computability did not necessarily have to do with numbers or numerical symbols, i.e. that linguistic structures themselves were mathematical. Eventually, languages involving recursion on lists, such as LISP (lists-processing), would be introduced and used. When the abstract setting was defined, it would be introduced in textbooks [Brainerd and Landweber, 1974]. With its huge numbers and awful, awesome factorizations, arithmetization nowadays is found only in old-fashioned logic textbooks.[10] Since theory and metatheory coincide, the use of numbers is rather considered as a complication, perhaps a theoretically important one, but confusing in practice. Indeed, as Bourbaki would put it, any single proposition can be interpreted in two ways, as a metamathematical sentence and as a proposition of formalized logic.

Today, in the theory and practice of programming, programs are linguistic texts and all mathematical operations on them are performed with no need for arithmetization. Paradoxes also are included as special technical tools and serve for undecidability results. In order to prove an impossibility, a contradiction must be reached, for which all paradoxes (e.g. Berry's [Machtey and Young, 1978]) are welcome.

The first who provided a mathematical version of Berry's paradox was Beppo Levi [1908]. He considered B to be the number of symbols needed for the expression of statements including letters, punctuation, and mathematical symbols (such as 'l' for exponentiation), and claimed that it can safely be supposed that B > 40, so that $\beta \geq 2$ where β is defined as the number of figures in B's decimal representation. He then considered an enumeration of all definitions and, in particular, the definition 'the number of place B|B' in the given enumeration. This definition has length $22 + 2\beta$, so that:

$$\text{place of 'the number of place B|B'} < B|(22 + 2\beta) < B|B.$$

Just as Richard and Finsler had been, Levi was cavalier with the notion of 'definition'. But he was slightly ahead in the mathematical dissection of linguistic objects. Recently, his argument has been formalized by Boolos [1989]. Levi's argument can be seen as forerunner of the phenomena of incompleteness due to definitions much shorter that what is defined, and of developments in information theory due to Gregory Chaitin [1987].

So commonplace nowadays is the large, and increasingly so, portion of mathematics connected to languages that it might be difficult, without a historical reconstruction, to perceive how laborious was its constitution. Long-term trends in abstract set theory, structuralism, and axiomatization came to converge in its acceptance. But it is one thing to say that structures are the most meaningful mathematical objects, but quite another to be able to discern a structure in an entity hitherto excluded from mathematical treatment, precisely because it represented a precondition for doing mathematics at all—the very language used for mathematics. This was Gödel's achievement and success. By a combination of soft and hard mathematics, he played with paradoxes, but also went through painstaking arithmetical calculations, a rewarding endeavor that was necessary in order to overcome resistance and scepticism.

NOTES

1. They are not true, properly speaking, but only implied by the consistency of the system.
2. For Peirce, see Myhill [1950].
3. Weyl [1918] and again in [1949]. For the climate of hopes and expectations surrounding Gödel's work, see Lolli [1992, 1995] and Girard [1983, 1989].

4. It is curious that Gödel should mention Richard first and, then, 'the Liar;' it is the latter which usually is associated with his proof, while the analogy with Richard's antinomy is far from obvious. See [van Heijenoort 1967, 439] for a partly successful effort at bringing Richard's argument together with Gödel's proof.
5. Technically, such a well-ordering cannot be Δ_2^1.
6. A more detailed history is told in Lolli [1985].
7. A correct interpretation of Bernstein's inequality merely proves that the continuum cannot be the first cardinal coming as a limit after the first uncountable cardinal, the second, the third and so on (for every natural number). This is the only clear-cut result that can be proved about the continuum in Zermelo-Frankel theory.
8. It was clear to Schönflies [1913, 176, n.2] that the notion of finite definability is not *definit*, in his report for the German Mathematical Society.
9. † means 'undefined;' this is the trick needed for using Richard's matrix machinery for a presentation of Gödel's proof.
10. Moreover, unless there is an interest for weak theories and smaller classes of functions (for complexity theory), the part on representability can usually be skipped. This does not mean that many technical points solved by Gödel, such as the proof that addition and multiplication sufficed in order to code sequences, are not interested in themselves or for coding theory.

REFERENCES

Bernstein, F., [1905a] 'Über die Reihe der Transfiniten Ordnungszahlen', *Mathematische Annalen*, 60, 187–195.

[1905b] 'Zum Kontinuum-problem', *Mathematische Annalen*, 60, 463–464.

[1905c] 'Die Theorie der reelen Zahlen', *Jahresbericht der deutschen Mathematiker-Vereinigung*, 14, 1905, 447–449.

Boolos, G., [1989] 'A New Proof of the Gödel Incompleteness Theorem', *Notices AMS*, 36, n° 4, 388–390.

Borel, E. *et al.*, [1905] 'Cinq Lettres sur la théorie des ensembles', *Bulletin de la Société mathématique de France*, 33, 261–273.

Bourbaki, N., [1960] *Elements d'histoire des mathématiques*, Paris, Hermann.

Brainerd W.S., Landweber, L.H., [1974] *Theory of Computation*, New York, Wiley.

Chaitin, G.J., [1987] *Information, Randomness and Incompleteness*, New York, World Scientific.

Finsler, P., [1926] 'Formale Beweise und die Entscheidbarkeit', *Mathematische Zeitschrift*, 25, 676–682.

Girard, J.Y., [1983] 'Sur le théorème d'incomplétude de Gödel', in Bouleau, N. (ed.), *Cinq conférences sur l'indécidibilité*, Paris, Presses des Ponts et Chaussées, 25–48.

[1989] 'Le champ du signe' in *Le théorème de Gödel*, Paris, Seuil, 141–171.

Gödel, K., [1930] 'Über formal unentscheidbare Sätze der *Principia mathematica* und verwandter Systeme', 1931, (Engl. transl. in *Collected Works I*, 145–195).

Gödel, K., [1986–] *Collected Works*, New York, Oxford University Press.

Hessenberg, G., [1906] *Grundbegriffe der Mengenlehre*, Göttingen, Vandenhoeck und Ruprecht.

König, J., [1904] 'Zum Kontinuum-problem',*Verhandlungen des Dritten Internationalen Mathematiker-Kongresses in Heidelberg 1904*, Leipzig, Teubner, 1905, 144–147.

[1905a] 'Zum Kontinuum-problem', *Mathematische Annalen*, 60, 177–180.

[1905b] 'Über die Grundlagen der Mengenlehre und das Kontinuum-problem', *Mathematische Annalen*, 61, 156–160.

[1907] 'Über die Grundlagen der Mengenlehre und das Kontinuum-problem' (Zweite Mitteilung), *Mathematische Annalen*, 63, 217–221.

Kreisel, G., [1988] 'Review of Gödel's *Collected Works*', *Notre Dame Journal of Formal Logic*, 29, 160–181.

Kuhn T., [1962] *The Structure of Scientific Revolutions*, Chicago, University of Chicago Press.

Levi, B., [1908] 'Antinomie logiche?', *Annali matematica pura e applicata* (3), 15, 187–216.

Lolli, G., [1985] *Le ragioni fisiche e le dimostrazioni matematiche*, Bologna, Il Mulino, (ch. IV: Da Zermelo a Zermelo), 175–239.

[1992] *Incompletezza. Saggio su Kurt Gödel*, Bologna, Il Mulino.

[1995] 'Completeness', AILA *Preprints*, Milan, n. 19.

Machtey, M., Young, P., [1978] *An Introduction to the General Theory of Algorithms*, New York, Elsevier.

Moore, G.H., [1982] *Zermelo's Axiom of Choice*, Berlin, Springer.

Myhill, J., [1950] 'Some Philosophical Implications of Mathematical Logic', *The Review of Metaphysics*, 6, n° 2, 165–198.

Peano, G., [1906] 'Additione a Super Theorema de Cantor-Bernstein', *Revista de Mathematica*, 8, 1902-6, n° 5, 143–157.

[1930] 'Algebra de Gramatica' in *Opere Scelte*, vol. II, Roma, Cremonese, 1958, 503–515.

Ramsey, F.P., [1925] 'The Foundations of Mathematics' in *The Foundations of Mathematics and Other Logical Essays*, London, Paul, Trench, Trubner, 1931.

Richard, J., [1903] *Sur la philosophie des mathématiques*, Paris.

[1905] 'Les principes des mathématiques et le problème des ensembles', *Revue générale des sciences pures et appliquées*, 16, 541–542.

[1907] 'Sur un paradoxe de la théorie des ensembles et sur l'axiome Zermelo', *L'Enseignement Mathématique*, 9, 94–98.

Schönflies, A., [1913] *Entwickelung der Mengenlehre und ihrer Anwendungen*, Leipzig, Teubner.

Skolem, Th., [1913] 'Review of Weyl [1910]', *Jahrbuch über die Fortschritte der Mathematik*, 41, 89–90.

[1930] 'Einige Bemerkungen zu der Abhandlung von E. Zermelo 'Über den Begriff der Definitheit in der Axiomatik'', *Fundamenta Mathematicae*, 15, 337–341.

van Heijenoort, J., [1967] *From Frege to Gödel*, Cambridge, Harvard University Press.

Wang, H., [1974] *From Mathematics to Philosophy*, New York, Routledge and Kegan Paul.

[1987] *Reflections on Kurt Gödel*, Cambridge, MIT Press.

Weyl, H., [1910] 'Über die definitionen der mathematischen Grundbegriffen', *Math. naturwiss. Blätter*, 7, 93–95 ; 109–113.

[1918] *Das Kontinuum*, Leizig, Veit.

[1949] *Philosophy of Mathematics and Natural Science*, Princeton, Princeton University Press.

Zermelo, E., [1908a] 'Neuer Beweis für die Möglichkeit einer Wohlordnung', *Mathematische Annalen*, 65, 107–128.

[1908b] 'Untersuchungen über die Grundlagen der Mengenlehre I', *Mathematische Annalen*, 65, 261–281.

[1929] 'Über den Begriff der Definitheit in der Axiomatik', *Fundamenta Mathematicae*, 14, 339–344.

Chapter 11

AN IMAGE CONFLICT IN MATHEMATICS AFTER 1945

Amy Dahan Dalmedico

INTRODUCTION

Was World War II an important rupture for twentieth century mathematics? According to the traditional historiography of mathematics, it may seem surprising to assign a status that is *a priori* epistemological to such a political event, however major it may have been. No historical event has hitherto been submitted to such an appraisal. What does it mean to confront mathematics and its development with a precise historical moment that is above all characterized by its political dimensions? A special topic in the work of several historians of science, the period of the French Revolution has similarly been studied, mostly in regard to the transformation of the social and institutional conditions for scientific life. These historians have emphasized the importance of the constitution of a true scientific community and of related institutions set up by the Revolution. As for mathematical disciplines, however, the history of analysis from Lagrange to Cauchy is still often written without the Revolution being explicitly mentioned; in the case of geometry, given Monge's political role in the foundation of the École polytechnique where descriptive geometry was taught for the first time, the intimate link of its destiny with the revolutionary event has, on the contrary, been widely recognized.

Concerning World War II, historians of physics were the first to reorient their ways of writing the history of their field and to stop treating the war period as a *parenthesis* in the history of the development of various physical theories (particle physics, field theory, thermodynamics, statistical physics, etc.). On the contrary, they started to study *mutations* in the physicists' practices and their new interactions with numerous partners (the military, engineers, scientists from other domains, politicians, etc.). In the introduction to the book *Science in the Twentieth Century*, J. Krige and D. Pestre have written: 'whereas in the former case [WWI], most scientists returned to their laboratories once the conflict was over, now the intimate link between science, the state, and the civil management of civil society remained in place ... Seen in terms of the relationship between science and the state, and the military in particular, World War II did not punctuate two periods of (relative) peace.' In particular, these historians have stressed the crucial shift in the physicist's *cultural role* in American society. Other historians have moreover emphasized the *pragmatic turn* taken by scientific activity, henceforth seeking mainly operative, predictive results able to be applied, rather than the ideal of fundamental, coherent, knowledge of phenomena and laws of nature.

In a way, mathematicians did not escape this major historical movement: new branches arose, original problems were raised, new practices emerged, and the mathematician's figure as viewed by his peers was partly modified. In this chapter, I wish to show that an additional element specifically characterizes the field of mathematics and its community. Indeed, the Second World War initiated what I shall call an 'image war' or a 'representation war' concerning what mathematics was about, what it dealt with, and how. Over the course of the 1950s and 1960s, this 'war' was progressively developed until the balance of power began to shift perceptibly at the end of the 1970s and during the 1980s. This 'war' was focused mainly on the cleavage between pure and applied mathematics, and on the tacit hierarchy—of concepts as much as of values—informing these categories of 'pure' and 'applied.'

At the turn of the twentieth century, all mathematicians obviously did not share the same practice nor the same conception of mathematics. As figureheads of the two most powerful national communities, Poincaré and Hilbert for example symbolized very distinct viewpoints. In his own words, Poincaré was above all concerned by 'problems that arise by themselves, and not problems that we raise ourselves [des *problèmes qui se posent, et non des problèmes qu'on se pose*].' By this, he sought to privilege important questions stemming from the natural sciences (dynamical systems, celestial mechanics), and to downplay artificial problems constructed by mathematicians. In order to solve these 'objective' problems, however, Poincaré forged indispensable abstract tools and founded entirely new disciplines such as topology and algebraic topology. Concerning Hilbert, his eminence derives from several achievements (invariant theory, number theory, geometry, Hilbert spaces, etc.); but, above all, he taught mathematicians how *to think axiomatically*. His enterprise concerning the axiomatic foundations of geometry soon started to represent an ideal for all mathematics. With the 23 problems of his celebrated talk in 1900, he suggested a vision and a choice about what was important and 'deep' in mathematics. Progressively, this vision and this choice would be transformed into a true 'hierarchy' of disciplines, and would for several decades position the trilogy—algebra, algebraic geometry, and number theory—at the pinnacle of mathematics. First constructed solely for algebra by Hilbert's students, and especially by E. Noether and B. van der Waerden, the axiomatic, structural image became the only prestigious one for the whole of modern mathematics. Nevertheless, Hilbert knew how to strike a deal with mathematicians opposed to him by taste and inclination, in particular Felix Klein.[1] The world's mathematical center until 1933, Göttingen united around its Faculty of Science several research institutes in physics, applied mathematics and mechanics, electrical technology, geophysics, etc. While in the first decades of the century the vogue for abstraction and the hegemony of Hilbert's school were on the rise in the European world of mathematics and in the United States, they were far from being dominant.

Until the 1930s, the American academic mathematicians' milieu was strongly marked by a tendency towards abstraction and very few of its members interacted with the industrial world or showed an interest in technical questions. True, connections between mathematics and physics (and notably relativity theory) provided the restricted elite of the American Mathematical Society (Oswald Veblen, George G. Birkhoff, Marston Morse) with possible strategies and discourses for increasing its prestige and funding, of which the foundation of the Institute for Advanced Study at Princeton was the most outstanding example. But this 'ideology'[2] was rejected by most of the professional mathematical community which remained inwardly focused, and uninterested in physics or technology. A few notable exceptions, however, deserve a mention: University of Wisconsin applied mathematician Warren Weaver who directed the Natural Science Division of the Rockefeller Foundation after 1934 and co-authored with Max Mason a book on the electromagnetic field; scientific director R.H. Kent at the Ballistic Research Laboratory at the Aberdeen Proving Ground after 1920; MIT professor Vannevar Bush, a reputed specialist in electric circuit theory and a pioneer in the application of advanced mathematical techniques to the transmission of energy, who collaborated with Norbert Wiener, and a few others. In a 1941 report, Thornton Fry, Mathematical Research Director at the Bell Telephone Laboratories, noted that the mathematical needs of industrial enterprises were essentially being fulfilled by physicists and engineers and that it was about time that applied mathematicians be trained directly. It will not come as a surprise that precisely these personalities would be prominent when science was mobilized for war.

THE WORLD WAR II RUPTURE

The very important immigration of German and Eastern European scientists into the United States, the rise of Nazism and the imminence of world conflict triggered a spectacular change of mind. The urgent need to train applied mathematicians became evident, and various institutional initiatives were taken: appointment of a 'War Preparedness Committee' by the AMS, organization of a summer school at Brown University addressing questions concerning partial differential equations and continuous medium mechanics, constitution of networks and small groups of applied mathematicians, such as the one around Richard Courant in New York City, etc. With the United States' entry into the war, Vannevar Bush established an agency, the Applied Mathematics Panel (AMP) headed by Warren Weaver, whose role was intended to be that of a mathematicians' organization providing other scientists engaged in collaboration with the military—or even the military itself—with mathematical help. Since a significant number of mathematicians were Jewish émigrés obliged to flee Nazism, they immediately accepted to work in project-oriented research contributing to the war effort and to collaborate with the military. In addition to several government laboratories of the civil administration and the armed forces (National Bureau of Standards, Ballistic Laboratory at Aberdeen, the various laboratories of the Navy, etc.), important laboratories set up in university centers (including notably the Radiation Laboratory at MIT) became places of active cooperation. For two and a half years, Weaver's agency supervised an effort to mobilize nearly 300 people including very prestigious mathematicians (John von Neumann, Richard Courant, Jerzy Neyman, Garrett Birkhoff, Oswald Veblen, etc.), producing hundreds of technical reports, and spending over three million dollars. With a program of applied mechanics at Brown University (Richardson and Prager), with J. Neyman's statistical laboratory at Berkeley, with Richard Courant's New York University group tackling mainly submarine sound transmission and shock wave behavior, the AMP promoted an institutionalization of applied mathematics. Working by contracts, the panel established a practice which, persisting after the war, modified several mathematicians' habits and states of mind by getting them much closer to other scientists and offering them a very wide spectrum of research topics. Joining talents and competences running from the pure to the applied, from fundamental research to operations and actions, teams on several occasions had the opportunity of collaborating. In the state of emergency, with the need for immediate results, the rigor and proof demanded by fundamental mathematics were often sacrificed to the benefit of numerical explorations, forms of experimentation or rough approximation hitherto considered to be the exclusive territory of physicists and engineers.

The Rise of New Fields: Partial Differential Equations, Probability and Statistics, 'Cyborg Sciences'

Three domains in particular would benefit from this powerful impulse. The first of them was the field of partial differential equations (PDE), numerical analysis, and algorithm theory. Questions of wave propagation and shock wave behavior in air and water were indeed central to the understanding of charges, rockets, and firing. From the Aeronautical Bureau, a demand for the conception of jet-propelled airplanes spurred numerous researches on gas dynamics, supersonic flow, explosion theory, etc. From Hermann Weyl to John von Neumann, via Courant, Karl Friedrichs, or van Kármán, a range of mathematicians from the purest to the most applied partook in the effort. Let us also mention Garrett Birkhoff's work at Harvard on submarine ballistics, as well as Saunders MacLane's, Hassler Whitney's and the Columbia group's work on air ballistics. Around Prager and Tamarkin at Brown University problems in continuous medium mechanics mobilized people like Feller, Bergman, Lipman Bers, etc. In 1944, the New York group prepared a classified textbook which prefigured Courant and Friedrichs's *Supersonic Flow and Shock Waves* published in 1948. Not allowed to remain theoretical and concerned only with solubility conditions, all of these studies were required to achieve numerical solutions because of the military and political context.

Secondly, probability theory and statistics witnessed a spectacular rise in the global architecture of mathematics. An important part of the research triggered by the aerial war dealt with the study of fragmentation: it relied on the probabilistic study of the damage caused by anti-aircraft fire on one airplane or a group of them. Combining theoretical analyses with statistical models, and problems of bombardment with the control of quality, Columbia statisticians in particular studied various firing configurations in air battles. These theoretical developments were picked up in Abraham Wald's postwar treatises on sequential analysis and statistics. In addition, understanding the statistical properties of noise and detection of a noisy signal were central to several of the undertakings of the Radiation Laboratory at MIT, where mathematicians—Nobert Wiener in the first place, but also Marc Kac and others—actively collaborated with physicists, engineers, and biologists. In the study of differential and integro-differential equations, the statistical approach was also developed within the framework of research on nuclear reactions undertaken for the Manhattan Project; S. Ulam and J. von Neumann studied Monte-Carlo simulations, as well as possible applications of probability theory to hydrodynamic computations.

Another field of research located at the intersection of operations research, game theory, and decision-making mathematics must be mentioned. During the conflict, operations research dealt with technological research, optimal strategies for convoys, optimal disposition of radar stations, etc. In the United States, Philip Morse's group at the Radiation Laboratory was the most important. He considered himself more as a 'consumer' of mathematics—and at least at the beginning a rather rudimentary kind of mathematics—than as a producer of new knowledge. Significantly, however, Morse was in 1947 invited to give the highly prestigious 21st Gibbs Lecture of the AMS, on which occasion he stressed the potentialities of operations research for the economy and industry. Together with the 1948 new edition of von Neumann and Morgenstern's famous *Theory of Games and Economic Behavior*, this conference launched the institutional development of decision-making mathematics.

Cooperation between mathematicians and the military or civil powers certainly did not evolve without tensions, rivalries, and conflicts. Ultimately some very talented mathematical minds, such as Nobert Wiener, whose expertise and suggestions were in high demand, were judged too eccentric by politicians or found it impossible to adapt to the constraints of guided research. Relations soured between military officers and a few mathematicians, even if deeply engaged in the effort: J. Neyman was loath to 'sell' them his most original theoretical statistical results; even J. von Neumann who was solicited on all sides tended to establish his own hierarchy of priorities. Examples of administrative inflexibility were moreover manifest: irritated by Byzantine squabbles within the Air Force, Weaver disappointedly renounced his ambition to coordinate the major tactical study of B-29s. Marston Morse, President of the AMS, and Marshall Stone, blamed Weaver and Bush for underestimating the potential role of mathematicians and for concentrating on physicists, chemists, and engineers. The ensouing resentment produced fractures in the mathematical community that would not be mended until long after the war. [Owens, 1989].

A New Paradigmatic Figure for Mathematicians: John von Neumann

In the course of the war, however, a new role model for mathematicians emerged—the figure of a mathematician acting outside academia, collaborating with other scientists (physicists, engineers, or economists), and integrated in complex social networks to which belonged also political and administrative decision-makers, businessmen, and military officers. Comparable to some physicists, a few mathematicians can be mentioned, who among others embodied this new figure: Wiener, Ulam, Goldstine, Garrett Birkhoff, etc. An extreme and paradigmatic symbol, however, was J. von Neumann. Socially engaged, he contributed to some strategic and technological decisions made by the United States, belonged to circles close to political power, and promoted the idea that the entire world was within the mathematician's domain of intervention. In particular, von Neumann's conceptions and practices led him to a new articulation of scientific disciplines, which shattered firmly established cleavages and hierarchies between the 'pure' and 'applied.' He blurred the accepted boundary between what belonged to mathematics and what did not, and, what would previously have been counted amongst topics in disciplines like as engineering science, the physical sciences, or economics.

Begun in the 1940s, von Neumann's most significant recomposition of interests concerned hydrodynamics, computer science, and numerical analysis. He judged hydrodynamics crucial to physics and mathematics but in need of new computational methods and means, whose development he then took on: under his impulse, the Electrical Computer Project (built at the Institute for Advanced Study) and the Numerical Meteorology Project were launched at Princeton. Von Neumann's field of intervention ranged from the theoretical level to practical implementation; his energy was invested as much in abstract analysis and the modeling of complex problems as it was in their technical implementation and the direction of the collective project necessary to their resolution. Von Neumann's scientific stature was so great, and his mathematical insight so unanimously recognized, that his engagement on the side of applied mathematics conferred on it a new dignity and prestige, which it had never enjoyed before.

THE NINETEEN-FIFTIES AND SIXTIES

In the aftermath of World War II, a redistribution of scientific forces occurred internationally. By the sheer size of its mathematical community, the breadth of its scientific coverage, and its dynamic university and research system, the United States became the foremost mathematical power in the world. Entirely new domains of intervention and enormous opportunities for interacting with more and more scientific and technological sectors were suddenly open to mathematicians. Testifying to increasing interest in applied mathematics, several organizations were established in the United States: The Association for Computing Machineries (1947), The Industrial Mathematical Society (1949), The Operations Research Society of America (1952), The Society for Industrial and Applied Mathematics (1952), The Institute for the Management Sciences (1953), and so on. In addition to university professors, they brought together people working in industry, government agencies, and the military. One should note, however, that the professionalization of applied mathematics would take place outside the American Mathematical Society.

Indeed on the academic level the mathematical community was reluctant to move into applied mathematical domains, notwithstanding the important stimulus they had received from the war. True, building on experience gained in the war, a few centers were developed in the United States and acquired excellent reputations in various applied domains: New York University's Courant Institute in the study of partial differential equations and numerical analysis; Brown University in the study of differential equations, dynamical systems, and control theory; Berkeley and Stanford in statistics, etc. At the research as well as educational levels, institutionalizing efforts in favor of applied mathematics were clearly successful in a few, well-circumscribed places. But, mere islands of resistance, they hardly affected the major tendencies in the country.

The case of Princeton University and the Institute for Advance Study, for example, would deserve a major study which is unfortunately still lacking.[3] Indeed, enjoying a tradition of excellence in pure mathematics, the Institute under John von Neumann's lead was the first to be endowed with a computer. Distinguished in the 1950s by its open spirit, Princeton University's mathematics department welcomed young men wishing to explore new areas of mathematics: game theory with John Nash, experimental computer economics with Martin Shubik, artificial intelligence with Marvin Minsky, etc. As a symbol of this openness, two of the department chairs, Solomon Lefschetz and his student and successor Albert Tucker, who were both initially pure mathematicians specializing in algebraic topology, moved in original directions (Dahan [1994]). Dynamical systems with Lefschetz, dynamic programming with Tucker, mathematical logic and combinatorics as a way to solve computer science problems with Alonzo Church—there is a long list of new mathematical sectors in which Princeton mathematicians would play a crucial role.

Nevertheless, on the whole, the American, as well as international, mathematical communities witnessed a forceful return of, let us say, structural pure mathematics along with the correlative sidestepping of applied mathematics. Several mathematicians have explained this evolution as due solely to the internal dynamical evolution of mathematics. For them, abstraction and formalization constituted an obligatory passage point in the progress of most domains. And it hardly bothered them whether both of these processes went along with an outgrowth of the formal, algebraic side of theories and the relative self-closure of the professional mathematical community.

In my opinion, the internal development of mathematics alone cannot account for what was the outcome of situations arising, and choices made, on political, institutional, and intellectual levels. It cannot explain what resulted from the actors' tacit *value hierarchy* and *disciplinary representation*. In the rest of this chapter, only the diffuse question of the cultural image of mathematics will be tackled.

THE CULTURAL IMAGE OF PURE MATHEMATICS

A characterization of the *cultural image* of pure mathematics in the 1950s and 1960s cannot be unique because images were not uniform; it cannot even be reduced to those produced either by American mathematicians or by Bourbaki's group and the French mathematical school. We must moreover keep in mind the properly French, *vs.* American, features of this image. Nevertheless, we think it is relevant to elucidate some of the main features of this image and understand how these features were inscribed in the spirit of the time. Some differences and nuances do exist but even the point where these nuances lie reveals main tendencies. In the 1980s, a shift in the dominant image goes along with very critical judgments in the mathematical community with respect to the preceding period. Even if occurring *post facto*, free expression of opinions helps to the historian to capture the cultural image of mathematics in earlier years.

American Mathematicians' Rebellion against Utilitarian Ideology

In 1957, the president of the American Mathematical Society Marshall Stone was surprisingly invited to give the 31st Gibbs Lecture. Entitled 'Mathematics and the Future of Science,' his talk is an instructive introduction to American views about what mathematics should be and how it should be considered.[4] His mathematical interest having always been directed toward the pure branches of the discipline, Stone shared with all mathematicians of the time—applied mathematicians or, let us say, 'cyborg mathematicians' (John von Neumann, Norbert Wiener, A. Wald, G. Dantzig, Philip Morse)—a tremendous faith in the growing importance of mathematical thought for the future of science. But while for von Neumann, this faith rested on the idea that mathematics could be applied to the whole world (nature, economics, society, human behavior, the brain, etc.), the possibility of making both deterministic and statistical predictions also being used to distinguish scientific from nonscientific disciplines, Stone, by contrast, expressed his faith in mathematics in a simple syllogism: 'science is reasoning; reasoning is mathematics; and therefore, science IS mathematics.' Hence, science was reduced to those disciplines in which mathematical reasoning played a predominant or crucial part. Fields such as geology or meteorology in which observation played an essential role could therefore hardly deserve to qualify as science! Opposing attitudes about applications of mathematics, Stone explained, resulted from a deep philosophical cleavage in American culture concerning the place of individuals in society. Due to the conflict between liberal and utilitarian conceptions of education, this cleavage opposed two standpoints: one valuing whatever developed the individual's intellectual and spiritual capabilities and the other prizing whatever worked or led to useful results.[5] 'For mathematics, which is at once the pure and untrammeled creation of the mind,' Stone emphasized, 'the adoption of a strictly utilitarian standard could lead only to disaster.' No doubt he shared National Science Foundation director Alain Waterman's conception of mathematics according to which this discipline, 'in a sense, bridges the gap, real or imaginary, which exists between the sciences and the humanities.'

According to Stone, an essential prerequisite for the extraordinary flowering of pure mathematics since 1900 was that 'mathematicians have recognized and acted upon the fact that mathematics is *not* closely bound to the material world or to physical reality—if, indeed, it is bound at all!' Hence, 'utility alone,' he declared, 'is not a proper measure of value' and could be, if strictly applied, dangerous and false as a measure. Pure mathematicians could not accept 'reference to action' as the sole criterion by which their work was to be judged. The sole use of the axiomatic method, however powerful or characteristic, could not capture the essence of this intellectual movement. Axiomatization had to be combined with an *ideology—the desire to free mathematical theory from a dependence on physical necessity*. Only this combination made possible 'the dissection of mathematical concepts into their elementary components, the recombination of these components into new constructs of intrinsic interest, the critical evaluation of alternative approaches to the important mathematical theories, the unification of hitherto unconnected branches of mathematics—that best expresses the spirit of modern pure mathematics.'

In fact, a paradoxical situation existed in the US: the utilitarian, pragmatic bent of American ideology and culture coexisted with strong tendencies towards abstraction and generality in American mathematics. How can we account for this paradox? American mathematicians' response to broad trends in the development of their discipline, Stone wrote, had been significantly affected by specifically American circumstances. The highly pragmatic character of industrialists and businessmen—an outgrowth of the pioneers' experience—prevented mathematics from being called on to play anything but the most modest utilitarian role during the critical period during the late nineteenth century and first part of the twentieth century. By chance, the development of great mathematical centers in the US almost exactly coincided with the flowering of pure mathematics. From this cultural lag, a complete freedom accrued to American mathematicians. Neither strong academic traditions nor previous higher mathematics prevented them from enjoying a singular independence from utilitarian demands and the freedom to direct their efforts towards central themes in modern mathematics.

Many years later, Peter Lax ventured a similar analysis of what he called 'the American tide of purity,' to which he personally strongly objected:

The bold proposal to cut the lifeline between mathematics and the physical world was put forth only in the 20th century, mainly by the Bourbaki group. Besides being wrong headed, this raises profound philosophical problems about value judgments in mathematics. The question 'What is good mathematics' becomes a matter of *a priori* aesthetic judgment, and mathematics becomes an art form ... Next to Bourbaki, the greatest champions of abstraction in mathematics came from the American community. This predilection for the abstract might very well have been a rebellion against the great tradition in the United States for the practical and pragmatic; the postwar vogue for Abstract Expressionism was another such rebellion. [Lax, 1986].

In this era of specialization, much of the best applied mathematics work was done by men who considered themselves not mathematicians at all but rather physicists, chemists, or biologists. According to pure mathematicians, a majority of scientists in America fell too easily into a pragmatic, utilitarian trap when dealing with mathematics. Ultimately, they generally considered it as a useful tool about which one needed to know no more than immediately useful features. In this way, communication was to a serious extent broken down between pure mathematics and many branches of applied mathematics. Moreover, mathematical instruction and training at the secondary level had been only slightly affected by the modernization of mathematics. The most serious obstacle to a modernization of the mathematical curriculum, Stone concluded, was the utilitarian spirit that pervaded secondary education.

Elitist Poetization of Pure Mathematics

Bourbaki's enterprise obviously had nothing to do with rebellion against a dominant utilitarian culture. For its first generation of members, the project was to reformulate mathematics on definitive foundations. As a new Euclid, Bourbaki would for thousands of years to come establish strict disciplinary standards. He was inscribed in a certain *Zeitgeist* of the interwar years: to lay down definitive solutions—some would even use the fateful expression 'a final solution'—to theoretical questions: 'There are good reasons to hate that sentence,' Pierre Cartier acknowledged, 'but people thought that we could reach a final solution.' In his historical notes, Bourbaki thus always expressed the conviction that mathematical developments were a series of stages unavoidably leading to the present, and ultimate, state—the axiomatic, structural conception—which would endure forever in the future [Corry, 1997].

This *eternalism*, let us say, also characterized the spirit of Hilbert's metamathematical program. Defeated by Gödel's work, this program was indeed intended to provide a 'final solution to the consistency problem' of arithmetic axioms.[6] Clearly, this attitude was also shared by the Vienna Circle project, which aimed at putting an end to metaphysics and to base the scientific conception of the world on the foundations of logical language analysis.

This spirit is obviously insufficient to characterize the Bourbaki group completely after World War II. Moreover, Bourbaki itself exhibited a certain degree of heterogeneity, as is shown by Cartier's list of four different generations in the group:

1°) The founding-fathers: André Weil, Henri Cartan, Claude Chevalley, Jean Delsarte, and Jean Dieudonné, the last of whom frequently expressed the definitiveness of solutions to certain questions, rigor for example, provided by modern mathematics;

2°) Those who joined the group during the Second World War or in its immediate aftermath, such as Laurent Schwartz, Jean-Pierre Serre, Jacques Koszul, Jacques Dixmier, Roger Godement, or Samuel Eilenberg. Among them, Serre emerged as the natural leader since he was the only one with a truly universal understanding of mathematics;

3°) Armand Borel, Alexander Grothendieck, Claude Bruhat, Pierre Cartier himself, Serge Lang, and John Tate: a generation which, according to Cartier, was 'more and more pragmatic' (or less and less dogmatic). Trained within the new axiomatic tradition, which obviously appeared excellent to them, these mathematicians may have felt that they had nothing to prove anymore.

4°) The next generation was formed by Grothendieck's pupils at the moment when himself left the group [Cartier, 1998].

Not only were there quarrels between different generations but also within them. Moreover, if the Bourbaki group was small and well-delimited, the *séminaire Bourbaki* was much more open, and it is quite impossible to determine precisely the exact borders of Bourbaki's following.

The cultural image promoted by Bourbaki was fundamentally elitist. For the most part, group members were exceptionally gifted individuals who enjoyed a broad humanist culture. Their interests were many: philosophy for Chevalley, classic Greek and Indian civilizations for Weil who knew Sanskrit, and so on. Most of them played musical instruments at an honorable level. In this vein, André Weil [1960] has described wonderfully the pleasure he derived from mathematics:

> Nothing, as every mathematician knows, is more fecund than these obscure analogies, these troubling reflections from one theory to another, these furtive caresses, these unexplainable scrambles [*brouilleries inexplicables*], nothing gives more pleasure to the researcher. One day, illusions are dissipated; presentiments become certainties; twin theories reveal their

common source before they fade away; as is taught by the Gita, knowledge is reached at the same time as indifference. Metaphysics has become mathematics, and is ready to form the topic of a treatise whose cold beauty would be incapable of moving us.[7]

Similarly, Laurent Schwartz has expressed in his autobiography the esthetic enjoyment of mathematical discovery [Schwartz, 1997]. In many respects, these mathematicians seemed closer to poets and literary creators than practical and experimental scientists.

They moreover considered that deep, important mathematics was produced by a very small number of individuals. True, lesser mathematicians could have a role to play in clearing the ground, in acting as sounding boxes, in teaching, etc. But only the production of significant results really counted.

The Bourbaki group recruited a few talented American mathematicians (Eilenberg, Tate, Lang) and many of its influential members spent extensive periods of time in the United States. While almost all spent at least one year there, Chevalley stayed from the end of WWII to 1955, and Weil and Armand Borel (who was Swiss) settled down indefinitely in Princeton. Between France and the United States, quite distant political and cultural contexts nonetheless fostered an axis around which the cultural image of pure mathematics was forged. In these two countries as well, this image was most clearly defined. Links were especially strong between both mathematical communities.

Bias Against Physics

In the 1920s, connections between mathematics and physics provided American mathematicians with opportunities to increase both their prestige and funding. At the time, mathematical physics and in particular relativity theory offered them the possibility, at least by means of rhetorical strategies, to escape their marginality, for instance in the creation of the Princeton Institute for Advanced Study [Butler Feffer, 1997]. Later, during the war years, mathematicians and theoretical physicists had several opportunities to collaborate. But, as already mentioned, the *ideology* of freedom from physical reality and independence from physics or other sciences established a tight barrier around pure mathematicians in the United States.

In France also, Bourbakists shared 'a strong bias against physics' [Cartier, 1998]. Visiting Göttingen in 1926, at the very moment when quantum theory was experiencing its strongest developments with

Heisenberg's elaboration of matrix mechanics, André Weil seemed to have completely ignored quantum mechanics. In the obituary he and Chevalley wrote for Hermann Weyl, both Bourbakis failed to mention two of Weyl's major books on general relativity and quantum mechanics. R. Hermann wrote:

> The most curious aspect of the Bourbaki story is that they started at the time of the initial flowering of quantum mechanics, reached full speed at the time when Einstein's geometric theory of gravitation was finally being understood, when elementary particle physics started its proliferation of accelerators and Lie groups across the countryside, and when many aspects of the 'core' mathematics with which they were concerned were being integrated into engineering and economics via system, control, and optimization theory and yet not a trace of such developments has ever appeared in their pages [1986, p. 32].

It seems obvious that several barriers were erected between mathematics and theoretical physics. Many pure mathematicians recognized that mathematical successes in the field of physics were impressive and spectacular, but in their opinion they were mainly based on nineteenth century mathematics. Although the geometrization of physics underwent crucial steps with general relativity, although Hermann Weyl exhibited the gauge invariance principle of physical laws, and although Élie Cartan studied from a mathematical standpoint the spinor notion, geometers generally resisted topics like spinor fields or differential operators acting on them. In 1954, two physicists working in the US, C.N. Yang and R. Mills suggested a model for the study of strong interactions in terms of a gauge theory with a very simple invariant group. But it was not until the 1970s that mathematicians began to recognize in the Yang-Mills theory an exact coincidence with modern differential geometry. A striking parallel between concepts elaborated separately by theoretical physicists and mathematicians was then established by Yang and Wu.[8]

The use of mathematical statistics has also been crucial in elementary particle physics, in the mechanisms governing heredity, in theoretical biology and genetics, in the social sciences, etc. Originally investigated by mathematicians S. Ulam and N. Metropolis for computer simulations of the H-bomb, Monte-Carlo methods have progressively acquired major theoretical importance. Notwithstanding its wide success, these methods drew very little attention from mathematicians.[9] In Marshall Stone's opinion, all sciences embodying the essence of inductive reasoning were based on a single principle: 'A sufficiently improbable event may be ignored.' He added: 'why the real world should be amenable to such a rule is a philosophical question no more—and no less—mysterious than the problem of why it should be amenable to logic.' These questions were but two aspects of the ultimate problem of the '*connection between mind and matter*.' The distinction between inductive and deductive reasoning offered no help in defining what was meant by science. In any case, he noticed, through measure theory and modern mathematical statistics, 'virtually all the detailed procedures of inductive reasoning are deductive in character'.

This last explanation should be considered along with two more statements which have flourished since they were first made. The first is Wigner's expression about 'the unreasonable effectiveness of mathematics' [1960] and the second Bourbaki's passage on the 'miraculous adaptation of mathematical structures to reality' [Bourbaki, 1948, p. 46]. Here are three expressions of the same philosophical conviction, the same *ideology* which could be summarized in the same terms: world and nature can, and must, be described in abstract mathematical structures, an adequacy which reflects the deep connection between mind and matter. But this absolute conviction conferred to mathematicians the legitimacy to turn their back on concrete, real-world problems and to be exclusively concerned with the internal development of mathematics.

Monistic Unity of Mathematics

In discourses and texts published in the 1950s, two features characterized modern mathematics: (1) the necessary *unity* of mathematics; and (2) the axiomatic method which allowed this unity to be constructed. Bourbaki strongly advocated a *structural* point of view for this unity, and in accordance with Hilbert's views set theory was thought to provide the needed general framework. But, as several people have since proved, category theory—not set theory—is perhaps a more flexible tool for logical foundations. Pleading for an *organic*, rather than structural, unity of mathematics, Cartier for instance recently wrote: 'Categories can offer a general philosophical foundation—that is the encyclopedic, or the taxonomic part—and a quite adequate tool to be used in *mathematical situations*. That set theory and structures are, by contrast, more rigid, can be seen by reading the final chapter in Bourbaki's set theory, with a monstruous endeavor to formulate categories without categories' [1998]. Still, developed by Samuel

Eilenberg, who belonged to the group, and by Saunders MacLane who was pretty close to it, category theory was by and large an offspring of Bourbaki's.

As was underscored by Leo Corry, the structural conception truly refers to a particular way of *doing* mathematics, to a tacit knowledge of the mathematicians' daily practice, and to a hierarchical conception of the mathematical corpus, but this character can only be defined in non-formal terms in relation to a discourse bearing *on* mathematics. The notion of 'mother-structure' and the description of mathematics as a hierarchy of structures, he has noted, do not internally result from any theory, but merely appear in non-technical popularization articles (see Corry, [1992]). In my opinion, this conception of mathematics can be appropriately grasped through an examination of the way some treatises were conceived. Concerned with mathematical domains not primitive for an architectural conception, these treatises were at the crossroads of several branches and involved method or intuition transfers from one to another.

Let us focus on the exemplary case of homological algebra. In the 1940s and 1950s in several mathematical topics, including algebraic topology and algebraic geometry, new foundations and internal results were systematically developed. Homological algebra first emerged from algebraic topology, but its domain of application was soon extended to several other fields including algebraic geometry. Homological algebra deals with both the homology of algebraic systems and the algebraic aspects of homology theory. The first topic leads to homological and cohomological theories of groups of associative algebras and of Lie algebras; the second to exact sequences and spectral sequences and the manipulation of functors of chain complexes. This algebra had been qualified, if indulgently, as '*abstract nonsense*' by the topologist Norman Steenrod [Lang, 1995, p. 340].

The first treatise on the topic was Cartan and Eilenberg's *Homological Algebra* which intensely mobilized categories. It is interesting to read MacLane's review of this book: 'The authors' approach can best be described in *philosophical* terms and as *monistic*: everything is unified.' MacLane explained this sentence with an example: 'Consider for instance the homology of groups; in view of its application to class field theory and to topology, this topic is central in homological algebra. In this book the homology of groups appears as a special case of the homology of monoids, which in turn is a special case of the homology of supplemented algebras, again a case of the homology of augmented algebras, which is an instance of a torsion product, which at your choice is an instance of a derived functor or an iterated satellite functor'. [MacLane, 1956]. He went on: 'Historically, each monistic doctrine is resolved by a subsequent pluralism. So it was here. When the authors started to write, it was true that all known cases of homology of algebraic systems (groups, algebras, and Lie algebras) could be neatly subsumed under the resolution, Tor and Ext pattern. When they finished writing this was no longer so—and this because of the authors' own separate efforts elsewhere! ... Perhaps Mathematics now moves so fast—and in part because of the vigorous unifying contributions such as of this book—that no unification of Mathematics can be up to date.' MacLane also criticized the accumulation of brilliant, promising ideas whose usefulness was obvious to no one at the time. As a result, he believed that uninitiated readers could hardly hope to understand the book. This was, he concluded, an unfortunate confusion between research paper and treatise.

The same impression of dogmatic, monistic tonality is present in A. Mattuk's book review of Chevalley's *Fundamental Concepts of Algebra*, which Mattuk considered as 'tight, unified, direct, severe ... Relentlessly and uncompromisingly, it pursues its end.' He added: 'The unity is *monolithic*. Gone is the discursive rambling of previous texts. This one marches unswerving and to its own music ... [The] book [is] *abominably hard* for a beginner, *unreasonably hard*, I should say' [Mattuk, 1957]. The author Claude Chevalley had warned: 'This is an exercise in *rectitude of thought*, of which it would be futile to disguise the austerity.' Mattuk countered: 'The voice that we hear resounding is that of an old Testament prophet, but the mental attitude is more like a tenth grade Latin teacher's wreaking with the old theory of formal disciplines.' A few years earlier, Weil had devoted a long technical, enthusiastic review of Chevalley's *Introduction to the Theory of Algebraic Functions of One Variable*. His friend however had to recognize that: 'It appears that the author has somewhat overstated his claims and has been too partial to the method dearest to his algebraic heart. Who would throw the first stone at him? It is rather with relief that one observes such signs of human frailty in this *severely dehumanized book* ' [Weil, 1951].

The reception of a book written by an American mathematician who was very close to Bourbaki provides us with a third example and enriches our global perception of their style in, and conception of, mathematics. Even among mathematicians working in the same pure, abstract domains, polemics pitted one against the other. In 1962, Serge Lang, who belonged to Bourbaki and shared many aspects of his spirit and conception of mathematics, published *Diophantine Geometry*. A historical authority on the topic, Mordell reviewed it in a piece which became famous for its very negative reaction to mathematical trends in the 1950s and 1960s. Mordell wrote: 'A general question that immediately suggests itself to a reader is what object an author has in mind when writing a book. Some have the true teacher's spirit or even a missionary spirit, wishing to introduce their subject to a wide circle of readers in the most attractive way ... Lang is not such an author'. After having quoted Lang's declaration: 'One writes an advanced monograph for one's self because one wants to give permanent form to one's vision of some beautiful part of mathematics, not otherwise accessible,' Mordell went on:

> Much of the book is *practically unreadable* unless one is familiar with among other Bourbaki, the author's books on algebraic geometry and Abelian varieties, and Weil's *Foundations of Algebraic Geometry* and is prepared occasionally to go to the original sources for proofs of some theorems needed in the present volume ... The reader will require the patience of Job, the courage of Achilles, and the strength of Hercules to understand the proofs of some of the essential theorems. He will realize that some of the proofs will be above his head [Mordell, 1964].

A German mathematician deeply involved in number theory, Siegel, sent Mordell a letter of approval that is also worth quoting:

> The whole style of the author contradicts the sense of simplicity and honesty which we admire in the works of the masters in number theory—Lagrange, Gauss, or on a smaller scale, Hardy [and] Landau ... Unfortunately there are many 'fellow-travelers' who have disgraced a large part of algebra and function theory. However until now, number theory had not been touched. These people remind me of the impudent behavior of the national socialists who sang '*Wir werden weiter marschieren, bis alles in Scheiben zerfällt.*' I am afraid that mathematics will *perish* before the end of this century if the present trend for *senseless abstraction*—as I call it: *the theory of the empty set*—cannot be stopped.[10]

In Lang's opinion, Mordell and Siegel were great mathematicians, but failed to understand the accomplishments of this period, in particular the relations between number theory and algebraic geometry which would later come to be known as number field case and function field. More profoundly they were missing the conceptual *unification* of topology, complex differential geometry, and algebraic geometry which went on in the seventies and beyond with the work of Alexander Grothendieck, and Pierre Deligne. According to Lang, this was the reason why algebraic geometry was largely at a dead end in England and Germany in spite of Atiyah's and Hirzebrook's work, and had been mainly developed in France and the United States.

The structural, monistic point of view had very important consequences in the way of presenting mathematics in rhetorical style, and in the extremely elitist conception of mathematical books or treatises. Moreover the required structural unity increased the dogmatic view of mathematics and arbitrarily excluded large branches of the discipline: applied mathematics whose method was clearly not axiomatic, combinatorics, concrete geometry, several branches of analysis among which differential equations, statistics, logic, etc. This vision strengthened a hierarchical conception of the mathematical corpus that privileged algebraic branches and internal mathematical dynamics.

Conception of Analysis: A Confrontational Matter

The domain of mathematical analysis is crucial. It is one of the oldest branch of mathematics, especially linked to the study of nature, physics, and engineering science. Various conceptions of analysis and what its teaching should be strongly opposed those of pure and applied mathematicians. In the 1940s and 1950s, the emphasis put by the former on functional analysis was absolutely enormous. For Bourbaki, this was justified by the general state of confusion in mathematics at the time. In fact, except for Laurent Schwartz, none of its members was really an analyst. Bourbaki labored towards a conception in which algebra, analysis, and topology would form a single unified domain giving rise to vast syntheses at increasing levels of abstraction. Traditional branches of analysis were considered bleak and limited in their ambitions. When he tackled nonlinear oscillations, Solomon Lefschetz noticed that differential equation theory was deemed the most boring topic possible. L. Carleson has described the reigning state of mind regarding classical analysis: 'There was a period, in the 1940s and 1950s when classical analysis was considered dead and the hope for the future of analysis was considered to be in the abstract branches, specializing in generalization.' Writing in 1978, he went on: 'As is now apparent, the rumor of the death of classical analysis was greatly exaggerated and during the 1960s and 1970s the field has been one of the most successful in all mathematics' [Carleson, 1978, p. 53].

About Coddington & Levinson's *Theory of Differential Equations*, which he very much appreciated, Richard Bellman (who had joined the Rand Corporation in the late 1940s) wrote:

> It has become fashionable of late, in various mathematical centers, to present the fundamental tools of analysis, real and complex variable theory in an *increasingly abstract manner* to those most defense-less, namely fledgling graduate students. In the process, motivation for the introduction of new concepts has been on the whole by-passed as an atrophied relic of those early pioneer days when mathematicians were forced to consort with astronomers and physicists, and indeed, in some cases were indistinguishable from them ... It is doubtful, whether this species can perpetuate itself. Most likely, mathematical education will continue along the same simple logical principles that have guided the greatest scientists of the past, from the simple to the complex, from the concrete to the abstract [Bellman, 1956].

This book review was a testimony to the irritation provoked by the fashion of the abstract mathematics, exclusively oriented towards itself. But abstraction was offered by pure mathematicians as the indispensable price one had to pay for achieving greatest theoretical simplification and deepest unification.

As mathematicians opposed to these trends observed, both constant distancing from concrete empirical sources and an increasing esthetic conception of mathematics always went together. Among scores of examples, let us mention two positions. In 1956, von Neumann wrote: 'Mathematical ideas originate in empirics; once they are conceived, the subject begins to live a peculiar life of its own ... As a mathematical discipline travels far from its empirical source, ... it becomes more and more pure aestheticizing; ... at a great distance from its empirical source, or after much abstract inbreeding, a mathematical subject is in danger of degeneration' [1956]. Criticizing the representation of mathematics as artwork, Paul Halmos violently denounced what he called the substituting of 'mathology to matophysics,' i.e. the tendency to generate problems and research areas by perfecting the elegance and symmetry of axiomatic formulations at the expense of increasing power of action on the world and the analysis of broad fields of classical problems and mathematical physics [Halmos, 1968].

At a conference on the evolution of modern mathematics in 1970, lively polemics on abstraction and axiomatization focused on analysis. While for some mathematicians like Browder and Dieudonné, this attitude remained fully justified, others like Garrett Birkhoff held that it exhibited a tendency to function solely for itself and had become excessive. In the long run, the latter asked, was not the computer more important than functional analysis? Was not the Lax-Richtmyer theorem concerning linear differential equations one of the major theorems of numerical analysis? Was the most important thing in analysis, not so much Hilbert or Banach spaces, but the notion of *norm* leading to error measures? etc. At the time, such iconoclastic claims shocked pure mathematicians. [Birkhoff, 1975].

In summary, it can be said that two concurrent images of mathematics collided:

— on the one hand, the image of pure mathematics developed above all 'in the honor of the human spirit', whose methodology *par excellence* was axiomatic and structural. Progressing via the internal dynamics of their problems at the interface of various branches, mathematics was a collection of deep theories produced by very few exceptional minds and constituted a structurally unified corpus to which one referred as artwork with a rhetoric of elegance and esthetics;

— on the other hand, the image of the mathematics which is applied, mathematics stemming from the study of nature, of technical problems, and of human affairs (war, weapon technology, statistics, economics, management, etc.). This was less prestigious mathematics, less rigorous mathematics as regards its approaches and methods (numerical analysis, approximations, modeling, etc.), less noble and less universal mathematics produced by the mass of the proletarians of science, and subordinate to social interests and conflicts.

In fact, the odds being clearly in their advantage until the end of the 1970s, pure mathematicians gave shape to this opposition. Applied mathematicians' self-identity was defined negatively in relation to the former. Except for bastions like the Courant Institute, the California Institute of Technology and a few others, applied mathematics was often developed outside the university world, in engineering schools, foundations, military agencies, or industry (the Bell Laboratories). Albeit establishing their own professional organizations, applied mathematicians remained relatively on the margins of the international mathematics community.

Over the course of the 1970s, the general landscape of mathematics was progressively modified as a result of new economic, technological, and cultural contexts in contemporary societies. In France, pure mathematics suffered from its elitist image. Paradoxically, since pure mathematicians always retained a leftist progressive political image, students' rebellious protests in 1968 upset their status. Whole domains which had remained dormant for decades, were reclaimed, unknown fields of research, linked in particular with the computer, randomness, and experimental mathematics, were reopened: probability theory and Brownian motions, dynamical systems, fractals, combinatorics, code theory, etc. Widely used in the aeronautic, space, nuclear, and medical imagery industry, modeling and simulation called for a development of new mathematical methods.

Bringing Mathematics Down to Earth

In 1984, Edward D. David chaired an American committee which produced a report titled 'Renewing U.S. Mathematics: Critical Resource for the Future.'[12] Looking at the explosion in the applications of mathematics due to the computer since the 1970s, the David report underscored the vital character of mathematics for science, technology, and contemporary mathematics. It had a major influence in increasing federal funding for mathematical research in subsequent years. Not surprisingly, this new conscience was very favorably welcomed by applied mathematicians [Lax, 1986].

A few years later, in a report titled 'The Endless Frontier Meets Today's Realities,' Richard H. Herman indicted the American mathematical community. The title of his report was a clear allusion to Vanevar Bush's 1945 'Science, the Endless Frontier.' Herman noted that media and government accused mathematicians of lacking social responsibility. A loss of public confidence, he wrote, was now perceptible. Society needed to re-negotiate its contract with the scientific community. Mathematicians could not go on 'cloning' themselves, self-reproducing identically. While appeals to responsibility had periodically taken place in the past, mathematicians could not ignore, this time, the social demands which concerned them [Herman, 1993].

Troubled by their isolation and eager to improve their social image, mathematicians increasingly tried to put forward a more open image of mathematics entertaining multiple interactions with other disciplines, the world, and human needs. The American Mathematical Society widely echoed David's and Herman's reports. In France, a large conference on the future of mathematics ('*Mathématiques à venir*') jointly organized by the Société mathématique de France and the Société de mathématiques appliquées et industrielles (SMAI) was witness to the receding of the 'tide of purity.' The actors' disciplinary ideological representations and the implicit philosophy sketched above also gave room to other representations privileging other values: ties with political power, capacity to earn money, entrepreneurial dynamism, pragmatic and operational character of results, etc. A mathematician was thus forced to note with nostalgia that 'the psychological climate has changed, there is less of an emotional element, a sharp decline in the poetization of pure mathematics' [Fomenko, 1986, p. 8].

While the increasingly pressing theme of mathematicians' social responsibility tended to bring mathematics back to earth, questions emerged in the community, which a few years earlier would been deemed totally incongruous. While few mathematicians shared Vladimir Arnold's violently provocative pronouncements about the 'criminal Bourbakizers and algebraizers of mathematics' [Arnold, 1995], many more wished to counter the esoteric image of their discipline and the myth of its inaccessibility.

Debates about Proofs, New Results, and Mathematical Activity

A first issue concerned the status of demonstration in mathematical production. Demonstrations of a new type that were heavily dependent on the computer, such as the four-color theorem established by K. Appel and W. Haken, gave rise to questions on the nature of the proof in mathematics: how could a human mind grasp a demonstration that it could not follow since it filled nearly 400 pages and distinguished close to 1500 configurations by means of very long automatic procedures? Authors moreover acknowledged having found, and then corrected, dozens and dozens of errors: what kind of conviction could one therefore have that the theorem was thus proved? [Appel & Haken, 1986]. From computer proofs, the debate was soon extended to other demonstrations either quite long (e.g. the exhaustive classification of simple finite groups) or mobilizing an extremely complex architecture of conjectures stemming from various domains, as for example Edward Witten's work on knot theory and string theory. Since impossible to follow by a single reader, numerous proofs done without the help of a computer were susceptible to the very same critiques as Appel and Haken's. After all, countered Haken, if logicians accepted that an 'unassailable' demonstration may be executable on a Türing machine, mathematicians could hardly demand higher certainty standards.

In 1990 a controversy shook the mathematical milieu. Initiated by an article of A. Jaffe and F. Quinn, the controversy was followed by a large number of replies to which the two mathematicians in turn reacted. Here, a different starting point concerned norms of rigorous, axiomatic-deductive demonstrations that mathematicians had to defend as regards new theoretical practices (speculative reasoning, physical intuitions, etc.) which were gaining ground as a result of theoretical physicists' entry in the mathematical community [Jaffe & Quinn, 1990, p. 171–172].[13] In fact, replies to this article, the majority of which were issued by prestigious pure mathematicians (W. Thurston, M. Atiyah, A. Borel, R. Thom, etc.) extended beyond the sole point of rigor and concerned the values of contemporary mathematical activity in general. They illustrated that the cultural image of mathematics described above, if it has been accurately sketched, was widely considered as out of date and that the social and institutional loci in which these practices were inscribed had been extraordinarily diversified. A strict defense of standards linked to a previous cultural image of mathematics now seemed hopeless.

The computer's place was already prominent in mathematical research; evolution in the exploration of results and conjectures by means of this tool was irreversible. The founders of the journal *Experimental Mathematics* declared: mathematicians 'are interested not only in theorems and proofs but also in the way in which they have been or can be reached ... The role of the computer in suggesting conjectures and enriching our understanding of abstract concepts by means of examples and visualization is a healthy and welcome development.'[14] A large number of conjectures emerging from extensive computer usage had sometimes been used for years before they could be rigorously proved, as was the case with the topological properties of the Lorenz attractor.

Similarly, W. Thurston has recently emphasized the multiple and collective aspects of mathematical activity. Adding that it should therefore not be judged on the sole criterion of new theorems obtained, he wrote:

> Mathematical knowledge and understanding were embedded in the minds and in the *social fabric* of the community of people thinking about a particular topic ... In any field, there is a strong *social standard* of validity and truth ... One used to analyze the motivation to do mathematics in

terms of a common currency that many mathematicians believe in: credit for theorems. This has a negative effect on mathematical progress. We must recognize and value a far broader range of activity ... Soccer can serve as a metaphor. There might only be one or two goals during a soccer game made by one or two persons. That does not mean that the efforts of all the others are wasted. We do not judge players on a soccer team only by whether they personally make a goal; we judge a team by its function as a team [Thurston, 1994].

A 'Sociological Turn'

The metaphor of the soccer team was a reference to the social which became surprisingly prominent under the mathematicians' pen. Apart from Thurston and among many others, we may cite Vladimir Arnold: 'The difference between pure and applied mathematics is social rather than scientific. A pure mathematician is paid for making mathematical discoveries. An applied mathematician is paid for the solution of given problems' [Arnold, 1997]. Similarly René Thom claimed that 'rigor can be no more than a local and sociological criterion.' When checking Andrew Wiles's proof of Fermat's theorem, several mathematicians recognized that the social and institutional dimensions of the confidence granted to some of them were at least as decisive as the rigor of the verification they could perform.

Even in the philosophy of mathematics, T. Tymoczo argued for a conception in which the notion of the mathematical *community*, rather than mere abstract isolated individuals, ought enter. Mathematics being a public, collective affair in which there was, says he, no commonly shared ideal of rationality and truth, the certainty of proofs in a theory today was laden with a collective dimension [Tymocko, 1986]. Thus, in practice, the rigor of the axiomatic-deductive character of mathematics functioned less as a reality than as an idealized myth. Promoted in the 1960s by philosophers and historians of science marching in Thomas Kuhn's steps, the 'sociological turn' was catching up with mathematicians who hitherto had seemed the least susceptible to it!

At the beginning of the twenty-first century, the value shift and hierarchy reversal occurring in mathematics is hardly independent of changes in the general image of science. While in the 1950s and 1960s, the first term in the oppositions mobilized in the 'image wars' in mathematics—pure vs. applied, abstract vs. concrete, structural vs. procedural, fundamental vs. useful, universal vs. specific, general vs.

operative—had always been privileged and charged positively, in the subsequent period the second term was always emphasized [Dahan, forthcoming b].

While *structure* was the emblematic term of the 1960s, *model* has now taken its place. In the physical sciences, climatology, engineering science, economics, and the social sciences, the practice of model-building has gradually dominated the terrain. It is today absolutely massive and intricately bound up with numerical experimentation and simulation. In some parts of the mathematical community, this practice naturally gives rise to new concerns: those who study, with the help of the computer, supersonic flow dynamics, plamas in fusion, or shock waves, all those who model a nuclear reaction or a human heart in order to test an explosion velocity or the validity of an artificial heart, what theorems have they precisely and clearly proved? Can we consider that all these people share the same profession? Can such diverse mathematical practices still be inscribed in a unified domain? At the Berlin International Congress of Mathematicians in August 1998, the old opposition between the pure and the applied—still widely shared in the community—has been formulated in quite different terms: 'mathematicians who build models versus those who prove theorems.' [Mumford, [1998]. But the respect enjoyed by the former is now definitely at least as high as that of the latter.

NOTES

* Translated by David Aubin.
1. See D. Rowe's paper in this volume.
2. We use this term in the same sense as Loren Butler Feffer [1997]; i.e.: we use it here without negative connotations, to encompass the prevailing values common to members of a group or of a scientific community.
3. About Princeton, several narratives are particularly lively. See (Nasar [1998], pp. 66–103; Mahoney [1997]).
4. Stone [1957]. Traditionally, Gibbs lecturers had been applied mathematicians.
5. Stone mentioned the report of the National Research Council's Committee on Training and Research in Applied Mathematics, on which he served; see F.J. Weyl [1955].
6. Hilbert's words are quoted and discussed by J-Y.Girard. See Nagel and *al* [1989], p. 155.
7. 'Rien n'est plus fécond, tous les mathématiciens le savent, que ces obscures analogies, ces troubles reflets d'une théorie à une autre, ces furtives caresses, ces brouilleries inexplicables; rien ne donne plus de plaisir au chercheur. Un jour vient où l'illusion se dissipe; le pressentiment se change en certitude; les théories jumelles révèlent leur source commune avant de disparaître; comme l'enseigne la Gita on atteint à la connaissance et à l'indifférence en même temps. La métaphysique est devenue mathématique, prête à former la matière d'un traité dont la beauté froide nesaurait plus nous émouvoir'.
8. See Wu & Yang [1975] This example is related by J.-P. Bourguignon [1989].
9. Dieudonné [1991] for example explicitly underlined that it would be a concession to computational fashion to consider Monte-Carlo methods and fast Fourier transforms as major mathematical results.

10. Siegel's letter seemed to have widely circulated at the time, but it has only been recently published by Serge Lang himself. See Lang [1995].

11. The first occurrence of this expression appeared in a letter from Jacobi to Legendre, July 2, 1830. Jacobi, Gesammelte Werke, vol. 1, Berlin 1884, p. 454. It reappeared in 1943 in André Weil, 'L'avenir des mathématiques,' *Les grands courants de la pensée mathématique*, ed. F. Le Lionnais (Paris: Blanchard, 1948). It has then been often picked up by various people in the 1960s and 1970s until it became the title of one of Jean Dieudonné's book.

12. David was President of Edward David Inc., and served as science advisor to President Nixon. See David [1985].

13. The whole debate has been published in the July 1993 and April 1994 issues of the *Bulletin of the American Mathematical Society*.

14. This journal was founded in 1991, see Epstein & Levy [1995].

REFERENCES

Albers, D.J., Alexanderson, G.L. (eds.), [1985] *Mathematical people. Profiles and interviews*, Boston, Birkhäuser.

Andler, M., [1994] 'Les mathématiques à l'École normale supérieure au XXe siècle: une esquisse', in Sirinelli, J.F., (ed.) *École normale supérieure, le livre du bicentenaire*, Paris, PUF, 1994, 351–404.

Appel, K., Haken, W., [1986] *Mathematical Intelligencer*, vol. 8, n° 1.

Armatte, M., [1996] 'Mathématiques modernes et sciences humaines', in Belhoste, B., Gispert, H. & Hulin, N., (eds), *Les sciences au lycée: un siècle de réformes des mathématiques et de la physique*, Paris, Vuibert.

Arnold, V., [1995] 'Will mathematics survive?', *Mathematical Intelligencer*, vol. 17, n° 3.

[1997] 'An Interview' by S.H. Lui, *Notices of the AMS*, vol. 44, n° 4.

Aspray, W., [1988] 'The Emergence of Princeton as a World Center for Mathematical Research, 1896–1939', in *History and Philosophy of Modern Mathematics*, ed. William Aspray and Philip Kitcher, Minnesota Studies in the Philosophy of Science, vol. 11, Minneapolis: University of Minnesota, 346–66.

[1990] *John von Neumann and the Origins of Modern Computing*, Cambridge (MA), MIT Press.

Aubin, D., [1997] 'The Withering Immortality of Nicolas Bourbaki: A Cultural Connector at the Confluence of Mathematics, Structuralism and the Oulipo in France', *Science in Context*, 10, 2, 297–342.

Batterson, S., [1996] 'The mathematical work of Stephen Smale', *Notices of the AMS*, vol. 43, n° 12.

Belhoste, B., Dahan-Dalmedico, A., Pestre, D., Picon, A., (eds.), [1995] *La France des X. Deux siècles d'histoire*, Paris, Economica.

Bellman, R., [1956], Book Review of Caddington & Levinson's *Theory of Differential Equations*, *BAMS*, vol. 62, n° 2, 185.

[1984] *The eye of the hurricane. An autobiography*, Singapore, World Scientific.

Bers, L., [1988] *The migration of European mathematicans to America*, in Duren & *al.*, [1988–89], vol. 1, 231–243.

Birkhoff, G., (ed.), [1975] 'Proceedings of the American workshop on the Evolution of Mathematics', *Historia Mathematica*, 2.

[1977] 'Applied mathematics and its future', in Thomson (R.W.) (ed.), *Science and technology in America*, NBS, publ. 465.

Bottazzini, U., [1990] *Il flauto di Hilbert. Storia della matematica moderna e contemporanea*, Torino, Utet Libreria.

Bourbaki, N., [1948] 'L'architecture mathématique', in Le Lionnais, (ed.) [1948–1962], 35–47.

[1969] *Éléments d'histoire des mathématiques*, 2nd ed., Paris, Herman.

Bourguignon, J.P., [1989] 'Géométrie et Physique' in *Démarches mathématiques, L'Encyclopédie Philosophique*, Paris, PUF.

Butler Feffer, L., [1997] 'Mathematical Physics and the Planning of American Mathematics: Ideology and Institutions', *Historia Mathematica*, 24, 66–85.

Carleson, L., [1978] 'The work of Charles Fefferman', in *Proceeding of the International Congress of Mathematicians*, Helsinki, 1978, 1, 53.

Cartier, P., [1998] 'The continuing silence of Bourbaki', *Mathematical Intelligencer*, vol. 20, 1, 22–28.

Corry, L., [1992] 'Nicolas Bourbaki and the Concept of Mathematical Structure', *Synthèse*, 92, 315–348.

[1997] 'The origins of Eternal Truth in Modern Mathematics: Hilbert to Bourbaki and Beyond', *Science in Context*, 10, 2, 253–296.

Dahan Dalmedico, A., [1994] 'Rénover sans se renier. L'École polytechnique de 1945 à nos jours', in Belhoste, B., Dahan-Dalmedico, A., & Picon, A., (eds.), *La Formation polytechnicienne, 1794–1994*, Paris, Dunod, 299–332.

[1994] 'La renaissance des systèmes dynamiques aux États-Unis après la deuxième guerre mondiale: l'action de Solomon Lefschetz', *Supplemento ai Rendiconti del Circolo Matematico di Palermo*, II, 34, 133–166.

[1995] 'Polytechnique et l'École française de mathématiques appliquées', in Belhoste & al., [1995] 283–295.

[1996] 'Le difficile héritage de Henri Poincaré en systèmes dynamiques', in Greffe, J., Heinzmann, G., & Lorenz, K., (eds.), *Henri Poincaré, science et philosophie*, Berlin, Akademie Verlag & Paris, Blanchard, 13–33.

[1996] 'L'essor des mathématiques appliquées aux États-Unis: l'impact de la Seconde Guerre mondiale', *Revue d'histoire des mathématiques*, 2, 149–213.

[1997] 'Mathematics in the 20th century', in Krige, J. & Pestre, D., (eds.), *Science in the 20th century*, London, Harwood Academic Publishers.

[1999] 'Pur versus appliqué? Un point de vue d'historien sur une 'guerre d'images'', *La Gazette des mathématiciens*, Publication de la Société Mathématique de France, n° 80, 31–46.

[forthcoming a] 'Chaos, Disorder and Mixing: A new 'Fin de siècle' Image of Science? ', to appear in *Growing Explanations*, N. Wise (ed.), Princeton University Press.

[forthcoming b] 'History and Epistemology of Models: Meteorology (1946–1963) as a Case Studies', *Archive for History of Exact Sciences*.

Dieudonné, J., [1977] *Panorama des mathématiques pures*, Paris, Gauthier–Villars.

[1982] *Notices of the AMS*, 20, 618–623.

[1987] *Pour l'honneur de l'esprit humain. Les mathématiques aujourd'hui*, Paris, Hachette.

[1991] Review of Halmos' paper [1990], *Mathematical Review*, [1991] 91 j: 01032.

Duren, P., Askey, R.A., Merzbach, U.C., (eds), [1988–89] *A Century of Mathematics in America*, vol. 1 (1988), vol. II (1989), vol. III (1989) Providence, American Mathematical Society.

Epstein, D. and Levy, S., [1995] 'Experimentation and Proof in Mathematics', *Notices of the AMS*, vol. 42, n° 670–674.

Fomenko, A.T., [1986] 'Mathematics and the External World' an interview with A.T. Fomenko, *Mathematical Intelligences*, vol. 8, n° 2, 8–17.

Forman, P., [1987] 'Behind quantum electronics: national security as basis for physical research in the United States, 1940–1960', *Historical Studies in the Physical and Biological Sciences*, 18, 149–229.

Forman, P. & Sánchez-Ron [1996] *National military establishments and the advancement of science and technology*, Dordrecht, Kluwer, Boston Studies in the Philosophy of Science, vol. 180.

Guedj, D., [1985] 'Interview of C. Chevalley', *Mathematical Intelligencer*, vol. 7, n° 2.

Halmos, P., [1968] 'Mathematics as a creative art', *American Scientist*, p. 375–389.

[1990] 'Has progress in Mathematics slowed down?', *The Encyclopedic Dictionary of Mathematics*, 561–588. With a condensed version presented in August 1990 at the Columbus Meeting of the Mathematical Association of America.

Heims, S.J., [1980] *John von Neumann and Norbert Wiener. From mathematics to the technologies of life and death*, Cambridge (MA), MIT-Press.

Herman, R.H., [1993] 'The Endless Frontier Meets Today's Reality', *Notices of the AMS*, vol. 40, n° 1, 6–8.

Hermann, R., [1986] 'Mathematics and Bourbaki' *Mathematical Intelligencer*, vol. 8, n°1, 32–33.

Hunger Parshall, K., Rowe, David E., [1994] *The Emergence of the Americain Mathematical Research Community, 1876–1900: J. Sylvester, Felix Klein, and E.H. Moore*, Providence, American Mathematical Society and London, London Mathematical Society.

Jaffe, A. and Quinn, F., [1993–1994] 'Theoretical Mathematics: Toward a cultural synthesis of mathematics and theoretical physics', *BAMS*, vol. 29, n° 1 (July 1993) and vol. 30, n° 2 (April 1994).

Kármán, T. von, Edson, L., [1967] *The wind and beyond*, Boston, Little, Brown.

Lang, S., [1970] Book review of Mordell's *Diophantine Equations*, *BAMS*, 76, 1230–34.

[1995] 'Mordell's Review, Siegel's Letter to Mordell, *Diophantine geometry* and 20th century Mathematics', *Notices of the AMS*, vol. 42, III, 339–350.

Lax, P., [1977] 'The bomb, sputnik, computers, and european mathematics', in Tarwater [1977], 129–135.

[1986] 'Mathematics and its Application' *Mathematical Intelligencer*, vol. 8, n° 4, 14–17.

[1988] 'The flowering of applied mathematics in America', in Duren & *al*, 1988–89, vol. II, 455–466.

Le Lionnais, F., (ed.), [1948] *Les Grands courants de la pensée mathématique*, Marseille, Cahiers du Sud; 2e éd., Paris, Blanchard, 1962.

Leray, J., [1972] 'La Mathématique et ses Applications', conférence à l'Académie Lincei de Rome. Dossier J. Leray à l'Académie des Sciences de Paris.

Mac Lane, S., [1956] Book review of Cartan's & Eidenberg's *Homological Algebra*, *BAMS*, vol. 62, n° 6, 615–624.

[1989] 'The Applied Mathematics Group at Columbia in World War II', in Duren & *al*, [1988–89], vol. III, 495–515.

Mahoney, M., [1997] 'Computer science: The Search for a Mathematical Theory', *Science in the Twentieth Century*, J. Krige & D. Pestre (eds), London, Harwood Academic Publishers.

Maria, M. de, (ed.), [1989] *The Restructuring of Physical Sciences in Europe and the United States, 1945–1960*, Singapore, World Scientific.

Mattuk, A., [1957] Book Review of C. Chevalley's *Fundamental Concepts of Algebra*, *BAMS*, vol. 63, n° 6, 412–417.

Mehrtens, H., [1996] 'Mathematics and war: Germany, 1900–1945', in Forman & Sánchez-Ron, 87–134.

Mendelsohn, E., Smith, M.R., Weingart, P., (eds.), [1988] *Science, Technology and the Military*, Dordrecht, Kluwer, 2 vol.

Mordell, L.J., [1964] 'Book Review of S. Lang's Diophantine Geometry', *BAMS*, 570, 491–498.

Morse, P.M., [1948] 'Mathematical problems in operations research', *Bull. Amer. Math. Soc.*, 54, 602–621.

Mumford, D., [1998] 'Trends in the Profession of Mathematics', *Berlin Intelligencer*, International Congress of Mathematicians Berlin.

Nagel, E., Newman, J., Gödel, K., and Girard, J.-Y., [1989] *Le theoreme de Gödel*, Paris, le Serif.

Nasar, S., [1998] *A Beautiful Mind. A Biography of John Forkes Nash*, New York, Simon & Schuster.

Neumann (von), J., [1956] 'The mathematician', in J.R. Newman, *The World of Mathematics*, vol. IV, 2059–2063, New York, Simon and Schuster.

Owens, L., [1989] 'Mathematicians at War: Warren Weaver and the Applied Mathematics Panel, 1942–1945', in *The History of Modern Mathematics*, ed. David E. Rowe and John McCleary, vol. 2, Boston, Academic Press, 287–306.

Pestre, D., [1984] *Physique et physiciens en France, 1918–1940*, Paris, Éditions des Archives contemporaines.

[1992] 'Les physiciens dans les sociétés occidentales de l'après-guerre. Une mutation des pratiques techniques et des comportements sociaux et culturels', *Revue d'histoire moderne et contemporaine*, 39, 56–72.

Pickering, A., [1985] 'Pragmatism in particle physics, scientific and military interest in the post-war United States', *History of Science Society*, Annual Meeting, 31 octobre-3 novembre 1985, Bloomington (Indiana).

Reingold, N.L., (ed.), [1991] *Science, American style*, New Brunswick and London, Rutgers University Press.

Rowe, D., [1989] 'Klein, Hilbert and the Göttingen mathematical tradition', *Osiris*, II, 5, 186–213.

Schubring, G., [1981] 'The Conception of Pure Mathematics as an Instrument in the Professionalization of Mathematics', in *Social History of Nineteenth Century Mathematics*, Mehrtens, H., Bos, H. and Schneider I. (ed.), Boston, Birkhauser Verlag, 111–134.

Schwartz, L., [1997] *Un mathématicien aux prises avec le siècle*, Paris, ed. Odile Jacob.

Schweber, S.S., [1986] 'The Empiricist Temper Regnant, Theoretical Physics in the United States, 1920–1950', *Historical Studies in the Physical Sciences*, 17, 55–98.

[1988] 'The mutual embrace of science and the military: ONR and the growth of physics in the United States after World War II', in Mendelsohn & *al*, vol. 1, 3–45.

Schweber, S.S., Fortun, M., [1993] 'Scientists and the legagy of World War II: the case of operations research', *Social Studies of Science*, 23, 595–642.

Stone, M., [1957] 'Mathematics and the Future of Science', 31° Gibbs Lecture, the 27th December 1956, *BAMS*, vol. 63, 61–76.

Tarwater, D., (ed.), [1977] *The Bicentennial Tribute to American Mathematics, 1776-1976*, Washington, Mathematical Association of America.

Thurston, W.P., [1994] 'On proof and progress in Mathematics', *BAMS*, vol. 30, n° 2.

Tymocko, T., [1988] 'Making rooms for Mathematicians in the Philosophy of Mathematics', *Mathematical Intelligencer*, vol. 8, n° 3, 44–49.

Ulam, S., [1976] *Adventures of a mathematician*, New York, Scribner.

Wald, A., [1947] *Sequential analysis*, New York, Wiley.

Weil, A., [1948] 'L'avenir des mathématiques', in Le Lionnais, [1948-1962], 307–321.

[1951] 'Book Review on C. Chevalley's Introduction to the theory of algebraic functions of one variable', *BAMS*, vol. 57, 384–398.

[1960] 'De la métaphysique aux mathématiques'

[1991] *Souvenirs d'apprentissage*, Bâle, Birkhäuser.

Wigner, E.P., [1960] 'The unreasonable effectiveness of mathematics in the natural sciences', *Communications on Pure and Applied Mathematics*, 13, 1–14.

Wu, T.T., Yang, C.N., [1975], 'Concept of non integrable factors and global formulation of gauge fields', *Physical Review*, D 12, 3845–57.

Chapter 12

FROM CATASTROPHE TO CHAOS: THE MODELING PRACTICES OF APPLIED TOPOLOGISTS

David Aubin

Sociologiquement, on peut dire que [la théorie des catastrophes] a fait ... un naufrage subtil, parce que la plupart des notions que j'ai introduites ... ont pénétré dans le bagage ordinaire des modélisateurs. Alors, il est vrai que, dans un sens, les ambitions de la théorie ont fait naufrage, mais la pratique, elle, a réussi. [Thom, 1991, 47].

During World War II, American mathematician George D. Birkhoff contended that 'topology deserves to obtain a more prominent position in physical theories than it has yet obtained' [1943, 310]. Because of its impact on various parts of mathematics, topology undoubtedly was among the greatest successes of the twentieth century. But, prior to the early 1970s, and despite Birkhoff's wish, topology had generally retained the image of an abstruse pursuit that had found little concrete application elsewhere. Or to put it more accurately, very few leading topologists paid any attention to the concrete.[1]

By the early 1990s, a compelling alternative had emerged, and this image had become obsolete. For instance, British topologist E.C. Zeeman characterized the history of topology as a succession of hegemonic approaches branching off to establish new subspecialties. After the 1890s when it untangled itself from its 'applied origins,' topology was 'analytic' from 1900 to 1920, 'geometric' in the 1920s and 1930s, and then 'algebraic.' But, the 1960s saw the resurgence of a 'geometric' standpoint, and the 1970s were labeled 'differential.' The wheel having turned a full circle, Zeeman saw in the 1980s the triumph of 'applications.'[2] Now, 'applied topology' is hard to find as a standard classification in mathematics. In more than fifty years, the *Mathematical Review* has recorded only one single use of the expression in the title of an article.[3] Clearly, the image of topology seems to have shifted recently.

Although part of larger mutations in the image of mathematics, this shift is best understood as resulting from the successful adaptation of notions and practices coming from topology to the modeling of certain phenomena in the physical, life, and social sciences. This, I claim, was the result a systematic effort undertaken by those I will call 'applied topologists.' From the late 1950s onwards, they pursued vast programs, well adapted to the then dominant ideology of pure mathematics [Aubin, 1997]. Sometime in the 1960s, they turned their attention to the real world. Exploiting the newest mathematical technologies of their arsenal, they forged *modeling practices* capable of providing, or so they wished, theoretical explanations for phenomena badly understood by means of conventional approaches.[4] But in so doing, they retained the Bourbakist ideal of exhibiting the deepest structures of the world.

Diversely known as the theories of catastrophes, dynamical systems, and deterministic chaos, these modeling practices benefited from the crucial catalytic role played by the Institut des hautes études scientifiques (IHÉS), at Bures-sur-Yvette near Paris [Aubin, 1998a; 1998b]. Hired there in 1963, French mathematician René Thom welcomed notable visitors, such as E.C. Zeeman and the prodigious Berkeley mathematician Steve Smale. With students and followers, they forged a small community which promoted new methods for modeling, as well as the mathematical technologies needed for the task. However, following the media frenzy that greeted catastrophe theory, a backlash challenged the legitimacy of their approaches outside pure mathematics. No matter how 'applied,' topologists had trouble convincing communities of specialists of the fruitfulness of their methods. Was the world actually structured by topological concepts, or did they merely provide a language for grasping it? Applied topologists' responses to skepticism varied greatly: Smale preferred to base his models on well-established mathematization processes; Zeeman tried to convince large audiences that catastrophe theory could actually be used to generate differential-equation models; and Thom embarked on a grandiose enterprise intended to revolutionize the philosophy of science.

Ultimately, the success of some of the modeling practices promoted by applied topologists depended on the willingness of other specialists to claim them as their own. For this mediation, the IHÉS also provided a well-suited environment. Inspired by Thom's and Smale's ideas, IHÉS physicist David Ruelle published, together with Dutch topologist Floris Takens, a seminal article in 1971, which in many ways launched the study of chaotic dynamics. An untypical physicist because of his emphasis on a mathematical rigor that bore Bourbaki's stamp, Ruelle served as mediator between applied topologists and physicists. This was by no means an obvious process. While topological modeling practices were widely picked up by specialists in various disciplines, this was done at the expense of substantial alterations and a loss of rigor. As a tradeoff, physicists exploited experiments and numerical simulations to bypass mathematical failures and philosophical pretense.

To make these modeling practices explicit, it has been necessary to select, as representative of particular variations, a few heterogeneous models, all of which constituted conscious and appealing attempts at extending the reach of topological tools and practices (Table 1). Not all of them, however, were promised a brilliant future; with the benefit of hindsight some might appear as rash acts of bravado soon to be harshly dismissed. By using these particular models as signposts, an overview of the evolution and diversification of topological modeling practices will be presented, and an important feature will be emphasized. This was the way topological approaches called into question the dominance of differential equations in modeling. In other words, they challenged the widespread assumption that theoretical accounts of natural phenomena were a matter of writing down, and if possible solving, the right equation. Topological technologies provided a language suited for a description of nature where differential equations were displaced in favor of much less important ontological commitments to fundamental laws.

This story therefore goes counter to the traditional historiography of 'application' in several ways: applied topologists' construction of mathematical frameworks appears as being intimately intertwined with their forging modeling practices applicable to concrete problems; the role of mediators is emphasized; and the specialists' successful 'applications' of topological practices are clearly seen as 'adaptations' betraying initial goals while offering original means of implementing abstract, philosophical undertakings. As a result, differential topology ceased to be just an abstract branch of mathematics to become a reservoir of tools and practices to be used in those cases when equations were hard to come by or solve.

THE ABSTRACT ROOTS OF APPLIED TOPOLOGY

Having been awarded their Ph.D.s in the 1950s, René Thom and Steve Smale have both emphasized the special conditions in which they first approached topology. While the former modestly acknowledged that it was his luck to have joined the field just when 'a river ... flooded the domain,' the later simply wrote that he 'was born into the "Golden Age of Topology"' [Thom, 1983, 21]; [Smale, 1990, 28]. A decade earlier,

TABLE 1 SCHEMATIC REPRESENTATION OF FEATURES THAT BEST ENCAPSULATE OF THE MODELING PRACTICES EXHIBITED BY SEVEN TYPES OF TOPOLOGICAL MODELS, NAMELY, [ZEEMAN 1965; THOM 1969; THOM 1973; ZEEMAN 1973; SMALE 1973; RUELLE AND TAKENS 1971]; AND THE POMEAU-MANNEVILLE INTERMITTENT MODEL FOR THE ONSET OF TURBULENCE.

	ZEEMAN TOPOLOGY OF THE BRAIN	THOM BIOLOGICAL MODELS	THOM LINGUISTICS	ZEEMAN NERVE IMPULSE & HEARTBEAT	SMALE ECONOMICS	RUELLE-TAKENS	POMEAU-MANNEVILLE
SUBSTRATE, THE EXPERIMENTAL "CHAOS"	NEURON ELECTROCHEMISTRY.	PHYSICS AND CHEMISTRY.	NONE (PHONETICS?)	BIOCHEMISTRY.	COMMODITY SPACE IN A PURE EXCHANGE ECONOMY.	NAVIER-STOKES EQUATION, TURBULENCE.	NAVIER-STOKES EQUATION, TURBULENCE.
DIFFERENTIAL EQUATIONS	CONTAIN TOO MUCH INFORMATION.	NOT IMPORTANT.	JUST A WAY OF SPEAKING.	GOAL OF THE MODEL.	NONE.	NOT SOLUBLE, AN APPROXIMATION?	NOT SOLUBLE, AN APPROXIMATION?
IDEALIZATION	THOUGHT CUBE.	GRADIENT FUNCTIONS.	ELEMENTARY SENTENCES.	QUALITIES. "SIMPLEST" MODELS.	TAKEN FROM DEBREU.	LOW-DIMENSIONAL ATTRACTORS.	ITERATED FUNCTION.
VARIABLES IN THE MODEL	INDIVIDUAL STATE OF NEURONS (0 OR 1).	HARDLY ANY, JUST A WAY OF SPEAKING.	HARDLY ANY. JUST A WAY OF SPEAKING.	DERIVED FROM SIMPLEST MODEL	TAKEN FROM DEBREU	REDUCTION OF VELOCITY VECTOR FIELD. NOT WELL IDENTIFIED.	REDUCTION OF VELOCITY VECTOR FIELD. NOT WELL IDENTIFIED.
USE OF TOPOLOGY	ORIGINAL: TOLERANCE SPACE.	ORIGINAL BUT NOT FORMAL: ATTRACTOR, CATASTROPHE, UNIVERSAL UNFOLDING.	ORIGINAL BUT NOT FORMAL: SAME PLUS ARCHETYPAL MORPHOLOGY.	"THOM'S THEOREM."	ORIGINAL. NEW THEOREMS.	GENERICITY, STRUCTURAL STABILITY. STRANGE ATTRACTORS.	LANGUAGE, NOT RIGOROUS.
EXPERIMENTS.	NOT POSSIBLE, BUT DESIRABLE.	IMPOSSIBLE IN PRINCIPLE.	IMPOSSIBLE IN PRINCIPLE.	DESIRABLE. ENCOURAGED. DECISIVE.	NONE.	NOT CONSIDERED AT FIRST, BUT POSSIBLE AND DECISIVE.	CONCEIVED WITH EXPERIMENTS IN MIND.
USE OF EXISTING LITERATURE.	MINIMAL.	WADDINGTON DELBRÜCK. SOURCE OF INFORMAL CONCEPTS.	FORMAL LINGUISTICS. SOURCE OF INFORMAL CONCEPTS.	IMPORTANT. PERSONAL CONTACTS.	IMPORTANT: DEBREU. PERSONAL CONTACT.	LANDAU, LERAY, HOPF. APPENDIX = AFTERTHOUGHT	WELL KNOWN AND USED.
WHAT DOES THE MODEL EXPLAIN?	WHY DO WE SEE.	MORPHOGENESIS OF BIOLOGICAL FORMS.	UNIVERSALITY OF GRAMMATICAL CATEGORIES.	DYNAMIC COMPLEXITY OF PHENOMENA.	DYNAMIC APPROACH OF PARETO OPTIMUM.	NATURE OF TURBULENCE.	"BURSTS" IN THE ONSET OF TURBULENCE.

G.D. Birkhoff [1942] confidently suggested that topology would soon 'greatly increase in scope and significance.' Having witnessed 'a kind of culmination in the abstract phase' of the history of the field that had performed the 'essential task of giving topological ideas their appropriate abstract setting,' he believed that the future would show the usefulness of topology for dynamics.

The field indeed skyrocketed in the postwar years, but in a direction hardly anticipated by Birkhoff. Following MacLane and Eilenberg's 1945 axiomatization of homology theory, extraordinary technical developments in algebraic topology ignited an 'explosion'—an internal explosion, but also an external one, extending the reach of topology 'by the creation of methods applicable to new domains of the concrete' [Lichnérowicz, 1955]. The main motor driving the explosion was the introduction of powerful algebraic tools for the study of topology and geometry. While vast fields of research opened up, extensions to concrete problems followed traditional patterns, application remaining outside the province of leading mathematicians. As put by Lichnérowicz, the next great 'explosion' took place in the direction of algebraic geometry: 'I dare say it was detopologized and partly transmuted into a purely abstract geometry.' Still, the systematic algebraic attack by the new generation was resented by some old-guard mathematicians who contended that 'while we wrote algebraic GEOMETRY they make it ALGEBRAIC geometry with all that it implies' [Lefschetz, 1986, 3]. Similarly, Marston Morse criticized algebraic supremacy and disdain for applications: 'Forever the foundations and never the Cathedral' [quoted in Bott, 1980, 908].

For those who would become applied topologists, these older traditions (found in the work of Whitney, Morse, and later Lefschetz, Birkhoff, and Poincaré) provided resources to complement the algebraic dominance. But, in the context in which they produced their first results, Smale, Zeeman, and Thom approached topological problems from purely internal motives, without paying attention to applications. In so doing, they achieved great successes acknowledged by their peers. In 1960, Smale proved the higher-dimensional Poincaré conjecture to which Zeeman contributed a different proof: for this the former was awarded a Fields Medal in 1966. Earlier, in 1958, Thom had already received the Medal for his work on cobordism. However, albeit educated by Bourbaki mathematicians, Thom, like Smale and Zeeman for that matter, tackled topological problems from a geometric, more than algebraic, standpoint. When topology was in a 'stage of vigorous ... algebraicization,' Heinz Hopf wrote when presenting Thom his Medal, lurked the danger of 'totally ignoring the geometrical content of topological problems.'

> In regard to this danger, I find that Thom's accomplishments have something that is extraordinarily encouraging and pleasing. While Thom masters and naturally uses modern mathematical methods and while he sees the algebraic side of his problems, his fundamental ideas ... are of a perfectly geometric-*anschaulich* nature [Hopf 1960, lxiii–lxiv].

By the late 1950s, Thom and Smale were moving away from their earlier concerns and embarking on ambitious programs. Reviving interest in topological notions such as genericity and structural stability, they endeavored to construct global classification schemes for familiar entities: real functions and differential equations, respectively. At the end of decade, Thom's project had partly been completed with the list of the seven elementary catastrophes [Thom, 1975], while Smale's, despite bold conjectures and great advances, was facing grave difficulties [Smale, 1969/70]. As a result of constant interactions, however, both were then definitively turning away from abstract pursuits and engaging in problems of modeling.[5]

THE EMERGENCE OF 'TOPOLOGICAL MODELS'

In 1969, René Thom proposed noteworthy 'topological models,' with photographs of caustics and plaster models, in the journal *Topology*. Concerned with the problem of explaining 'the stability ... of the global spatio-temporal structure *in terms of the organization of the structure itself*,' he thought of his models in terms of a 'striking analogy between this fundamental problem of theoretical Biology and the main problem considered by the mathematical theory of Topology, which is to reconstruct a global form ... out of all its local properties' [Thom, 1969, 313]. For him, it was not only the tools of topology which could be applied to biology, but the very nature of both endeavors which suggested that involved interactions could be productive for both disciplines.

Obviously, this dramatic extension of the meaning of *model* is but a small part of the larger story of mathematical modeling, which to a large extent remains to be told by historians of science.[6] The idea came to Thom from his encounter with an original topologist, Christopher Zeeman, who had already spoken of *topological models* in an article by which Thom had been 'singularly fascinated.' According to von Neumann, 'the sciences do not try to explain, they hardly even try to interpret, they mainly make models' [quoted in Dahan Dalmedico, 1996, 179]. Contrary to this pragmatism, Thom and Zeeman believed their models to be more than descriptions or computational techniques, but *forms of explanation.*

Topology of the Brain: The Irrelevance of Differential Equations

The intelligibility of the experimental 'chaos' depended on mathematics' very 'power to simplify and explain,' Zeeman [1964a] claimed in accordance with the credo of a mathematician raised in Bourbaki's heydays. More than any other topologist of his time, however, he seriously worked out models, in particular 'to try and explain the relationship between mind and brain' [Zeeman, 1965, 277]. His principal goal was to provide explanations for such problems as: How does the brain perceive an image from nervous impulses? To tackle this question Zeeman used algebraic topology, because it was *'well adapted to ignore local variations and capture global properties'* [277]. In biophysics, one usually started with electrochemical properties expressed as measurable quantities, then one derived and if possible solved differential equations which they obeyed. But, as Zeeman explained, models built with such equations 'frequently give the impression of being too detailed neurologically and oversimplified in the large.'

> To pursue an analogy, think of blowing up a balloon into a funny shape. The local behaviour of the rubber material is described accurately by differential equations, but globally the equations become either very complicated or else inadequate, whilst the topology remains very simple [287].

Granted that mind mechanisms relied on interactions of neurons, algebraic topology, like a net catching global, relevant features in a sea of local, irrelevant complexity, provided original tools enhancing the understanding of the workings of the brain.

Zeeman introduced 'a simple model of the brain' by organizing neurons into a cube of 10 billion dimensions called 'the thought cube.' To work with such a complicated construct, topological techniques were called to the rescue. He devised the notion of *tolerance spaces,* accounting for pairs of distinct 'states of neural activity ... so close that they "feel the same" and consequently give rise to the "same thought"'

[282]. Modeling the brain, he was led to original and 'precise' mathematics.

Some features however made Zeeman's model unattractive to biologists. Even if he claimed that it was 'based on the well known anatomical structure of the brain,' his model was too crude for them [291–2]. It was moreover difficult to test in the laboratory.

> The results are expressed in geometrical language, and are *qualitative* rather than *quantitative*. This means that so far the theory ... has attempted to *explain* phenomena rather than *predict* the measurements that experiment would obtain [277].

As is well known, while Thom mathematically defined *catastrophes*, Zeeman introduced the catchy phrase of *catastrophe theory*.[7] Extending its range, he audaciously developed scores of famous (and infamous) models [Zeeman, 1977]. Far from resulting from his mere reading of an 'underground' copy of Thom's manuscript, Zeeman's interest stemmed from constant contacts between the two men, thanks to which emerged a modeling practice that topologists could call their own.

By the same token, exploiting topological tools to make sense of the brain without paying attention to biochemical processes, Zeeman invented modeling procedures exhibiting many of the features which became trademarks of catastrophe theory. In his model, the substrate (the neurons) was replaced by an idealization with crude dynamics: the thought cube. The model variables were more or less realistic, and differential equations deemed uninteresting because containing more local information than needed. Topological technologies could filter out irrelevant information in favor of meaningful wholes. Experimental confirmation would be difficult, but perhaps not impossible; the goal of modeling was to explain rather than to predict. These features, which would all explicitly or implicitly inform later attempts at forging topological modeling practices, were more systematically expounded by Thom.

Topological Models in Biology: Dynamics without Equations

From 1966 onward, Thom publicly embarked on the ambitious adventure of catastrophe theory. Although mentioning 'topological models,' he casually used the term to refer to abstract mathematical constructs. He distinguished two kinds of models. On the one hand, he called *differential models* those given by a dynamical system: $dx_i/dt = X_i(x,\tau,t)$, where

τ represented external parameters. Classic and common, these models suffered from two well-known problems: X was rarely obvious to find especially in the nonphysical sciences, and solutions were often impossible to compute formally.

On the other hand, Thom defined a new type of mathematical model. Inspired by British embryologist C.H. Waddington, he provided a mathematical definition for his informal notion of *chreods*, or stable pathways of development. For Thom, an 'experimental morphology' defined a 'catastrophe set' of discontinuities dividing the substrate into regions controlled by chreods. This decomposition, he wrote, could 'be considered as *a kind of generalized m-dimensional language*: I propose to call it a "*semantic model*".' He identified two kinds of problems to be considered given such a model:

1) To classify all types of chreods, and to understand the nature of the dynamic processes which insure their stability.

2) ... Generally, there are some associations of chreods which appear more frequently than others. One may speak, in that case, of a *multi-dimensional syntax* directing the semantic model. The problem is then to describe this syntax [Thom, 1969, 321–322].

Like Zeeman's, these were problems well suited for topology. Thom referred to Poincaré's qualitative dynamics, and more crucially to Andronov and Pontrjagin's notion of *structural stability*. Recent developments due to Smale [1967] having weakened its usefulness, Thom defined *attractors* and restricted his study to systems which had a finite set of structurally stable attractors. The decomposition of the substrate in *basins of attractors* characterized 'entirely the dynamical behaviour of the system.' Thus were topological tools brought to bear on the 'semantic' problems raised by Thom. 'In such a model, the fundamental phenomenon to be studied is the destruction of a structurally stable attractor by variation of the vector field' [Thom, 1969, 323]. To use the tools of topological dynamics, he assumed that, like for traditional models, natural processes were described by vector fields. Inspired by Andronov, Thom allowed the equation to vary, and instead of studying solutions, he focused on global topological features. But, contrary to Andronov, the ultimate validity of idealized differential equations was irrelevant to Thom's concerns.[8]

Thom's models were no more amenable to experimental control than Zeeman's. His method merely provided an '*art of models*.' Faced with the 'need ... to classify the [empirical] data,' modelers could use his proposals as a substitute to 'pure chance and lucky guess' and achieve 'a *qualitative understanding* of the process studied.'[9] However, the most important consequence of Thom's proposals lay not in his strenuous efforts at constructing a philosophy of science, but rather in the introduction of modeling practices taken up by some topologists, as the following examination of three models shows.

APPLIED TOPOLOGY? MODELING PRACTICES AT BAHIA, 1971

In 1971, the University of Bahia, Brazil, held a symposium on dynamical systems that provided an occasion for applied topologists to proclaim vocally that one could put 'under the sway of the mathematician a vast array of phenomena thus far considered beyond his reach' [Peixoto, 1973, xiii]. Although most participants dealt with pure mathematics, Smale, Thom, and Zeeman presented models based on differential topology. Thom's article aimed at providing a geometric interpretation of language and its grammatical categories. In his paper, Zeeman intended to 'abstract the main dynamical qualities of the heartbeat and nerve impulse, and then build the simplest mathematical model with these qualities.' By far the most mathematically involved, Smale's contribution attempted to put time back into the equations of equilibrium economics. A comparison of these three articles makes points of convergence and divergence appear explicitly.

The Bahia Models: Topological Modeling at Work

Of the three, Thom's [1973] article was the most verbose. Were it not for its mathematical metaphors, it would have seemed closer to a philosophical paper. As it is difficult to consider this paper as an actual attempt at mathematical modeling, its legacy still being a matter of important dispute, it will not be discussed in detail. However, one should note that, while often claiming that his practice was independent of the substrate, Thom made clear that it was rather *without* substrate. In his purely topological modeling, no variable nor equation was involved. In the absence of precise substrata lay the most extreme difference between Thom's practice and that of other applied topologists.

Catastrophe Theory à la Zeeman: Deriving Equations from Topology

A perfect exemplar of the modeling practice most often associated with catastrophe theory, Zeeman's Bahia paper was an *exposé* of the way in which this theory provided 'not only a better conceptual understanding, ... but also explicit equations for testing experimentally.' He took seriously the idea that global analysis modeled *qualities*. For the heartbeat and nerve impulse, he contended, three main dynamical qualities were displayed: '(I) stable equilibrium; (II) threshold, for triggering an action; (III) return to equilibrium' [Zeeman, 1973, 684]. Clearly, these were not simple physiological descriptions, but idealizations of biochemical processes, crucially informed by the types of behavior best suited for qualitative dynamics. Equilibrium meant 'look for attractors', and threshold that a catastrophe was involved. Starting from the three qualities, Zeeman derived the simplest mathematical model displaying such features:

$$\varepsilon \dot{x} = -(x^3 - x + b), \quad \dot{b} = x - x_0,$$

where dots denoted derivation with respect to time [699].

At this point catastrophe theory explicitly entered Zeeman's modeling practice. Using general arguments for deriving his 'simplest' model, he needed 'Thom's deep uniqueness theorem' to argue that these models were indeed the right ones. 'Let us pause for a moment to consider what we are doing,' Zeeman wrote:

> The topologist regards polynomials as rather special, and tends to turn his nose up at so crude a criterion of simplicity ... So perhaps we ought to consider all possible surfaces [models]. Now comes the truly astonishing fact: when we do consider all surfaces, not only is this particular surface the simplest example, but in a certain sense it is ... the unique example. Herein lies the punch of the deep and beautiful catastrophe theory [704].

A common strategy among applied topologists consisted in substituting mathematical justifications for metaphysical assumptions, namely in this case, the postulate of simplicity.[10] As Zeeman interpreted it, Thom's theorem implied that if the dynamics were postulated to depend on a generic potential, then the simplest model represented 'the most complicated thing that could happen locally ... The theorem is the key mathematical fact behind our whole approach' [706]. His actual use of topology was therefore limited but crucial.

Next, by confronting his model with observations, experiments, and empirical models, Zeeman wished to interpret its variables in terms of physical parameters, but very loosely: one variable, for example, being identified as 'chemical control ... possibly membrane potential' [712–3]. In striking contrast to his previous attitude when modeling the brain, differential equations remained the *goal* of Zeeman's modeling practice, using topological qualities to get them, rather than observations of empirical quantities. Using radical means to derive applied mathematicians' classical objects, he laid himself open to the harshest critiques [esp. Zahler and Sussmann 1977].

Topologizing the Mathematical: The Case of Economics

Before his 1970 visit to the IHÉS, Steve Smale had scarcely dared to publish articles devoted to mathematical modeling. Inspired by Thom, and wondering 'whether I should explicitly direct my work toward socially-positive goals,' he turned to applications [Smale, 1972, 3]. At Bahia, albeit providing economic justifications, he nevertheless expressed his problem in uncompromising mathematical terms:

> One is given real differentiable functions $u_i: W \to R$ defined on a manifold W, say $i = 1, \ldots, m$. What is the nature of curves $\varphi: R \to W$ with the derivative $(d/dt)(u_i o\varphi)(t)$ positive for all i, t [Smale, 1973, 532].

Contrary to Thom and Zeeman, Smale chose to rely on a rigorous, axiomatized treatment—provided by [Debreu, 1959]—of the domain he was dealing with. Entering a field already using sophisticated mathematical techniques, which he translated into a topological framework, he was making them 'attractive to the modern mathematician, ... brought up in the purist, Bourbakist style of education' [Smale, 1980, 100]. Economics provided him with interesting mathematical problems where he could use dynamical systems theory. Thus, as opposed to Thom and Zeeman who did not prove anything, the main body of Smale's paper was a list of theorems proved for the sake of pure mathematics. Building on an already well-mathematized discipline and not turning his back on the specialists' previous work, Smale promoted modeling practices that could be more easily adapted to established practices.[11]

Since Smale made no actual attempt at building economic models, his Bahia article provides a poor example of his modeling practice. Like Thom and Zeeman, however, he saw in topology a reservoir of techniques for the modeling of phenomena in biology, mechanics, electronics, etc. Abstracting topological features from known models, his modeling practice consisted in topologizing extent models and accounting for their dynamic behavior. This led him to specify assumptions hidden in models, understand some of their consequences, and modify hypotheses when needed. Later, Smale promptly recognized the importance of the Ruelle–Takens model, and crucially mediated between mathematicians and physicists.

A New Status for Differential Equations?

'In a large number of cases, a kinship of structures in extremely diverse domains has been noticed. This allows today's mathematician, without becoming an expert in a branch that is not the object of his study, to understand its essential [features].' In 1968, this was how director Léon Motchane explained the fruitfulness with which many mathematicians orbiting the IHÉS were tackling concrete problems. Their effort was systematic and concerted. Designed to promote the advancement of 'fundamental research' but sponsored by industry, the Institut of Bures-sur-Yvette provided a fertile ground for the development of topological modeling practices. Beyond applied topologists' personal motives, one must note how well this undertaking fitted with the ideology of research promoted by Motchane. Indeed, by insisting on the independence of research while emphasizing potential concrete benefits for industry, he favored abstract frameworks which paid attention to the outside world. Vehemently autonomous with respect to traditional practices in the sciences, applied topologists developed models intended as intelligible explanations rather than computing procedures for action [Aubin, 1998b]. By contingency, the IHÉS provided an impulse and some of the means for the emergence of a community of mathematicians willing to adapt their practice to the concrete. By 1971, the most eminent among them, Smale, Thom, and Zeeman each believed that, armed with their topological background, they could build *dynamical* models for all sorts of sciences, while keeping them at arm's length. Dynamics however had an unusual sense: it emphasized changes, but not forces responsible for change.

Clearly, it was not a unified modeling practice that emerged from applied topologists' work. Their respective attitudes differed markedly with respect to the sciences they intruded, with respect to extant models, experimental results, or literature. At first glance, the most important difference concerned the goal of modeling and the place they assigned to differential equations. As mentioned, a most striking aspect of Zeeman's modeling practice lay in the Copernican reversal he advocated. Instead of looking for equations obeyed by observable variables, he derived them through topological considerations; only as a second step did he identify his variables with observations. Still, the derivation of equations remained the goal of his practice. On the contrary, Thom's and Smale's papers made little use of substrata, variables, or equations. Consequently, by proposing mathematical descriptions in the absence of equations, the latter two opposed more drastically traditional practices than the former who merely proposed new means to achieve standard objectives.

To emphasize this opposition too strongly nonetheless is deceptive, since all shared a belief that topology served to constrain possible representations of phenomena. For Thom, more than a mere reservoir of metaphors, mathematics provided a way of thinking. For Smale, the mathematician proved theorems which disturbed the established consensus. Of course, Zeeman was the most explicit in expressing this belief, attracting the obvious criticism that he let mathematics dictate what reality should be like. As a consequence of this shared belief, applied topologists challenged the traditional role ascribed to differential equations in modeling. Although a dynamical substrate in terms of vector fields was always assumed, they paid little attention to usual ways of deriving equations. While Zeeman's Bahia models proposed new ways of achieving standard goals, other models, including his own of the brain, suggested methods not only to analyze global features without solving equations, but more importantly to provide explanations of phenomena when underlying dynamical equations were not precisely known. This last suggestion would be widely taken up by physicists.

TOPOLOGICAL PHYSICS? MODELING PRACTICES OF CHAOS

Remarkably, applied topologists' efforts, which if restricted to them only might have remained rather sterile, were expanded upon by some physicists. In this process, the IHÉS served as a crucial mediator. Its physics section had always been characterized by a focus on rigor, and some noticeable instances of interactions with mathematicians had already taken place [Zeeman, 1964b]; [Froissart, 1966]. That Ruelle and Takens's theory of turbulence relied on *concepts* introduced by Smale and Thom is clear simply by looking at their joint paper. But Ruelle's *modeling practice* also was shaped by his situation at the IHÉS, which was not only an opportunity for him to learn about recent theories, but showed him how to exploit them for model-building, as a comparison with the Bahia models will make clear.

Ruelle-Takens: A New Definition for Turbulence

Russian physicist Lev Landau [1944] and German mathematician Eberhard Hopf [1942; 1948] contended that when a fluid was submitted to increasing external stress, it went through a series of bifurcations, where appearing frequencies gave rise to quasiperiodic motions that seemed increasingly turbulent. Being in contact with applied topologists, Ruelle and Takens suggested, but did not show rigorously, that this bifurcation sequence stopped after the manifestation of three different modes because a 'strange attractor' appeared in a 'generic' manner. As the title of their paper indicated, what was at stake was the very nature of turbulence. *Aperiodic*—not quasiperiodic—motion was the definition they offered for it.

There was a striking feature in Ruelle and Takens's article which brought out the new status they assigned to differential equations. Indeed, they did not feel the need to write down the Navier-Stokes equations (NSE), the fundamental law for fluid flows, more explicitly than:

$$\frac{dv}{dt} = X_\mu(v).$$

'For our present purposes,' they added, 'it is not necessary to specify further ... X_μ' [Ruelle and Takens, 1971, 168]. A unique parameter depending on physical characteristics, μ represented external stress on the fluid (e.g. the Reynolds or Rayleigh number). The determination of critical values at which motion became turbulent had motivated studies for almost a century [Aubin, 1998a]. Not interested in particular critical values, Ruelle and Takens only looked at general features of motion as the parameter increased.

When $\mu = 0$, the fluid tended to rest; for small μ, it tended toward a stationary motion in which the velocity field remained constant. At a critical value μ_1, the system went though a Hopf bifurcation: the velocity field started to oscillate at a given frequency ω_1. In phase space, while for stationary flows a fixed point existed which was an attractor of the system, when the oscillatory mode appeared this point exploded into a closed curve. At a further critical value μ_2, a second bifurcation gave rise to a frequency ω_2, and so on. When μ increased sufficiently, 'the fluid motion becomes very complicated, irregular and chaotic, we have turbulence' [Ruelle and Takens, 1971, 168]. But how to describe this 'chaotic' flow? Based on their topological knowledge, Ruelle and Takens claimed that since the quasiperiodic motion was not generic for general dissipative systems, it had no chance of being observed. One had to look elsewhere for a 'mathematical explanation' of turbulence [Ruelle, 1972].

Topology in the Ruelle–Takens Model

Compared to the Bahia models, Ruelle and Takens's was closest to Smale's. Since fluid mechanics is one of the oldest mathematized disciplines, this might have been expected. Given by the velocity field of fluid flows and NSE, Ruelle and Takens's substrate was quite uncontroversial. But as in the above, identification of model variables with observable ones was not immediate, for the phase space of the velocity field was infinite-dimensional, a problem as far as dynamical systems theory was concerned. Thus, while the substratum was straightforward, the pertinence of reducing it to a low-dimensional manifold was not obvious.

When, using an example of Smale's (the famous horseshoe), they defined strange attractors and argued for their genericity, Ruelle and Takens made one crucial conceptual innovation in dynamical systems theory. But more importantly, they adapted a crucial part of applied topologists' modeling practices. Revived by Thom, the notion of genericity was the core of applied topologists' work both in mathematics and modeling. Still, Thom [1975, 35] was aware of its slipperiness, and Smale [1967, 748] made a welcome clarification when he restricted its use to properties of topological spaces rather than points. The use of genericity remained an art difficult to make rigorous. One of the weakest in Zeeman's use of catastrophe theory for modeling, this point remained a tricky matter for physicists, as we shall see.

Noting that Smale's horseshoe was stable under small perturbations, Ruelle and Takens concluded that 'the existence of such a "strange" attractor therefore is not a non-generic pathology' [Ruelle and Takens, 1971, 171]. By indicating that in the neighborhood of quasiperiodic motions in more than 3 dimensions a generic set of such attractors should exist, they felt entitled to pronounce that quasiperiodic motions could not physically occur. But one should note the tentativeness of their language:

> For $\mu > 0$ we know *very little* about the vector field X_μ. Therefore it is *reasonable* to study generic deformations from the situation at $\mu = 0$. In other words we shall ignore possibilities of deformations which are *in some sense* exceptional ... It *appears* ... that a three-dimensional viscous fluid conforms to the pattern of generic behavior which we discuss [Ruelle and Takens, 1971, 168, my emphasis].

By adapting topological modeling practices to the turbulence problem, Ruelle and Takens were in a situation similar to Smale's. A huge literature could be tapped into in order to argue for the plausibility of their model, but this had to be done in a new framework, which, in traditional views, still lacked a solid mathematical basis. Significantly, most references to this literature were included in an appendix written later and a note published in a following volume of the journal. Ruelle and Takens had come up with their model while trying to adapt topological practices to the turbulence problem, but still tried to root it in a mathematized tradition. More significantly still, the process by which their paper became a seminal one was hardly straightforward. The resistance they encountered was overcome only after specialists in the mathematical study of fluid dynamics could incorporate topological practices into their analytical framework, and after experimental results indicated that the Ruelle–Takens model accounted for observation more accurately than Hopf's and Landau's.

Topological Modeling by Physicists: The Example of Intermittence

Ruelle and Takens's proposal revolutionized the way theoretical physicists could deal with differential equations. This alternative modeling practice displaced earlier emphases put on specific models of nature in order to tackle classes of models. That they made no use of the specific form of NSE was symptomatic. A hydrodynamicist later wrote that this was 'a point of *philosophy*:' without arguing about the relevance of NSE, 'I ought to confess we can forget about them here' [Velarde, 1981, 210]. Without resolving the conundrum of the relationship between fundamental laws and observation, this practice made models cheap and dispensable, and focused rather on essential topological features of observed behaviors assimilated to structural, yet dynamical, characteristics of classes of models. In short, physicists were now allowed to stop looking for specific representations of nature, and acquired means of studying the consequences of the mode of representation itself.

Characteristic of applied topologists' modeling practices, this approach would exert an important appeal in the following decade. In the last section of this chapter, the discussion of another model for the onset of turbulence will underscore the fact that the extension of topological modeling practices to physics was done at the expense of the rigor that applied topologists wished to retain and in interaction with the results of experimentation and simulation. The model selected for discussion is known as the Pomeau-Manneville scenario. Developed in the late 1970s, it is representative of the tradeoffs accompanying the adaption of topological modeling practices to physics and other mathematized disciplines.

One of the rare theoretical physicists who paid early attention to the Ruelle–Takens model, Yves Pomeau, from the French Commissariat à l'énergie atomique (CEA), had arrived, by 1976, at a coherent picture of chaotic behavior. Inspired by Thom, Smale, and Ruelle, he adopted a highly mathematical language. But, Pomeau understood that merely to propose 'revolutionary' models was not enough. One needed to get one's hands dirty, compare theory with experiment, collaborate with people from previously separate disciplines, and forge a common language. In 1976, two CEA experimental physicists, Pierre Bergé and Monique Dubois had built a Rayleigh–Bénard system—a cellular cavity filled with fluid where a gradient of temperature induced convection. A singular phenomenon caught their attention: 'the velocity amplitude shows intermittent periodic oscillations versus time' [Bergé and Dubois, 1976]. Studying a computer model, Pomeau noticed similar intermittent flashes. Together with Paul Manneville, a young CEA theoretician, they came up with a series of bifurcations, different from the Ruelle–Takens scheme, leading to the onset of turbulence.

Pomeau and his collaborators faced similar data on systems that *a priori* had little to do with one another: time series exhibiting regular periodic behavior randomly and abruptly disrupted by erratic bursts. In trying to find a common cause for these behaviors, a dynamical systems approach proved useful. Disputing the claim that the Ruelle–Takens model was the only way to turbulence, they contended: 'theories based on genericity arguments [are] sufficiently versatile to allow for different possible transitions' [Bergé & *al.*, 1980, L341]. They considered a Poincaré map of a dynamical model $y_{n+1} = f(y_n, r)$, where for the control parameter r slightly below a critical value r_T, the curve f had two intersection points with the diagonal, while 'for $r > r_T$ the curve is lifted up and no longer crosses the [diagonal] so that a "channel" appears between them' (Fig. 1) [Pomeau and Manneville, 1980, 190]. This simple picture, they contended, was 'displaying generic features susceptible of *explaining* the experimental observations' [Bergé & *al.*, 1980, L343]. When the system passed through the channel, its behavior seemed regular. Leaving the channel, it explored chaotically other regions of phase space until it was again trapped into such a channel.

Pomeau and Manneville's modeling practice lay in between that of applied topologists and the traditional practice of physicists based on explicit dynamical systems. Like applied topologists, they started with

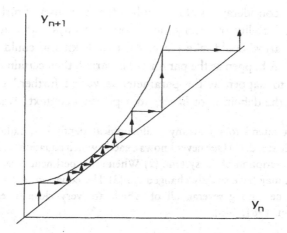

FIGURE 1 POINCARÉ MAP FROM THE LORENZ MODEL, WITH THE PARAMETER r SLIGHTLY ABOVE THE CRITICAL TURBULENT VALUE r_T. CORRESPONDING TO THE REGULAR FLUID MOTION, THE SLOW DRIFT THROUGH THE CHANNEL IS QUITE IMPERCEPTIBLE ON EXPERIMENTAL OR NUMERICAL TIME RECORDS. REDRAWN FROM [POMEAU AND MANNEVILLE 1980, FIG. 4]

features to be explained topologically, rather than by relying on precise differential equations; bifurcations again were interpreted as sources for change. Like them, they assumed that an 'unknown realistic dynamical system' existed, which 'should share some generic properties with already well studied models' [Bergé & *al.*, 1980, L342]. Properties of *other* systems provided an explanation for experimental behaviors stemming from a similar, but unknown, equation. Unlike applied topologists, however, they neglected to ground their model in rigorous mathematics. 'The reader must be warned that the discussion is made *in physical terms*. No proof is given' [Pomeau and Manneville, 1979, 331]. Their use of genericity arguments was especially loose. For them, a generic property was simply one often encountered when varying parameters in numerical simulations or laboratory experiments.

In their footsteps, Jean–Pierre Eckmann, from Geneva, reviewed the physicist's attitude with regard to topological methods, which constituted a '*language* for describing deterministic evolution equations.'[12] Although a rigorous classification was far from achieved, *experiments* could guide physicists to relevant bifurcations. Topology did not constrain reality. On the contrary, reality prescribed the topological notions that could be usefully mobilized for the understanding of phenomena. While Ruelle and Takens had argued that topological considerations alone implied that the onset of turbulence was due to the stability of strange attractors, the physicists countered that many 'scenarios' could take place. All that Eckmann could say was 'if certain things happen as the parameter is varied, then certain other things are likely to happen as the parameter is varied further.' Linked with genericity, the definition of *likely*, 'in a physical context,' was tricky.

> I do not intend to go to any philosophical depth but, rather, take a pragmatic stand. (1) One never knows exactly which equation ... is relevant for the description of the system. (2) When an experiment is repeated, the equation may have slightly changed ... (3) The equation under investigation is one among several, all of which are very close to each other [Eckmann, 1981, 646].

Inspired by Thom and Ruelle, this type of mathematization was 'answering new types of questions which are more or less independent of detailed knowledge of the dynamics of any given physical system' [Collet & *al.*, 1981, 1]. Its validity therefore did not depend on fundamental laws. Out of a mathematical failure, a modeling practice was built by physicists for physicists. While mathematicians had not succeeded in classifying generic bifurcations, Eckmann's scheme provided an *experiment-based* topology of dynamical systems.

CONCLUSION

In 1977, in the midst of raging controversies, René Thom granted that catastrophe 'theory could have the same fate as cybernetics or information theory ... a considerable sociological craze, which short of effective success ends up by falling on the distaff side' [Thom, 1977a, 681]. And to a large extent it did. To explain the strong rejection he encountered, Thom mustered sociological arguments. 'It was a corporate reaction: the whole community of applied mathematicians rose up against the theory' [Thom 1992, 45]. He went as far as suggesting that 'the interests of the computer industry [were] perhaps not entirely foreign to this affair' [Thom, 1977b, 196]. Fundamentally devoted to computation, they just had to oppose an essentially qualitative enterprise! It is not uncommon for scientists to ascribe the rejection of endeared theories to social factors. But, if we accept Thom's interpretation, how are we to account for the fact that, although sharing many practical features with catastrophe theory, the qualitative study of chaotic systems was rather welcomed by applied mathematicians and the computer industry alike?

In fact, Thom perceptively hinted at the different ways in which topological modeling could be put to use when he suggested that catastrophe theory followed a 'twofold way.'

> Either, starting from known scientific quantitative laws, ... you insert the CT formalism; ... this is the 'physical' way. Or, starting from a poorly understood experimental morphology, one postulates 'a priori' the validity of CT formalism, and one tries to reconstruct the underlying dynamics which generates this morphology: this is the 'metaphysical' way [Thom, 1976, 235].

Ridiculed by critics, this alternative amounted, in their view, to a choice between triviality and arbitrariness, if not nonsense. Using Thom's dichotomy, the models discussed above can be seen as treading the 'metaphysical' way in Thom's and Zeeman's cases, and following the 'physical' way for the others. The Ruelle–Takens model aptly showed that this latter way by no means implied triviality. Topological modeling practices provided means to account for behaviors without solving equations and could even be exploited to exhibit hitherto unsuspected phenomena.

As suggested by the case of fluid mechanics, corporate reactions were not directed against the goal of the modeling defended by applied topologists, so much as against its means. Even if especially exciting because they allowed new kinds of questions to be tackled, topological tools, people objected, had no right to constrain reality, at least for as long as the general mathematical framework was not completely proved rigorously, something that was not likely to happen quickly. Specialists opposed the perception that models could be built while neglecting established traditions, literature, experimental and numerical results, etc. But they acknowledged that tools coming out of applied topologists' work provided bases for a new hybrid language useful in the description—not the constriction—of reality.

By discussing the modeling practices of applied topologists and their adaptation by physicists, this chapter has demonstrated remarkable shifts in the practice of some topologists who imagined that they could extend their skills to the world of models, and in that of some physicists who started to adapt concepts and practices of this heretofore abstruse branch of mathematics, topology. In this process, of course, images were not left untouched. But was the reverse also true? How did global shifts in images of mathematics also contribute to the emergence of alternative practices for topologists and physicists alike? The puzzling issues raised by Dahan Dalmedico [1996] about the hegemony of pure mathematics in 1945–70, and the subsequent feeling that, after decades of Bourbakism, mathematics went back to the concrete [Houzel, 1979], are nicely illustrated by the story of applied topologists. Other case studies are needed to complete the picture.

ACKNOWLEDGMENTS

This paper has benefited from the guidance of M.N. Wise, A. Dahan Dalmedico, and M.S. Mahoney. I thank IHÉS and the Collège de France for allowing me to use their archives. It is a pleasure to acknowledge the kindness with which V.I. Arnol'd, P. Bergé, M. Dubois, J.P. Eckmann, P. Manneville, Y. Pomeau, M. Peixoto, D. Ruelle, and R. Thom answered my questions either personally or in writing. This work was supported by the Social Science and Humanities Research Council (SSHRC) of Canada and the American Philosophical Society.

NOTES

1. Classic histories of topology bear a strong Bourbakist stamp [Dieudonné 1989; 1994; Pont 1974]. Recent studies, devoted to earlier periods, have however tended to dispute the internalist view; see e.g. [Epple 1998; and Feffer 1997].
2. The transparencies of Zeeman's lecture delivered on October 27, 1993 at Berkeley, are reproduced on the World Wide Web: http://www.math.utsa.edu/ecz/l_ht.htm. By no means was he alone in seeing the traditional image of topology as having been radically challenged; see e.g., [Browder 1989; Jaffe and Quinn 1993–1994; Atiyah 1995].
3. [Henderson and Maunders 1969]. Reviewing a paper on 'Hilbert's Sixteenth Problem,' by George Wilson (*Topology*, 17 [1978]: 53–73), H.B. Griffiths also used the expression (with quotation marks). Cf. *Mathematical Review*, item 58#16684.
4. By *modeling practice*, I mean the actual activities in which scientists engage when they build models. They include tacit sets of assumptions, mathematical technologies, and frameworks for interpreting results; see [Aubin 1998a]. On 'theoretical technologies,' which include concepts, tools, and practice, see [Warwick 1992].
5. Applied topologists were sensitive to changes in the social and political status of mathematics. Growing demands for utility and concerns for the political responsibility of mathematicians were among the factors that directed their attention to issues of modeling [Aubin 1998a]. In addition, economic conditions may also have been involved, although as far as I was able to determine this was never mentioned by protagonists. In the U.S. the ever increasing production of new topologists in 1951–1975 was accompanied, in 1967–1974, by a funding cut of nearly 40% for research [Cohn 1986, 38].
6. One source for the emergence of 'a new paradigm for making mathematics useful, mathematical modelling,' has been examined in [Alberts 1994, 280]. For another viewpoint, see [Israel 1996].
7. For histories of catastrophe theory, see [Aubin forthcoming; Ekeland 1988; and Woodcock and Davis 1978].
8. Andronov's philosophical concerns leading to the concept of structural stability are found in the introduction to Andronov et al. [1966]. About Andronov, see [Diner 1992; and Dahan Dalmedico 1995].
9. [Thom 1969], 333–334. For extended discussions of topological models in biology, see [Thom 1975], ch. 9–11.

10. In modeling actual systems, many assumptions and idealizations always are involved. How one motivates them can vary. In some cases, philosophical arguments are used: e.g. nature selects the simplest equation consistent with hypotheses. Zeeman's gist was to justify with a theorem what hitherto had been merely postulated.

11. A deeper analysis of the reception, adaptation, and development of Smale's ideas in economics is called for, as it would provide fascinating insights into the processes by which topological practices entered the social sciences.

12. [Eckmann 1981], 643. Although this has not been attempted here, a linguistic analysis in terms of 'pidgins' and 'creoles' might be enlightening for the study of the history of chaos [Galison 1997].

REFERENCES

Alberts, G. [1994] 'On Connecting Socialism and Mathematics: Dirk Struik, Jan Burgers, and Jan Tinbergen', *Historia Mathematica* 21, 280–305.

Andronov, A.A., Witt, A.A., and Khaikin, S.E., [1966] (1937) *Theory of Oscillators*, trans. F. Immirzi, Oxford, Pergamond; Reading: Addison-Wesley.

Atiyah, M., [1995] 'Reflections on Geometry and Physics', *Surveys in Differential Geometry*, 2, *Proceedings of the Conference on Geometry and Topology Held at Harvard University, April 23–25, 1993, Sponsored by Lehigh University's Journal of Differential Topology*, Boston, International Press, 1–6.

Aubin, D., [1997] 'The Withering Immortality of Nicolas Bourbaki: A Cultural Connector at the Confluence of Mathematics, Structuralism, and the Oulipo', *Science in Context* 10(2), 297–342.

[1998a] *A Cultural History of Catastrophes and Chaos: Around the Institut des Hautes Etudes Scientifiques, France*, Ph.D. dissertation, Princeton University, UMI #9817022.

[1998b] 'Un pacte singulier entre mathématiques et industrie. L'enfance chaotique de l'Institut des Hautes Études Scientifique', *La Recherche*, n° 313, 98–103.

'Forms of Explanations in the Catastrophe Theory of René Thom: Topology, Morphogenesis, and Structuralism', in Wise, M.N., (ed.), *Growing Explanations: Historical Perspecitive on the Sciences of Complexity*, (to appear).

Bergé, P., Dubois, M., [1976] 'Time Dependent Velocity in Rayleigh-Bénard Convection: A Transition to Trubulence', *Optics Communications*, 19, 129–33.

Bergé, P., Dubois M., Manneville, P., Pomeau, Y., [1980] 'Intermittency in Rayleigh-Bénard Convection', *Journal de physique—Lettres*, 41, 341–45.

Birkhoff, G.D., [1942] 'Review of S. Lefschetz, *Algebraic topology* (Providence: AMS, 1942) and G.T. Whyburn, *Analytic Topology* (Providence: AMS, 1942)', *Science*, 96, 581–584; (repr. *Collected Papers*, 3, 839–842).

[1943] 'The Mathematical Nature of Physical Theories', *American Scientist*, 31, 281–310; (repr. *Papers*, 2, 890–919).

Bott, R., [1890] 'Marston Morse and his Mathematical Works', *Bulletin of the American Mathematical Society*, 3, 907–50.

Browder, W., [1989] 'Commentary on Topology', *A Century of Mathematics in America*, Part II, ed. P. Duren, Providence, American Mathematical Society, 347–351.

Cohn, S.F., [1986] 'The Effects of Funding Changes upon the Rate of Knowledge Growth in Algebraic and Differential Topology, 1955–75', *Social Studies of Sciences*, 16, 23–56.

Collet, P., Eckmann, J.P., Koch, H., [1981] 'Period Doubling Bifurcations for Families of Maps on R^n', *Journal of Statistical Physics*, 25, 1–14.

Dahan Dalmedico, A., [1995] 'Le difficile héritage de Henri Poincaré en systèmes dynamiques', *Sonderdruck aus Henri Poincaré: Science et philosophie, Congrès international de Nice, 1994*, Berlin, Akademie; Paris, Blanchard, 13–33.

[1996] 'L'essor des mathématiques appliquées aux Etats-Unis: l'impact de la Seconde Guerre mondiale', *Revue d'histoire des mathématiques*, 2, 149–213.

Debreu, G., [1959] *Theory of Value: An Axiomatic Analysis of Economic Equilibrium*, New Haven, Yale University Press.

Dieudonné, J., [1989] *A History of Algebraic and Differential Topology, 1900–1960*, Boston and Basel, Birkhäuser.

[1994] 'Une brève histoire de la topologie', in Pier, J.P., (ed.), *Development of Mathematics, 1900–1950*, Boston and Basel, Birkhäuser, 35–155.

Diner, S., [1992] 'Les voies du chaos déterministe dans l'école russe', in Dahan Dalmedico, A., Chabert, J-L., Chemla, K., (eds.), *Chaos et déterminisme*, Paris, Seuil, 331–70.

Eckmann, J-P., [1981] 'Roads to Turbulence in Dissipative Dynamical Systems', *Reviews of Modern Physics*, 53, 643–654.

Ekeland, I., [1988] *Calculus and the Unexpected*, Chicago, University of Chicago Press.

Epple, M., [1998] 'Topology, Matter, and Space I: Topological Notions in 19th-Century Natural Philosophy', *Archives for History of Exact Sciences*, 52, 297–392.

Feffer, L.B., [1997] 'Mathematical Physics and the Planning of American Mathematics: Ideology and Institutions', *Historia Mathematica*, 24, 66–85.

Froissart, M., [1966] 'Applications of Algebraic Topology to Physics', in Goodman, R., Segal, I., (eds.), *Mathematical Theory to Elementary Particles: Proceedings of the Conference Held at Dedham, Massachusetts, 1965*, Cambridge, MIT Press., 19–28.

Galison, P., [1997] *Image and Logic: A Material Culture of Microphysics*, Chicago, University of Chicago Press.

Henderson, J.C. de C., Maunders, E.A.W., [1969] 'A Problem in Applied Topology: On the Selection of Cycles for the Flexibility Analysis of Skeletal Structures', *Journal of the Institute of Mathematics and Applications*, 5, 254–269.

Hopf, E., [1942] 'Abzweigung einer periodischen Lösung von eine stationären Lösung eines Differentialsystems', *Berichten der Mathematisch-Physischen Klasse des Sächsischen Akademie der Wissenschaften zu Leipzig*, 94, 1–22; Eng. Transl., in Marsden, J.E., McCracken, M., (eds.), *The Hopf Bifurcation and Its Applications*, New York, Springer, 163–93.

[1948] 'A Mathematical Example Displaying Features of Turbulence', *Communications on Applied Mathematics* 1, 303–322.

Hopf, H., [1960] 'The work of R. Thom', *Proceedings of the International Congress of Mathematicians* [Edinburgh, August 1958], Cambridge, Cambridge University Press, lx–lxiv.

Houzel, C., [1979] 'Les mathématiciens retournent au concret', *La Recherche*, n° 100, 508–509.

Israel, G., [1996] *La Mathématisation du réel*, Paris, Seuil.

Jaffe, A., Quinn, F., [1993–1994] 'Theoretical Mathematics: Toward a Cultural Synthesis of Mathematics and Theoretical Physics', *Bulletin of the American Mathematical Society*, 29 (1993), 1–13; and the various responses in *ibid*. 30 (1994), 178–207.

Landau, L., [1944] 'On the Problem of Turbulence', *Doklady Akademi Nauk SSSR* 44, 311–314.

Lefschetz, S., [1986] 'A Page of Mathematical Autobiography', in Sundaraman, D., (ed.), *The Lefschetz Centennial Conference: Proceedings of the Conference Held December 10–14, 1984*, I, Providence, American Mathematical Society, 1–26.

Lichnérowicz, A., [1995] Rapport pour la création d'une chaire d'Algèbre et Géométrie, Assemblée des professeurs (27 November). Archives du Collège de France, Paris. Document #G-iv-m 20R.

Manneville, P., Pomeau, Y., [1980] 'Different Ways to Turbulence in Dissipative Dynamical Systems', *Physica* 1D, 219–226.

Peixoto, M., (ed.) [1973] *Dynamical Systems: Proceedings of a Symposium Held at the University of Bahia*. New York, Academic Press.

Pomeau Y., Manneville, P., [1979] 'Intermittency: A Generic Phenomenon at the Onset of Turbulence', in Laval, G., Grésillon, D., (eds.), *Intrinsic Stochasticity in Plasmas*, Orsay, Éditions de physique Courtaboeuf, 330–340.

[1980] 'Intermittent Transition to Turbulence in Dissipative Dynamical Systems', *Communications in Mathematical Physics*, 74, 189–197.

Pont, J.C., [1974] *La Topologie algébrique des origines à Poincaré*, Paris, Presses Universitaires de France.

Ruelle, D., [1972] 'Strange Attractors as a Mathematical Explanation of Turbulence' in Rosenblatt, M., van Atta, C., (eds.), *Statistical Models and Turbulence: Proceedings of the Symposium at La Jolla, 1971*, Berlin, Springer, 292–299.

Ruelle, D., Takens, F., [1971] 'On the Nature of Turbulence', *Communications in Mathematical Physics*, 20, 167–192; 23, 343–344.

Smale, S., [1967] 'Differentiable Dynamical Systems'. *Bulletin of the American Mathematical Society*, 73, 747–817. (repr. [Smale 1980], 1–82).

——— [1969/70] 'Stability and Genericity in Dynamical Systems', *Séminaire Bourbaki*, 22, (374), Berlin, Springer, Lecture Notes in Mathematics 180 (repr. [Smale 1980], 90–94).

——— [1972] 'Personal Perspectives on Mathematics and Mechanics', in Rice, S.A., *et al.*, (eds.), *Statistical Mechanics: New Concepts, New Problems, New Applications*, Chicago, Uinversity of Chicago Press, 3–12. (repr. [Smale 1980], 95–105).

——— [1973] 'Global Analysis and Economics I: Pareto Optimum and a Generalization of Morse Theory', in Peixoto [1973], 531–44.

——— [1980] *The Mathematics of Time: Essays on Dynamical Systems, Economic Processes, and Related Topics*, New York, Springer.

——— [1990] 'The Story of the Higher Dimensional Poincaré Conjecture (What Actually Happened on the Beaches of Rio)', *Mathematical Intelligencer*, 12(2), 44–51, (repr. In Hirsch, M.W., *et al.*, (eds.), *From Topology to Computation: Proceedings of the Smalefest*, New York, Springer, 1993, 27–40.

Thom, R., [1969] 'Topological Models in Biology', *Topology*, 8, 313–35.

——— [1973] 'Langage et catastrophes: éléments pour une sémantique topologique', in Peixoto [1973], 619–654; and in *Mathematical Models of Morphogenesis*, transl. Brooks, W.M., Rand, D., Chichester, Ellis Horwood, 1983, 214–243.

——— [1975] (1972), *Structural Stability and Morphogenesis: Outline of a General Theory of Models*, trans. D.H. Fowler, Reading: Benjamin.

——— [1976] 'The Two-Fold Way of Catastrophe Theory' in Hilton, P., (ed.), *Structural Stability, the Theory of Catastrophes, and Applications*, Berlin, Springer, 235–252.

——— [1977a] 'René Thom répond à Lévy-Leblond sur la théorie des catastrophes', *Critique*, 33, 675–681.

——— [1977b] 'Structural Stability, Catastrophe Theory, and Applied Mathematics', *SIAM Review*, 19, 189–201.

——— [1983] *Paraboles et catastrophes. Entretiens sur les mathématiques, la science et la philosophie réalisés par G. Giorello et S. Morini*, Paris, Flammarion; (orig. Italian ed. 1980).

——— [1991] *Prédire n'est pas expliquer*, interview by Émile Noël, Paris, Eshel.

Velarde, M.G., [1981] 'Steady States, Limit Cycles and the Onset of Turbulence: A Few Model Calculations and Exercises' in Riste, T., (ed.), *Nonlinear Phenomena at Phase Transitions and Instabilities*, New York, Plenum, 205–247.

Warwick, A., [1992] 'Cambridge Mathematics and Cavendish Physics: Cunningham, Campbell, and Einstein's Relativity 1905–1911', *Studies in the History and Philosophy of Science*, 23, 625–656; 24, 1–25.

Woodcock, A., Davis, M., [1978] *Catastrophe Theory*, New York, Dutton.

Zahler, R.S., Sussmann, H.J., [1977] 'Claims and Accomplishments of Applied Catastrophe Theory', *Nature*, 269, 759–763.

Zeeman, E.C., [1964a] 'Les mathématiques et la pensée créatrice', *Sciences et l'enseignement des sciences*, 5(34), 11–14.

——— [1964b] 'Causality Implies the Lorentz Group', *Journal of Mathematics Physics*, 5, 490–494.

——— [1965] 'Topology of the Brain', *Conference on Mathematics and Computer Science in Biology and Medicine (MRC, Oxford, July 1964)* , London, Her Majesty's Stationary Office, 277–311.

——— [1973] 'Differential Equations for the Heartbeat and Nerve Impulse' in Peixoto [1973], 683–741; and in Zeeman [1977], 81–140.

——— [1977] *Catastrophe Theory: Selected Papers, 1972–1977*, Reading: Addison-Wesley.

Chapter 13

BEYOND ONE-CASE STATISTICS: MATHEMATICS, MEDICINE, AND THE MANAGEMENT OF HEALTH AND DISEASE IN THE POSTWAR ERA

Jean-Paul Gaudillière

In 1961 Daniel Schwartz, a statistician studying the relationship between tobacco and cancer for the INH—the French medical research agency—wrote a popular paper on 'The Statistical Method in Medicine' published in a journal widely circulated within the medical profession. Schwartz scorned French physicians who thought that statistics had little to do with medical knowledge:

> The reasons explaining this backwardness are diverse. France is a leading country in pure mathematics. As a consequence applied mathematics have been neglected. The statistical method emerged in the 'Anglo-Saxon' countries and slowly penetrated our country.
>
> Our peculiar frame of mind as well as our education system favor the mathematics of certainty and not the mathematics of probability. The latter prevails in all aspects of life and in decision-making processes but they are never taught ...
>
> On top of cognitive patterns, cultural traits may well have played a part in the disdain for statistics. French people are highly individualistic ... This behavior dominates the medical world. The French doctor is a very good clinician and an outstanding care provider. In his perception, every patient is an individual rather than a figure in a table ... These values are highly recommended in medical practice but the price paid for their supremacy is a farfetched emphasis on the individual in scientific research.
>
> This standing is not inescapable. It is true that respect is due to the special relationship between a patient and his doctor. But it is not true that this respect should result in one-case statistics. [Schwratz, 1962, p. 1920]

In his essay Schwartz offered a multi-layered interpretation of the role of the statistician in medicine. His analysis of the problematic status of 'big numbers' in French medical circles highlighted the ambiguities of the expertise claimed by pioneers of medical statistics in the aftermath of World War II. Being neither clinicians nor biologists many statisticians sought the role of mathematically-trained mediators. They argued that medical judgment should be based on objective procedures, i.e. aggregated data, comparative studies, controlled samples, and signification tests. Professionals 'applying' mathematics or translating probabilities into rules of evaluation and algorithms were barely needed. Hence the notion highlighted by Schwartz that physicians—especially French physicians trained in the 'great clinical tradition'—quite naturally opposed the intervention of outsiders and any radical departure from accepted individual practices.

The relationship between mathematics, statistics, and medicine is hardly an uncharted territory in the history of science. Early attempts at gathering clinical cases and computing figures for medical decision making are usually traced back to the invention of the 'numerical method' by Pierre Louis in the early nineteenth century [Desrosières, 1993]; [Bynum, 1994]; [Matthews, 1995]. Louis' story also is one of medical resistance. His failure to promote the numerical method has been explained with the very same factors as those listed by Schwartz: clinical tradition, one patient-one doctor relationship, organization of the medical profession, or absence of public health bodies.

Because of this failure, the main roots of contemporary medical statistics usually are ascribed to the British 'biometric school' of the first half of the twentieth century. The careers of Galton, Pearson, Fisher, or Major Greenwood have been the topic of several studies [Gigerenzer, 1989]; [Mackenzie, 1981]; [Matthews, 1995]; [Porter, 1986]. One shared feature of these analyses is to locate innovation—whatever its motives, i.e. cognitive, cultural, disciplinary, professional, ideological—in the science of numbers. This perception inevitably results in viewing medicine as a field for diffusion and a site of application. Statistical tools like the regression technique invented by Galton as part of his studies of human inheritance are often viewed as tools which emerged OUT OF a biological *terrain* to live an independent life and to occupy all sorts of scientific niches [Porter, 1986]; [MacKenzie, 1981]; [Matthews, 1995]. Explanations for this dissemination process may be intellectual—a new way of seeing the world circulated with new representations and computational techniques or social, the profession of statistician emerged in the midst of institutional conquests. In both cases however inventors are described as mathematicians having little or no interaction with specialists working in the domain of applications.

This chapter will build on more recent analyses of the part played by statistics in public life [Porter, 1995]; [Schweber, 1996] in order to propose a more balanced account focusing on the invention of medical statistics as a process of interaction between actors involved in health and disease management and building computational tools for decision making. This perspective will be illustrated with two episodes in the history of medical statistics after the Second World War when an apparent 'trust in numbers' started to dominate entire segments of biomedicine.

THE RANDOMIZED CLINICAL TRIAL: A DEVICE AT THE INTERSECTION OF STATISTICS AND MEDICINE

In 1948 the British Medical Journal published the results of an experiment on the therapeutic uses of a new antibiotic called streptomycin which seemed to be a potent means of curing lung tuberculosis [Medical Research Council, 1948]. This paper is often described as the 'birth of the modern clinical trial' because what was at stake was a new way of evaluating therapies and the statistical method as well as the efficiency of streptomycin (something many clinicians were already convinced of).[1] Run by a special committee established by the British Medical Research Council, the 1947 streptomycin assay actually was a statistician's trial. It was designed by Austin Bradford Hill and tuberculosis specialists according to methodological choices which would later become the basic norms for standard objective assays of therapeutics [Meldrum, 1994]. The streptomycin trial was planned as a 'randomized and controlled' trial meaning that a) patients were randomly assigned to the assay group (receiving the antibiotic) or to the control group (receiving established treatments); b) when making decisions about admission to the trial, clinicians did not know this allocation; c) clinical evaluation of therapeutic progress was based on the measurement of tuberculosis spots on X-ray pictures of the patient's lungs; d) this examination was done by radiography specialists with no knowledge of the clinical status of the patient; e) final evaluation of the significance of these results was completed by statisticians.

The history of such randomized clinical trials (RCT) may be traced back to the biometricians' work and it seems to be a good example of 'applied statistics'. On the one hand, there was a direct lineage from Pearson to Bradford Hill via Fisher and Major Greenwood, the last being the first physician to study with biometricians and the first a statistician appointed by a medical research institution (the Lister Institute, London, where he participated in the evaluation of sera and vaccines; he later joined the National Institute for Medical Research at Hampstead) [Cox-Maximov, 1998]. On the other hand, it is not too difficult to argue for conceptual legacy, since the basic concepts grounding the choice of randomization can be traced back to R.A. Fisher's work.

Fisher's first job was at the Rothamsted Agricultural Research Station where he focused on problems of sampling and statistical significance [Fisher-Box, 1978]. He thus worked on the design of experiments [Fisher, 1935]. Fisher's main interest lay in the reliability of a given experimental result and the role of chance, for instance, in the result of trials comparing the yields of different varieties of cultivated plants. He argued that an observed difference in yield between two varieties planted in different fields might be due either to differences among seeds or to differences in soil, exposure, treatment, and other unknown factors. If so, how to decide on the results' meaning? Randomization was an invention targeted at taming chance in experimentation without having to depend on the practitioner's experience of terrain and agricultural practices actually involved. Chance was the enemy but it was to be defeated with more chance. Fisher built on existing practices in agricultural stations and first proposed to standardize treatments. Randomization was an additional step that consisted in dividing the plots into narrow strips and assigning the place of different varieties to be tested by use of chance mechanisms. The rational was that randomization would replace replication of experiments or that the use of large numbers and chance allocation would reduce the effects of unknown variations in experimental conditions. Randomization reduced what Fisher called the objective bias.

Fisher's argument had a significant impact on agricultural research in the 1930s. In the medical world however, it was of limited influence, although some statisticians dealing with biological and clinical problems started to refer to random allocation.[2] Austin Bradford Hill, a student of Greenwood at the MRC Statistical Committee was among them, and he referred to 'alternating' treatment assignments [Bradford Hill, 1937]. But it is only with the postwar antibiotic trials that he fully committed himself to the organization of randomized trials and chance allocation. Bradford Hill took over Fisher's procedure but developed a different rationale. Therapeutic evaluation in medicine is more complex than decision making in agriculture because two different biases are involved. The analog to Fisher's objective bias was the fact that the same disease in two different bodies could follow variable paths according to unknown differences in the terrain, i.e. body constitution, immunity, etc. But Bradford Hill and his followers focused on another—subjective—bias originating in the medical situation. Unlike agricultural researchers, physicians take care of people and they usually think that the treatment they prescribe works while patients usually trust their doctor. This conjunction of subjective hopes and trust does affect the selection of patients to be included in a study, the management of a treatment, and its final evaluation. As Bradford Hill put it in 1951 when summing up the motives for controlled clinical trials: 'Drugs are not ordered by doctors at random, but in relation to a patient's condition when he first comes under observation and also to the subsequent progress of the disease. The two groups are therefore remotely comparable ... The same objections must be made to the contrasting of volunteers for treatment with those who do not volunteer, or between those who accept and those who refuse ... The contrast of one physician, or one hospital, using a particular form of treatment, with another physician, or hospital, not adopting that treatment, or adopting it to a lesser degree, is fraught with the same difficulty ...' [Bradford Hill, 1951, p. 279]. These were the most compelling reasons for introducing randomization in medicine: to regulate and undermine the clinician's preferences and idiosyncrasies. In the statistician's eyes, chance should be used to narrow down both objective and subjective sources of variability.

This plea against the traditional investigator should not be undermined. Randomization did not only oppose the clinician's role but also the very sense of a doctor's duty and experience. As recalled by many observers, clinical trials did exist before the Second World War and sometimes included control groups [Ross, 1936]. The allocation of patients however was based on other assumptions. Clinicians were aware of the variability of patients and diseases but sought to control uncertainty by accumulating experience and controlling trial parameters. The right choice of patients to participate in a study was to be based on the doctor's intimate knowledge of the individual and the intricacies of the disease. A good assay group would therefore maximize the factors enhancing a positive response to the drug, while every patient would be matched by another individual in the control group. The latter was consciously selected by the investigator so as to show strong analogies with the former, at least with respect to the clinically most important features of the disease. Chance was not to be mastered with more chance but conquered with experience and medical knowledge. This approach was also taken to be an ethically responsible one: treatment would not be denied those who would in all medical knowledge (and not in all probability) benefit from them. The statisticians' ethos clearly was departing from these attitudes [Bradford Hill, 1963].

This foundation story has recently been refined by several historians of medicine. On the one hand, Desiree Cox-Maximov has stressed the pre-history of clinical trials and emphasized, for instance, the legacy of the Statistical Committee and the Therapeutic Trials Committee established in the 1930s by the British Medical Research Council [Cox- Maximov, 1998]. On the other hand, Harry Marks has discussed the work of American medical reformers and convincingly argued that the norms of the RCT were far from being universally accepted and applied in postwar medicine [Marks, 1997].

In-depth analyses however do not abolish the transformation typified by the reference to the MRC streptomycin trial. How is it then that methodological power shifted to the statisticians' hands? Cognitive history has it that the postwar expansion of randomized clinical trials was triggered by the need for objective means of evaluating medical treatments and by the technical efficiency of the statisticians' method [Himsworth, 1982]; [Lock, 1994]. One is nonetheless left with the question of knowing which factors created and made visible this 'need' for mathematically-certified objectivity and outside evaluation. In order to account for postwar changes, one must analyze controlled trials and their underlying notions as decision-making procedures rather than abstract methodological norms. In other words, one must look at the changes in medical practice during and after the Second World War. This leads one to stress other aspects of the MRC streptomycin trial.

In Hill's words: 'It is the gradual development of this [scientific] attitude of mind coupled with the concurrent introduction of one antibiotic, one modern drug, after another, that has led in the past few years to the highly organized and efficiently controlled therapeutic trial of new remedies' [Bradford Hill, 1951, p. 278]. In other words, it was of fundamental importance that the MRC trial took part in the testing of a chemical 'wonder drug', an early product of the 'antibiotic revolution'.[3] Postwar clinical research was actually dominated by tightened relationships among biological laboratories, pharmaceutical firms, and hospital wards [Starr, 1982]. A critical change in the role of the elite physician and the medical systems of most industrialized countries was associated with the dramatic expansion of chemotherapy. Following the development of penicillin and early antibiotics, chemical laboratories of the pharmaceutical industry started to produce thousands of compounds annualy having some structural analogy with known anti-microbial agents and therefore of putative therapeutic value. The same hope of developing 'magic chemical bullets' soon pervaded numerous domains other than infectious diseases caused by bacteria, with cancer above all. Clinicians faced the difficult task of making decisions regarding the claims of industrial chemists and doctors, or regarding the choice of protocols for treating a given disease. This shift obviously contributed to the creation of new 'demands' for evaluation procedures. The production of antibiotics however does not in itself explain why these demands took the form of 'highly organized', 'large scale' and 'cooperative' trials, and why these trials gave rise to an alliance between the medical elite and statisticians. Changes in medical organization rather than chemical breakthroughs have to be taken into account.

In his detailed analysis of American clinical reformers, Harry Marks points to the war-time medical mobilization as a turning point [Marks, 1997, chapter 4]. His perceptive description shows how the development of early antibiotics within the context of military medicine paved the way to new collaboration patterns, acculturated elite clinicians to the use of statistical tools, and contributed to sow the seeds of postwar changes. The war assays of penicillin and streptomycin are emblematic of this trend.

Although 'discovered' in the late 1920s in Fleming's laboratory at St. Mary's Hospital, London, penicillin did not become a therapeutic agent until the Second World War when its development was taken over by the US pharmaceutical industry and the Office for Scientific Research and Development (OSRD)—the federal agency established by President Roosevelt which coordinated the scientific war effort [Macfarlane, 1984]; [Hobby, 1985]. It is well known that the mobilization of American scientists for war was organized by the OSRD. From 1941 onwards its Committee for Medical Research (CMR) prompted studies on the mass production of penicillin and acted as a regulator and coordinator for a network composed of a few federal laboratories (for instance, the US Department of Agriculture Laboratory at Peoria), a dozen university teams, the War Production Board and large pharmaceutical firms (Merck and Pfizer in the leading roles). Initially CMR was also in charge of supplying the drug to civilian clinicians, and organized investigations of penicillin (and later of other antibiotics). The series of cooperative investigations launched within this context shows how organization drifted toward the randomized clinical trial.

The distribution of penicillin by CMR was handled by Chester Keefer, a Harvard professor of medicine. CMR first relied on classical 'trials': the drug was distributed to a handful of well known clinicians who determined who would be treated and how. CMR's role was to draw a list of targeted infections and gather information about the outcomes. As production was scaled up, and as clinicians started to use massive doses of penicillin, spectacular cures of meningitis and staphylococcal infections were reported. Although there were some tensions regarding the list of authorized pathologies, CMR's monopoly on the drug, as well as its reported wonderful effects, forced compliance among participating clinicians who feared being refused access to the 'magic bullet'. At this stage statistics played no, or a very limited, role with the presentation of 'all or no' figures.[4]

Organizational patterns started to change in 1943 as expansion of production prompted the Army, which envisioned massive uses of penicillin, to take over the drug. For the military, syphilis was a high-rank priority because of the incidence of the disease and the slow effects of arsenic treatments then in use. Syphilis was a complex and chronic disease with latent a-symptomatic phases whose diagnosis and follow-up were based on sophisticated bacteriological and immunological procedures. Systematic and organized comparison among cases was then considered necessary for evaluating the effects of penicillin on the disease. The first study organized by the military took advantage of the large number of enrolled men. Thousands of patients were given the drug and progress was monitored. Data collection however remained highly problematic, and the army officials ultimately requested the OSRD to organize a carefully controlled trial [Pillsbury, 1946]. Under CMR's guidance, syphilis investigators agreed to set up a cooperative trial based on one single treatment scheme, standardized laboratory tests, and homogenized data collection. Although the study was plagued with problems in assuring follow-ups, it contributed to make multicentric cooperative work and large numbers more visible and more normal.

As war ended and medicine returned to peacetime conditions, this style of operations was passed on to the evaluation of other antibiotics. In September 1945, when a team of researchers working at the Mayo Clinic in New York announced that streptomycin may benefit tuberculosis patients, the army and the pharmaceutical industry once again turned to governmental scientific bodies (this time the National Research Council) to resolve uncertainties about the drug and therapeutic priorities [Hinshwa, 1954]. The investigation was organized with the Veterans Administration which ran its own network of hospitals. Conducted within a bureaucratic organization by people who had all been involved or aware of penicillin studies, the VA streptomycin trial reinforced cooperative and planning ideals. Innovations focused on the need for a control group and for 'objective' outside analyses of X-ray pictures. The program could not be fulfilled (for instance, the assembling of a control group was dropped after a few months because investigators could not get Veterans Administration hospitals in the study to find enough patients) but the VA study clearly promoted the notion of cooperative AND controlled trials.[5] It was followed by another study of streptomycin in tuberculosis launched by the US Public Health Service. This time tight control was enforced. All decisions to admit patients were reviewed by a panel of senior investigators; a central statistical unit assigned patients to treatment or control groups; patients in the control group were not told that they participated in a study while clinicians' scruples were softened by introducing an appeal procedure for handling control patients whose disease worsened critically; finally evaluation followed the statisticians' recommendation for X-ray measurements rather than having clinicians hold clinical case conferences.[6]

The growing importance of statistics and methodological rules in this series of trials reflects forms of objectification rooted in the advent of a cooperative regime in medical research which was unequivocally facilitated by mobilization for war and the context of military medicine. Its success and expansion after the war, however, is inseparable from the changing organization of medicine and particularly from the birth of a hospital-based high-tech medicine which transformed the nature and procedures for clinical decision-making.

Placed within this series, the MRC streptomycin trial thus appears as just one more instance of coordinated clinical work organized under the umbrella of a scientific regulating body. The acceptance of randomization and related notions, as well as the new role of 'big number' specialists, may be viewed accordingly as decision-making tools meaningful within the large medical networks emerging during and after the war. To follow Ilana Löwy's analysis of the case of cancer research where the randomized clinical trial became a widely used norm in the 1950s and 1960s, the RCT was an 'organizational device' useful in domains where the routine use and evaluation of fast-moving chemotherapeutic innovations became part of the normal caring of patients [Löwy, 1996].

RULES OF INFERENCE: TOBACCO, CANCER, AND EPIDEMIOLOGY IN THE PUBLIC ARENA

In 1964 the US Surgeon General released an official statement on the correlation between smoking and cancer [US Department of Health,

1964]. This document was seen as bringing to a close the decade-long controversy on the causes of lung cancer, which had been debated among epidemiologists, public health officials, cancer specialists, and the tobacco industry [Brandt, 1982]; [Berlivet, 1995]. This debate, which catalyzed the disciplining of epidemiology, has been viewed as another key process in the postwar history of medical statistics.

The discussion on smoking and lung cancer was initiated by the publication of a British survey of lung cancer patients conducted by Bradford Hill and a student of his; Richard Doll. This first study compared 1,500 lung cancer patients and a control population assembled by matching every cancer case with a patient of similar age, and if possible taken care of in the same hospital. The study focused on the significantly higher number of smokers among cancer patients than among the control population, but remained quite elusive on the nature of the association between smoking and lung cancer [Doll, 1950].

For reasons that have still to be properly investigated, studies multiplied in the following decade with an increasing emphasis on statistical arguments and an imputation of causality claimed by organizers of the largest surveys, beginning with Bradford Hill and the British Medical Research Council. Opponents, like statistician R.A. Fisher or geneticist C.C. Little, repeatedly argued that a causal inference was impossible to make, that the control population was biased, or, more critically, that smoking was just a marker for a population with a high incidence of another 'third' (causal) factor, for example genetic predisposition or socioeconomic status [Fisher 1958a and b]. In response to critics there was in an increasing sophistication of techniques. The size of populations grew. Prospective studies complemented retrospective analyses. Control populations were selected according to additional variables: age, sex, medical history. In addition, new statistical tools were invented. These concentrated on procedures for analyzing the significance of multiple associations and comparing multiple control groups [Schwartz, 1969]. For instance, fourfold table analysis [Berkson, 1946] was developed into complex strategies for analyzing retrospective data and computing relative risks by means of an extension of chi- square testing to situations in which data could be subclassified according to variable sets of factors [Mantel, 1959].

Growing mathematical sophistication did not in itself put an end to the controversy [Brandt, 1982]; [Berlivet, 1995]; [Proctor, 1995]. Public support for—and administrative endorsement of—the notion of a causal relationship between smoking and lung cancer proved necessary. They crystallized with events like the above-mentioned publication of the US Surgeon General's 1964 report building on the comparison of two dozen studies including an American Cancer Society follow-up of one million people whose aim was to urge public action. A similar role was previously played by a 1957 report of the Medical Research Council. Warning the British government of public-health consequences of the causal relationship between smoking and lung cancer, this report was based on a survey comparing the incidence of lung cancer among smoking and non-smoking British physicians.

It is noteworthy that, in both instances, public expertise was associated with strong claims regarding the rules of inference and the nature of epidemiology. In 1964, the US report argued for its own series of criteria for causality, which emphasized large numbers and the notion of relative risk [Lilienfeld 1983]. In 1965 Bradford Hill summarized the methodological lessons of the tobacco debate with a series of nine heterogeneous criteria to be taken into account for similar issues: 1) the strength of the association (for instance the fact that 'prospective inquiries have shown that the death rate from cancer of the lung in cigarette smokers is nine to ten times the rate in non-smokers'); 2) the consistency of the observed association (accordingly the Advisory Committee to the US Surgeon-General 'found the association of smoking with cancer in 29 retrospective and 7 prospective inquiries'); 3) the specificity of the association; 4) the temporality of the association ('which is the cart and which is the horse'); 5) the biological gradient or dose-response curve; 6) the biological plausibility; 7) the coherence 'with known facts of the natural history and biology of the disease'; 8) experimental evidence; 9) analogy with other pathological situations [Bradford Hill, 1965]. Although Bradford Hill viewed methodological innovation as an important outcome of the lung cancer controversy, he warned his readers that 'none of my nine viewpoints can bring indisputable evidence for or against the cause-and-effect hypothesis and none can be required as a sine qua non ... Finally, in passing from association to causation I believe that in 'real life' we shall have to consider what flows from that decision. On scientific grounds we should do no such a thing ... But in another and more practical sense we may surely ask what is involved in our decision.' [Bradford Hill, 1965, p. 300].

Such conjunction of statistical and policy debates is emblematic of the rising notion of risk in the postwar public and medical cultures. Relative risk, attributable risk, or risk factors are epidemiological constructs, which during the two decades following the war emerged at the intersection of biological laboratories, statistical bureaus, hospitals, and public health institutions. They all combine evaluation, assessment, and collective management of pathological events. They all point to the invention of quantitative procedures for handling concerns about chronic and old-age diseases (cancer, obesity, hypertension, heart diseases, etc.), which rapidly took infections' place in the psyche of inhabitants of Western industrial countries. One major advantage of the notion of risk and associated computational techniques was that it replaced the 'infinite regress', typical of the search for isolated and final causes, by a juxtaposition of 'relative' causes, i.e. a list of factors which may encompass genes, hormones, behavioral traits, lifestyles, food, socioeconomic status, age, ethnicity, etc. In a mathematically more refined way, this form of knowledge recalled old hygienic inquiries more than the bacteriology-rooted surveys of the first half of the century. The emphasis placed on the notion of risk thus highlighted the fact that postwar epidemiology was constructed as a statistical specialty producing models and evidence for public health specialists.

This brings us back to Daniel Schwartz' career and the diverse meanings of hybridization between statistics and medicine. The rise of epidemiology was closely associated with the tobacco question in France, too. But the field took a different shape due to the specificity of local clinical and mathematical cultures. Little emphasis was put on 'risk' and on the public-health dimension of the controversy.

A graduate of the École Polytechnique, Schwartz began his research career as an engineer at SEITA, the French state-run tobacco company. As a mathematically-trained statistician, he knew a little about statistics in biological research, but nothing about statistics in medicine.[7] In 1954, Pierre Desnoix—then head of the cancer commission of the Institut National d'Hygiène (INH), the state medical research agency—solicited his help in launching a study of lung cancer and tobacco.[8] The study was later expanded to a small group of epidemiologists, first established at Desnoix' cancer research institute, which eventually became the first unit for medical statistics research in the country.[9] Ironically, this cancer study was initially supported by both the medical research agency and the state tobacco industry.[10]

By a not so strange twist in the story, the mathematically-trained French analysts of 'cancer and smoking' focused on methodological innovations. Their contribution was not to add decisive evidence by scaling up previous inquiries (something American and British investigators were actively doing) but to refine the design of control groups and significance tests [Schwartz, 1957, 1961]. Participating in the 'experimental culture' characterizing the French reconstruction of biological and medical research and the INH during the 1950s and 1960s, they did not try to further the public debate, but focused on developing new tools and advocating more systematic uses of statistics in clinical research [Schwartz, 1960].

It is important to note that this inovation apart, very little was done by the French administration to advance the tobacco question. As stressed by social historians, one reason for this was that public health had for decades been a rather marginal domain in French medicine [Murard & Zylberman, 1997]. Moreover, when thinking of major health problems in the country, doctors and officials in the Ministry listed factors affecting the size of the population on a large scale: child mortality, tuberculosis, and alcoholism. Smoking was a matter of lifestyle, not a scourge. Financial and bureaucratic interests in the production and sale of tobacco added to these factors to make sure that there would be little administrative pressure for expertise or public debates on tobacco and cancer. It is therefore not altogether surprising that the 'risk culture' played little role in the discourses of the 1950s and 1960s. It is only in the early 1970s, when Desnoix became head of the French National Cancer Commission, that tobacco, conceived of as a 'risk factor', surfaced as a problem for administrative intervention and public inquiry [Berlivet, 1998].

French statistical epidemiology therefore developed as a specialty that emphasized both the search for the (multiple) causes of diseases and research on the 'fundamental' problems of causal imputation and significance. In contrast to Bradford Hill, Schwartz rarely made 'cases for action', but repeatedly pleaded for the statisticians' irreplaceable role on the basis of his knowledge of the measurement of uncertainty [Schwartz, 1992].

CONCLUSION

Robert Musil's 'The Man Without Qualities' begins with a brilliant insight into the love of statistical data characterizing the twentieth century's understanding of modernity:

'A barometric low hung over the Atlantic. It moved eastward toward a high-pressure area over Russia without as yet showing any inclination to bypass this high in a northerly direction. The isotherms and isothers were functioning as they should. The air temperature was appropriate relative to the annual mean temperature and to the aperiodic monthly fluctuations of the temperature ... In a word that charecterizes the facts fairly accurately, even if it is a bit old-fashioned: It was a fine day in August 1913'. [R. Musil, 1995].

Of course Musil is wrong. Isotherms, annual means, and weather correlations are not equivalent to a nice summer day. His comment however reminds us that when dealing with statistical entities observers should not forget 'the fine day in August'. In the context of the historiography of the field, the warning may be translated into: one must not forget that statistics is a form of government. More precisely, statistics is a boundary field continuously creating decision-making procedures. This chapter argues this point by focusing on examples taken from the history of medical statistics. The origins and fate of randomized clinical trials as well as epidemiological debates about smoking and lung cancer show that major changes in the nature and images of statistics did not stem from mathematical breakthroughs, but emerged out of the work of statisticians operating within a rapidly changing medical arena. The rise of a new statistical culture, typified by randomization and relative risk was therefore rooted in the birth, after the Second World War, of a new form of medical organization focusing on cooperative work, chemotherapy, and state intervention. In other words, looking at statistical creativity from the viewpoint of the 'dead-end users' may occasionally shed new light on the dynamics of the domain.

NOTES

1. The introduction of the paper thus explained: 'The history of chemotherapeutic trials in tuberculosis is filled with errors due to empirical evaluation of drugs; the exaggerated claims made for gold treatment, persisting over 15 years, provide a spectacular example. It had become obvious that, in future, conclusions regarding the clinical effect of a new chemotherapeutic agent in tuberculosis could be considered valid only if based on adequately controlled clinical trials.'
2. Practitioner's historiography has merely focused on this point. For example see (Armitage, 1992).
3. On this conjunction, see the forthcoming work of Alan Yoshioka, *The British Clinical Trials of Streptomycin*, Ph.D. thesis, Imperial College, London.
4. As Bradford Hill later put it: 'No controls are essential to prove the value of a drug such as penicillin which quickly reveals dramatic effect in the treatment of the disease. Such dramatic effects occurring on a large scale and in many hands cannot be long overlooked. Unfortunately these undeniable producers of dramatic effect are the exception rather than the rule even in the halcyon days of the antibiotics.' (Bradford trial, 1951, p. 281).
5. W.B. Tucker, 'The Evolution of the Cooperative Studies in the Chemotherapy of Tuberculosis of the Veterans Administration and Armed Forces of the USA' *Advances in Tuberculosis Research* 10 (1960), 3–4. W.B. Tucker, 'Evaluation of Streptomycin Regimens in the Treatment of Tuberculosis: An Account of the Study of the Veterans Administration, Army and Navy' *American Review of Tuberculosis* 60 (1949), 745–746.

6. H. Marks, op. cit. for a detailed analysis of this trial.
7. Schwartz eventually started some work on the tobacco mosaic virus.
8. Daniel Schwartz, 'A quoi sert l'épidémiologie' seminar EHESS, December 11, 1996.
9. D. Schwartz, *Note sur le fonctionnement de l'unité de recherche statistique de l'Institut Gustave Roussy*, undated manuscript, presumably 1960, Archives INSERM.
10. According to Schwartz the latter was interested at finding the chemicals inducing lung cancer and eliminating these side-products through the cigarette production process.

REFERENCES

Armitage, P., [1992] 'Bradford Hill and the Randomized Controlled Trial', *Pharmaceutical Medicine*, 6, 23–67.
Berlivet, L., [1995] *Controverses en épidémiologie: Production et circulation de statistiques médicales*, Rapport de recherche, Mission Interministérielle à la Recherche, Paris, Mire.
Berlivet, L., [1998] *Une santé à risque. L'action publique de lutte contre l'alcoolisme et le tabagisme en France, 1954–1997*, thèse de Sciences Politiques, Université de Rennes 1.
Berkson, J., [1946] 'Limitations of the application of fourfold table analysis to hospital data', *Biometrics Bulletin*, 2, 47–53.
Bradford Hill, A., [1937] *Principles of Medical Statistics*, London, Lancet.
    [1951] 'The Clinical Trial', *British Medical Bulletin*, 1951, 7, 278–282.
    [1963] 'Medical Ethics and Controlled Trials', *British Medical Journal*, 20 April 1963.
    [1965] 'The Environment and the Disease: Association or Causation?', *Proceedings of the Royal Society of Medicine*, 58, 295–300.
Brandt, A., [1982] 'The Cigarette, Risk and American Culture', *Daedalus* Fall.
Bynum, W., [1994] *Science and the Practice of Medicine in the Nineteenth Century*, Cambridge, Cambridge University Press.
Cox-Maximov, D., [1998] *The Making of the Clinical Trial in Britain, 1900–1950*, Ph.D. thesis, Cambridge University.
Desrosières, A., [1993] *La Politique des grands nombres*, Paris, La Découverte.
Doll, R., Bradford Hill, A., [1950] 'Smoking and Carcinoma of the lung', *British Medical Journal*, 73, 739–748.
Fisher Box, J., [1978] *The Life of Scientist*, New York, John Wiley and Sons.
Fisher, R.A., [1935] *The Design of Experiments*, London, Oliver and Boyd.
    [1958] 'Lung Cancer and Smoking', *Nature*, 12 July 1958.
    [1958] 'Cancer and Smoking', *Nature*, 30 August 1958.
Gigerenzer, G. et al., [1989] *The Empire of Chance: How Probability Changed Science and Everyday Life*, Cambridge, Cambridge University Press.
Hinshwa, H.C., [1954] 'Historical Notes on Earliest Use of Streptomycin in Clinical Tuberculosis', *American Review of Tuberculosis*, 70, 9–14.
Himsworth, Sir H., [1982] 'Bradford Hill and Statistics in Medicine', *Statistics in Medicine*, 1, 301–303.
Hobby, G.L., [1985] *Penicillin: Meeting the Challenge*, New Haven, Yale University Press.
Lilienfeld, A.M., [1983] 'The Surgeon's General Epidemiologic Criteria for Causality', *Journal of Chronic Diseases*, 36, 837–845.
Lock, S., [1994] 'The Randomised Clinical Trial: A British Invention' in Lawrence, G., (ed.) *Technologies of Modern Medicine*, London, Science Museum.
Löwy, I., [1996] *Between Bench and Bedside*, Cambridge, Harvard University Press.
Macfarlane, G., [1984] *Alexander Fleming: The Man and the Myth*, Cambridge, Harvard University Press.
Mackenzie, D., [1981] *Statistics in Britain, 1865–1930. The Social Construction of Scientific Knowledge*, Edinburgh, Edinburgh University Press.
Mantel, N., Haenzl, W., [1959] 'Statistical Aspects of the Analysis of Data from Retrospective Studies of Disease', *Journal of the National Cancer Institute*, 22, 719–748.
Marks, H., [1997] *The Progress of Experiment: Science and Therapeutic Reform in the United States, 1900–1990*, Cambridge, Cambridge University Press.

Matthews, G., [1995] *The Quest for Certainty*, Princeton, Princeton University Press.

Medical Research Council Streptomycin in Tuberculosis Trials Committee, [1948] 'Streptomycin treatment of pulmonary tuberculosis', *British Medical Journal*, 30 October 1948.

Meldrum, M.L., [1994] *Departures from Design: The Randomized Clinical Trial in Historical Context*, Ph.D. thesis, State University of New York at Stony Brook.

Murard, L., Zylberman, P., [1997] *L'hygiène dans la République*, Paris, Fayard.

Musil, R., [1995] *The Man Without Qualities* (English translation by S. Wilkins), New York, Alfred A. Knopf.

Pillsbury, D.M., [1946] 'Penicillin Therapy of Early Syphilis in 14,000 patients: Follow-Up Examination of 792 Patients Six or More Months after Treatment', *American Journal of Syphilology and Dermatology*, 30, 134–135.

Porter, T., [1986] *The Rise of Statistical Thinking, 1820–1900*, Princeton, Princeton University Press.

[1995] *Trust in Numbers: The Pursuit of Objectivity in Science and Public Life*, Princeton, Princeton University Press.

Proctor, R., [1995] *Cancer Wars*, New York, Basic Books.

Ross, O.B., [1936] 'Use of Controls in Medical Research', *JAMA*, 145, 20–21.

Schwartz, D., Desnoix, P., [[1957]] 'L'enquête française sur l'étiologie du cancer broncho-pulmonaire. Rôle du tabac', *Semaine des hôpitaux de Paris*, 62, 424–437.

Schwartz, D., Flamant, R., Lellouch, J., Desnoix, P., [1961] 'Results of a French Survey on the Role of Tobacco, Particularly Inhalation, in Different Cancer Sites', *Journal of the National Cancer Institute*, 26, 1085–1108.

Schwartz, D., Flamant, R., Lellouch, J., Rouquette, C., [1960] *Les essais thérapeutiques cliniques*, Paris, Masson.

Schwartz, D., [1962] 'La méthode statistique en médecine: mode ou nécessité ?', *Gazette Médicale de France*, 68, 1919–1922.

[1969] *Méthodes statistiques à l'usage des médecins et des biologistes*, Paris, Flammarion.

[1992] *Le jeu de la science et du hasard: la statistique et le vivant*, Paris, Flammarion.

Schweber, L., [1996] *The Assertion of Disciplinary Claims in Demography and Vital Statistics: France and England*, Ph.D. thesis, Princeton University.

[1997] 'L'échec de la démographie en France au XIXème siècle', *Genèses*, 29 décembre 1997.

Starr, P., [1982] *The Social Transformation of American Medicine*, New York, Basic Books.

US Department of Health, Education and Welfare, [1964] *Surgeon's General Report on Smoking and Health*, Washington D.C., Government Printing Office.

INDEX

Printed in the United States
by Baker & Taylor Publisher Services